BASIC STATISTICS AN INTRODUCTION TO PROBLEM SOLVING WITH YOUR PERSONAL COMPUTER

BY JERRY W. O'DELL, PH.D.,
EASTERN MICHIGAN UNIVERSITY

TAB BOOKS Inc.
BLUE RIDGE SUMMIT, PA 17214

FIRST EDITION

FIRST PRINTING

Printed in the United States of America

Library of Congress Cataloging in Publication Data

O'Dell, Jerry W.
BASIC statistics, an introduction to problem solving with your personal computer.

Bibliography: p.
Includes index.
1. Statistics—Data processing. 2. Electronic digital computers—Programming. 3. Basic (Computer program language) I. Title.
QA276.4.034 1984 519.5'028'542 84-8804
ISBN 0-8306-0759-5
ISBN 0-8306-1759-0 (pbk.)

Contents

Introduction

I have waited many years for someone to write a book such as this one, so that I could use it in my classes. Alas, no one has, and thus this book.

The book is intended for anyone with very little background in mathematics; you'll need algebra to read it, but nothing beyond that. No real background in computers is necessary, but it would be a good idea to read the manuals that come with your machine!

Many people deserve acknowledgment for helping me with the writing of this book. This is especially the case for my wife, Alma, to whom the book is dedicated. She not only put up with the writing process, but also drew the cartoons. I have fallen in love completely with Minnie Computer! I should also like to thank Messrs. Abramowitz and Stegun. Without their book, *Handbook of Mathematical Functions* (1964), I certainly could not have calculated the tables. Government publications, such as theirs, are not copyrighted. They certainly deserve great credit for producing such an enormous source of valuable mathematical information.

1
Statistics and Computers

Statistics is one of the most fascinating branches of modern mathematics. The introduction of statistical analysis has revolutionized the social sciences, making possible the analysis of data which could not be dealt with before. The field of computers is perhaps even more interesting; everyone knows of the growing influence of computers in our culture. Computers handle our bank accounts, keep time for us, and even control our microwave ovens.

The combination of statistical analysis and computers is a doubly powerful one. Solution of statistical problems is often a painfully long and arduous task. When I was a senior in college, I had to carry out a rather simple statistical analysis. No computers were available in those days, so it was necessary to carry out the computations with a set of logarithm tables. The task required *two months* of agonizing labor, and when I was all through, the answer was wrong! A modern computer could solve the problem in well under a second of time, and with perfect accuracy. The use of computers has made possible the rapid solution of ever-larger statistical problems which would require several lifetimes of calculation from any one person. With modern technical developments, you can do all of the computer analyses shown in this book, in your home, on a microcomputer that costs well under $250.

Strangely, often the same sort of person finds both statistics and computers terrifying. Both subjects are thought by many to be topics to be avoided at all costs. Yet the use of a computer in statistics removes all the drudgery. Further, if approached correctly both statistics and computers are rather simple topics. Most people have an inherent, intuitive feel for both areas. Many students have been amazed to find just how much fun either topic can be, once the terror has been eliminated. Knowledge of a little computer programming makes it possible to work many, many statistics problems, thus increasing one's understanding of statistics. And the great utility of even a minimal amount of computer programming becomes thoroughly apparent at such a time, making people like computers.

It is the purpose of this book to teach both elementary statistics and elementary computer programming at the same time—two terrifying topics at once! As you'll learn very shortly, computer programming is a rather simple business. If you don't try to be a virtuoso, most people can pick up the elements very quickly. In the first part of the book, the emphasis will of necessity be on simple computer procedures, using statistical examples that most people know already. Then, as the chapters progress, and you have the programming under your belt, the emphasis will shift more and more to the statistics. By working through all the examples, learning will be easier than you thought possible.

Going through this book will take dedication on your part. Alas, you can't learn systematic topics like statistics or computer programming without daily work. Working through the book will also be fun if you're the sort of person who likes to learn things—and that's practically everybody. The book is designed to treat both topics as simply as possible. I hope that once the fear of either has dissipated the student will be able to read far more difficult books without difficulty.

STATISTICS DEFINED

Statistics is usually considered to be the branch of mathematics which deals with the organization of numerical data. According to the American Heritage dictionary, the term statistics comes from the Latin *statisticus*, meaning of state affairs. Apparently the Romans needed organized information about their people, just as we do, and thus the term. Of course, our census bureau collects just such statistical data. In general there are two broad areas of statistics:

Descriptive statistics has to do with simple description of data—one gets the average age of a group of people, finds how much a set of scores spreads out, measures the relationship between two groups, and so on. Most people have little difficulty with this area of statistics—they calculate their grade point average in school with no difficulty, and they would have no trouble in figuring out the average age of their children. The man on the street has no difficulty in calculating the range of scores in his bowling league. You won't have much trouble understanding descriptive statistics, and so these concepts are presented first, in Chapters 1 through 8.

Inferential statistics are used to make inferences about things that we don't know, or perhaps can't know. For example, say that we measure the average salary of a group of 30 people, and find it to be $25,000 per year. The question addressed by inferential statistics is that of whether the $25,000 figure can legitimately be extended to a larger group of people, or even to the population at large. Inferential statistics are inherently more confusing and more difficult than descriptive statistics, and so they'll be presented later in the book, in Chapters 9 through 14. They can be of tremendous value in practical situations. Thus, only the Census Bureau can know for sure the average salary of workers. But to learn this, they must measure 215,000,000 people, a prohibitively expensive task. Wouldn't it be nice if something about the average salary could be derived from a smaller sample of people? This is the sort of task for which inferential statistics is used. Since it attempts to generalize from data which hasn't been collected, inferential statistics is much more controversial than descriptive statistics. The mathematics in inferential statistics is more involved and, hence, a good deal more confusing to many people. People who have trouble with statistics generally have trouble with inferential procedures, not the descriptive ones.

WHAT SORT OF COMPUTER SHOULD I USE?

As mentioned above, the programs in this book can all be run on very small computers, even those costing under $300. They've all been tested on a TRS-80 computer made by Radio Shack (Fig. 1-1), a Commodore VIC-20 (Fig. 1-2), an Apple Computer (Fig. 1-3), and the Sharp PC-1500 Pocket Computer (Fig. 1-4). These are all tiny computers by commercial standards. The programs also can be run on many large computers that offer timesharing and the BASIC language. Of course, most people don't have access to a very large computer, so there isn't any dilemma. Assume for the moment, however, that one has the choice between working the problems on a large computer, or a tiny one. Which should you choose?

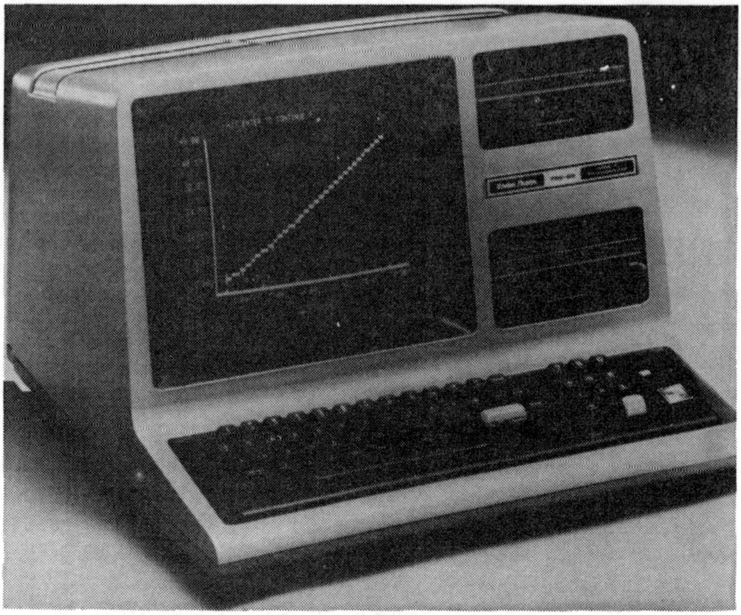

Fig. 1-1. The TRS-80 computer (courtesy Radio Shack).

A few years ago, I would unhesitatingly have said that the student should use a large computer. Large computers have almost limitless advantages: They have many, many languages available, they have awesome amounts of memory, they have every conceivable peripheral device connected to them—printer, plotters, disks, tapes, and on and on. Large computers are hundreds, thousands of times faster than microcomputers. They can solve problems of staggering complexity. They offer almost infinite flexibility. Why

Fig. 1-2. The Commodore VIC-20 computer.

Fig. 1-3. The Apple computer (courtesy Apple).

would anyone want to use a personal computer with the power of a large computer at his command?

One of the chief advantages of a small computer is convenience. For the past week, I've been trying to make contact with the large computer at our University, and all I've gotten are busy signals. Large computers are always busy, or, alternatively, always broken. During the main part of the school year, it is very difficult indeed to get hooked up to the computer. When you finally do get connected, the large computer can be a slow, frustrating beast. One

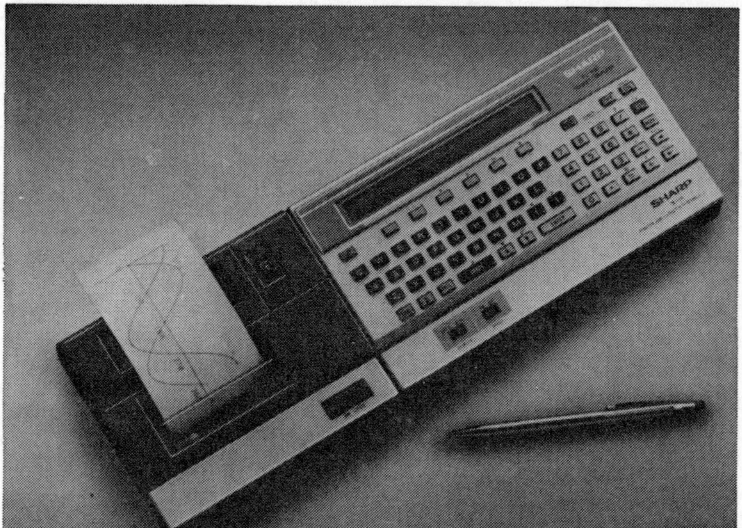

Fig. 1-4. The Sharp PC-1500 computer (courtesy Sharp).

sends a message to the computer, and the device thinks it over for a long time before answering. A friend of mine who used very, very large computers tells me that it frequently takes 30 minutes to get an answer to even the simplest question. Furthermore, large computers are very, very expensive to operate. In many institutions, people can't use the large computer simply because they can't afford it. Computer time for these monster machines often costs hundreds of dollars per hour, and does not include telephone costs, tape-mounting costs, and so on.

Microcomputers have none of these disadvantages. The operator shares them with no one else; they give him their undivided attention, and they're cheap. Frequently, a complete microcomputer costs less than the terminal used to talk to the large computer. Also, the microcomputer is not without virtues of its own. Modern microcomputers are reasonably fast machines, especially for tiny programs such as the ones in this book. They don't seem to break as often as the larger machines. While they don't have the memory that is available in the large computers, most microcomputers still have more than enough memory for most purposes. You really don't have to use 500,000 words of memory for every problem! Most computer languages are available for the more popular microcomputers.

Personally, I have a choice between virtually unlimited time on a large computer, and my microcomputer. Which do I choose? More and more, I find myself using the micro; I'm beginning to wonder whether I'll use the big computer at all once I get all of my programs converted. There's no hassle with the microcomputer, no system crashes (except the ones I make myself), and no people playing computer games while I'm trying to get useful things done.

At the same time, if you have access to a decent, large computer, all of the programs in this book will run just as well on it. Of course, certain very large problems can be solved only on a large, number-crunching computer, but a beginner doesn't need all of that power.

WHY BASIC?

Basic is the computer language used in this book, because it's the language that is supplied with most personal computers. Most of them have Microsoft BASIC, or its equivalent, built in. Believe it or not, the BASIC on many microcomputers is far superior to that found on many large computers.

I suspect that BASIC is as popular a computing language today as FORTRAN was a few years ago. Everyone who uses computers knows a little BASIC, and every computer large or small has BASIC available. You'll find that there are perhaps hundreds of other computing languages: PIL, APL, PL/I, Pascal, COBAL, ALGOL, SNOBOL, and on and on. Be forewarned: it is rather risky to learn any of these languages without knowing BASIC or FORTRAN because they are not universally available and because of certain limitations in them. Of course, all computing languages in the scientific area do more or less the same job. FORTRAN, for example, can do most of the same things that BASIC can do, and it's usually faster. But you can't get it on your personal computer without spending a lot of money.

The first computer language that I learned was MAD. One of my instructors told me that I'd regret this choice for a first language. He was right! It's lonely if you're the only person around who knows an obscure language. Worse than that, obscure languages have a way of disappearing from the face of the earth forever. When the University changed computers, my hard-learned first language was no more. The people who had learned the more common language had a tremendous advantage. The language that they learned, FORTRAN, is still going strong, while my first language is virtually forgotten.

BASIC was created at Dartmouth College in the middle 1960s by Professors John Kemeny and Thomas Kurtz. The people at Dartmouth had at that time what was a totally novel approach: they wanted everyone on campus to know computers and to use them. Computers at other institutions were elitist devices strictly limited to the most famous researchers. They were thought far too sophisticated and expensive for ordinary people.

Consequently, Kemeny and Kurtz tried to make up a language that would be friendly to the user and could be learned very, very rapidly. BASIC stands for Beginners' All purpose Symbolic Instruction Code. BASIC's creators succeeded admirably. BASIC has remained THE language for friendly computers since that time.

You can solve almost any problem with BASIC. The language is capable of solving awesomely large problems, however, most BASICs are rather slow in solving problems because they're what is called *interpreters*. Each program line must be interpreted by the main computer program each time it is encountered. Other languages, such as FORTRAN, use *compilers* to convert the program statements to the machine's own language before they do anything else. Therefore, the programs run much faster. Most people find compilers intimidating, and, frankly, for most problems, interpreters work just fine. After all, your microcomputer does about a million things a second. If speed of operation is really important BASIC compilers are available. BASIC will solve simple problems, such as those found in this book, in the twinkling of an eye.

THE VERSATILITY OF THE MICROCOMPUTER

The usefulness of the microcomputer is evident in the writing of this book. The book was written on my TRS-80 using a word processor. Almost all of the mathematical illustrations were made with the TRS-80 computer, a Radio Shack digital plotter, and an Epson printer. The statistical tables in the Appendices were calculated by the TRS-80. People who don't take small computers seriously are often surprised to find that a small computer could carry out these extremely complex computations at all, much less with great accuracy. In all, my micro saved me thousands of hours of work.

All of these tasks could have been carried out by any of the small computers listed earlier. Extra memory and other peripheral devices might have been needed on the other machines. The TRS-80 was used simply because I had added more features to it than to any of the others. Oddly enough, perhaps the most useful of the computers at my disposal was my tiny Sharp pocket-sized computer. I carried it with me everywhere for a few months, and almost all of the programs in the book were developed on it. You really can't carry a TRS-80 or an Apple to a faculty meeting, and peck away on it without notice, but you can with a pocket-sized computer.

I'M JUST AN AVERAGE COMPUTER

2
The Average

Everyone is familiar with averages. Sports fans continually calculate batting averages, students compute grade point averages, the government publishes yearly information on the average income of workers, and so on. Therefore, this statistic makes an excellent starting place. The sort of average that you're already familiar with has a technical name: it's called the *Mean.* It is the most widely used of the many types of averages in statistics and is by far the best, and so we'll limit ourselves to this one average. It makes an excellent example for use in teaching elementary programming.

Assume that we have the following IQ scores on a group of students:

SCORE
110
120
150
180

I'll bet you already know how to calculate the Average from this set of data. One simply adds up all the numbers, and then divides by the number of numbers. That is, it's 110+120+150+180=560. Then, 560 divided by 4 yields 140, which is the average IQ of that group of rather bright students. Students generally learn to calculate the Average in grade school.

Statisticians have an elegant way of writing the formula for the Average or Mean.

$$\overline{X} = \frac{\Sigma X}{N} \qquad \textbf{2-1}$$

To read this formula aloud, simply say "X-bar equals the sum of the Xs divided by N."

The X with the line over it, called X-bar is the statistical symbol for the Average. Statisticians often use many widely different symbols for the same thing, but they're pretty consistent about using X to denote the Average.

ΣX, otherwise known as sigma-X simply means add up all the Xs. The symbol X in mathematics always means an unknown, a mythical value that isn't known until actual values are plugged into a formula. In the little problem above, the Xs are 110, 120, 150, and 180. Now, in statistics, again and again we need a term to indicate that a set of numbers has to be added up. In the old days, several decades ago, people used to write

$$SX$$

to mean add up all the values. S meant sum. The problem with this system was simply that it was confusing. S and X are both letters of the alphabet, and many students got confused by thinking that S was also a variable and that SX meant "multiply S by X." Consequently, S was replaced by sigma.

$$\Sigma X$$

Σ is the Greek letter sigma, which corresponds directly to the letter S. But somehow ΣX looks a great deal less confusing than SX.

Finally, N stands for number, or the number of things in the Average. Statisticians are rather consistent in using N to refer to the number of things, I think that in all the hundreds of statistics books that I've read, I've come across only one book that didn't use N for that purpose. Unfortunately, that was one of the better books!

In short, ΣX/N simply means add up all the numbers and divide by the number of numbers. But you already knew that! When put in the form of a formula, even the simplest things become confusing to many people. Fortunately, most people have little difficulty in understanding the formula for the Average once they figure out what sigma means.

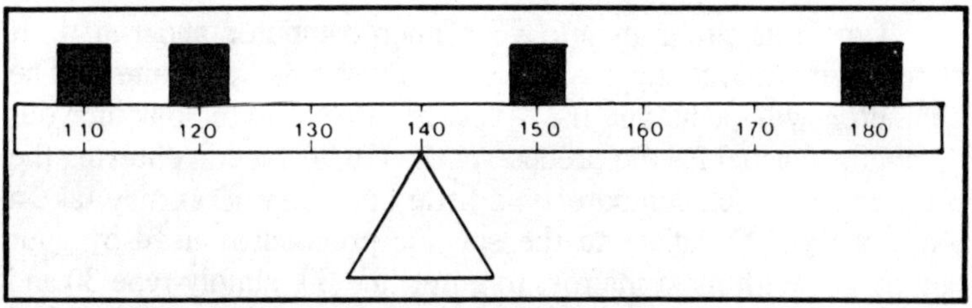

Fig. 2-1. The Average as the point of balance.

WHAT IS THE AVERAGE?

Everyone knows that the Average is sort of the center point or the focal point for a set of scores. Thus, in our group of students the person in the most typical position has an IQ score of 140. Thus, the Average provides a nice, short way of describing the data. Everyone knows that if a group of people have an average score of 140, they're a very bright group of people.

There is another interpretation of the Average or Mean that is often given. The Mean is roughly equivalent to the concept of center of gravity in physics. Assume that you have four one pound weights, each one of which will represent the scores of 110, 120, 150, and 180. Further, assume that these weights are placed on a weightless board so that they're spaced properly, with the distance corresponding to their IQ size. This arrangement is shown in Fig. 2-1.

Where will the board balance? Well, it will balance precisely at 140. The Average or Mean, then, corresponds point of balance in physics, or the center of gravity, as some call it.

The Mean is by far the most useful of the many, many Averages that statisticians have developed. Now, how can you calculate it with a microcomputer?

COMPUTER CALCULATION OF THE MEAN

There are many ways of calculating the Mean on your computer. As we go through several of them, you'll learn a great deal about how to use BASIC. Be sure to work all the examples on your computer. They're all short, and each one teaches you something that you'll need for the next one.

Probably the simplest method, and the least useful, is that of doing it more or less directly, as shown:

```
10 REM PROGRAM 2.1. SIMPLEST AVERAGE PROGRAM
20 S=110+120+150+180           :REM GET SUM
30 A=S/4                       :REM GET AVERAGE
40 PRINT A                     :REM PRINT AVERAGE
```

Type this program into your microcomputer and run it. It should run without any conversion on most small computers. The computer will print out the Average. Be sure to look in your computer manual for the precise method to be used in entering the program, for each computer is a little bit different. It may take a while for you to adjust to the specific procedures used by your computer. With most micros, to enter line 30, simply type 30 and A=S/4, and then press the Return key on the keyboard. Any program line in BASIC must begin with a number, called a line number. Before going any further, look at your computer manual to discover the proper way of getting programs into the computer. If you don't learn this now, you'll be hopelessly lost in the pages to come.

Figure 2-2 is a diagram of the program. Many people find flowcharts useful, and thus I will provide charts of this type

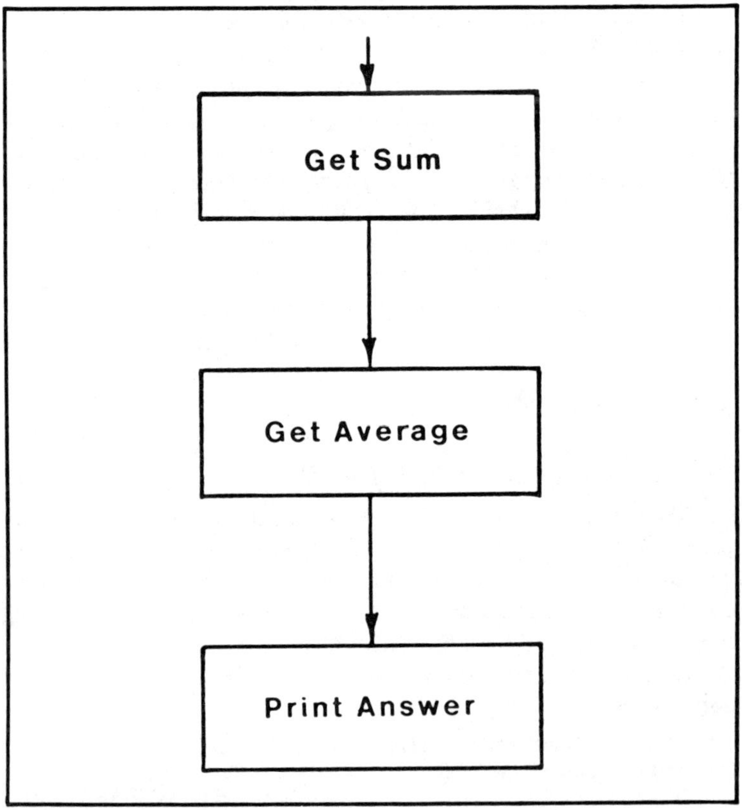

Fig. 2-2. Simple average flowchart.

whenever possible for the programs in this book. There are many, many sets of symbols which are useful in drawing flowcharts. I like them to be as simple as possible. You'll find no special symbols in mine. Also, they will be short. I've seen flowcharts that were twenty pages long! One tends to get lost in diagrams of that length.

Let's go through the program in detail. Line 10 says:

```
10 REM PROGRAM 2.1. SIMPLEST AVERAGE PROGRAM
```

Line 10 is what is called a remark line. Whenever BASIC sees the letters REM in a statement, it simply ignores the remainder of the statement. In short, Line 10 is merely a note to help the programmer remember the purpose of the program. As soon as BASIC encounters the REM it jumps to the next line. Actually, it isn't necessary for you to type the REM line at all in such a short program. But every program that I list will start out with just such a line so that I can identify it.

Line 20 gets the sum of the numbers. It says:

```
20 S=110+120+150+180     :REM GET SUM
```

What happens in Line 20 is the following; the computer adds up 110, 120, 150, and 180, and then puts the sum into a space called S. Thus, S will be 560 in this problem. I used S to stand for sum, but you could use any letter that you wanted. Thus, the sum of the numbers has been obtained in line 20, and put into variable S.

What about REM GET SUM? What does the rest of the line mean? That's easy. That's another remark. When BASIC sees a colon (:) in a program line, it assumes that another instruction is coming. You can put more than one statement on a line in most BASICS; that's not the case in many other computer languages. Then, right after the colon, REM GET SUM follows. This is another remark, just like the one in Line 10. Remarks of this sort are useful in remembering what a program line means at a later date. Right at the moment you probably have no difficulty in understanding what the line means, but a year from now, unless it has a comment on it, you'll probably forget its purpose completely. However, as a student with the book in front of you, you really don't have to put in all the remarks in short programs such as this one. At any rate, the computer will ignore REM GET SUM.

The next line gets the Average.

```
30 A=S/4     :REM GET AVERAGE
```

The sum, S, is divided by 4, and the result placed in variable A (/ means divided by). The sum of the numbers divided by the number or numbers is the Average! Therefore A is the Average. We now have the Average. Keep in mind that :REM GET AVERAGE is a comment and will be ignored by the computer.

The last line prints the average on the screen.

```
40 PRINT A      :REM PRINT AVERAGE
```

You could make things a bit fancier by changing line 40.

```
40  PRINT "THE AVERAGE IS",A   :REM PRINT AVG
```

Try it. You'll find that the letters between the double quotes are reproduced exactly, and then the value for A is printed out.

You see, you have successfully calculated an Average! It's not at all difficult. In fact, you could get the whole thing on one program line, as shown:

```
10   PRINT (110+120+150+180)/4
```

Try it. It works. It's a little harder to explain such a terse statement, however, and consequently we'll continue to present programs in as much detail as possible. On most computers you can simply type the following statement:

```
PRINT   (110+120+150+180)/4
```

The computer will display the answer on the screen without the need for a program at all. This is known as the *direct mode*, and is much like a calculator.

ANOTHER DIFFICULT WAY TO DO THE AVERAGE

The trouble with the preceding program is that it is very inflexible. It can calculate the Average on only four numbers; there's no easy way to change a program based on that scheme so that it will handle a group of any size. To get around that limitation, I'll present another program with precisely the same limitation! This program has possibilities for expansion:

```
10 REM PROGRAM 2.2. THE AVERAGE THE HARD WAY
20 S=0                          :REM CLEAR LOCATION
30 S=S+110                      :REM CUMULATE SUM
```

```
40 S=S+120                          :REM CUMULATE SUM
50 S=S+150                          :REM CUMULATE SUM
60 S=S+180                          :REM CUMULATE SUM
70 A=S/4                            :REM FIND AVERAGE
80 PRINT "THE AVERAGE IS",A :REM PRINT RESULT
```

Run the program. It yields the Average. This program looks a great deal more complex than the last program, but it really isn't. Figure 2-3 will help you to understand the program.

In line 20, variable S is set equal to zero. This is the location that will be used to build up the sum of the numbers. In actual practice, most BASICs will clear all the variables before you use them the first time, but I'll show this step explicitly for a while. My Sharp PC-1500 pocket-sized computer doesn't clear all the variables, and a special clear statement (5 CLEAR) must be added at the beginning of programs on that machine.

Line 30 says, simply, "Take the present value of variable S, add 110 to it, and put it back in the storage location for variable S." Before line 30 is executed, I know that S is 0. In line 30, then 110 is added to the 0 in S, and the sum put back as variable S. At the end of line 30, S will then be 110.

Why is this sort of statement confusing to many people? It's simply because mathematically, the statement doesn't make sense. You have to be obsessive to worry about things like this, but you see, in point of fact, S does not equal S+110! Obviously, S and S+110 are completely separate quantities. Fortunately, most people are not that picky and are able to accept the idea that S=S+110 means add 110 and S and put it back in S. I'll bet most readers weren't even aware that there was a problem.

Line 40 follows the same scheme as line 30. The value of S (110 at the beginning of the statement) is added to 120, and the result placed back in the space for variable S. At the end of line 40, then, S will have cumulated to 230. Line 50 does the same thing, adding 150 to the value of S, making 380. Line 60 adds 180 to the sum, making a grand total in of 560 in S. This is the same total that we got in the previous examples. Finally, line 70 calculates the Average, and line 80 prints it out.

"My heavens,"you say. "What a silly system. This program is much more complex than the last one, yet it does the same job. There is no advantage to such a complex method of doing things."

Ah, but there is! Look at Program 2-2 closely. Notice that lines 30 through 60 are all identical, except for the numbers. If we could figure out some way to change the new number to be added into the

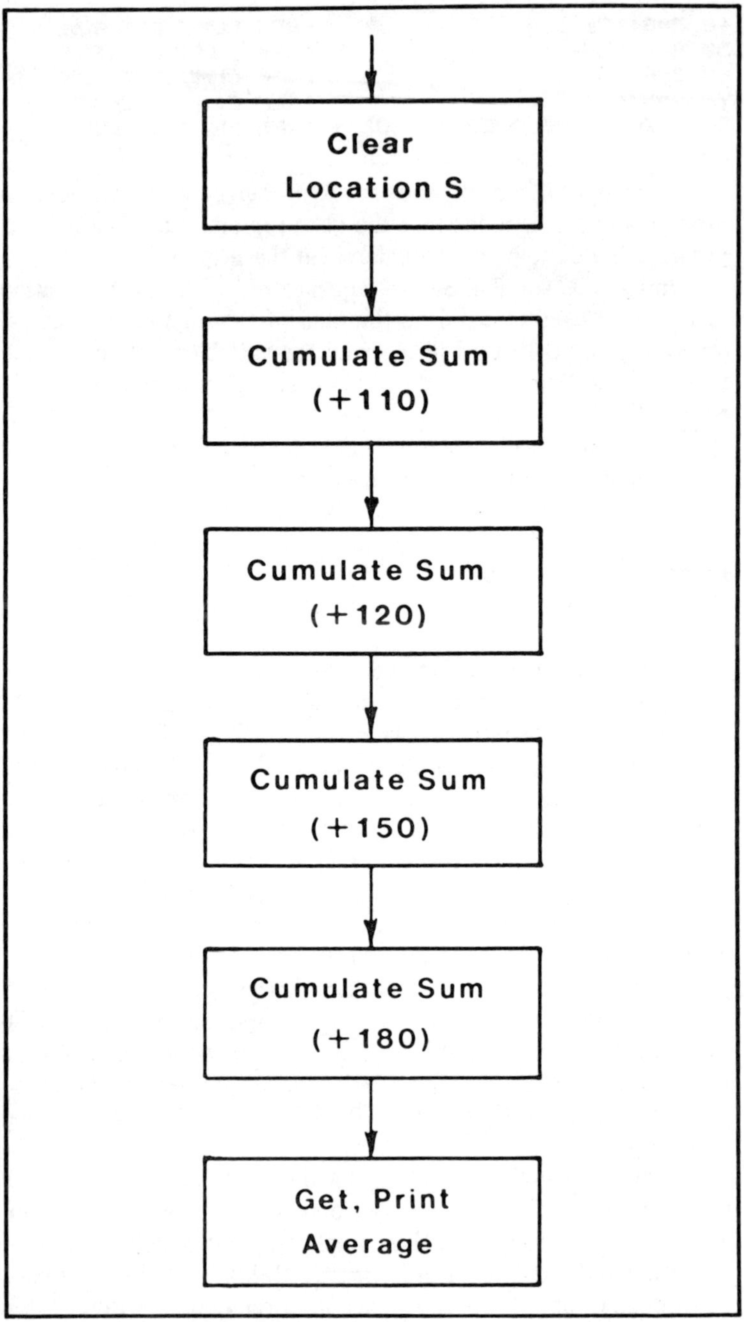

Fig. 2-3. Harder average flowchart.

Average each time, the statement "S=S+" could be used over and over again. Computer experts simply love situations in which programming elements are repeated. Such concepts lead to great savings in the amount of computer memory needed.

As an experiment, change line 80 to the following:

80 PRINT "SUM IS",S,"AVERAGE IS",A

The program will tell you both the Sum and the Average. One of the nice things about writing your own computer programs is the opportunity to change the output in any way that you want. It's not easy to do that with the canned or prewritten programs used by so many statisticians.

A STILL HARDER WAY TO DO THE AVERAGE!

The following program is closely related to the last one:

```
10 REM PROGRAM 2.3. AVERAGE WITH "READ"
20 S=0                          :REM CLEAR LOCATION
30 READ X                       :REM GET DATA
40 S=S+X                        :REM CUMULATE SUM
50 READ X                       :REM GET DATA
60 S=S+X                        :REM CUMULATE SUM
70 READ X                       :REM GET DATA
80 S=S+X                        :REM CUMULATE SUM
90 READ X                       :REM GET DATA
100 S=S+X                       :REM CUMULATE SUM
110 A=S/4                       :REM GET AVERAGE
120 PRINT "THE AVERAGE IS",A:REM PRINT RESULT
130 DATA 110,120,150,180        :REM DATA TO USE
```

This is basically the same program as the last one, although it doesn't look like it (Fig. 2-4). The new concept in the program is the READ statement.

30 READ X

When the computer encounters a READ, it jumps to the line starting with DATA. Then it picks up a value from the DATA statement.

130 DATA 110,120,150,180

When the first READ X statement occurs at line 30, the variable X will be set equal to 110. Then in line 40, just as in the last program, X will be added to S.

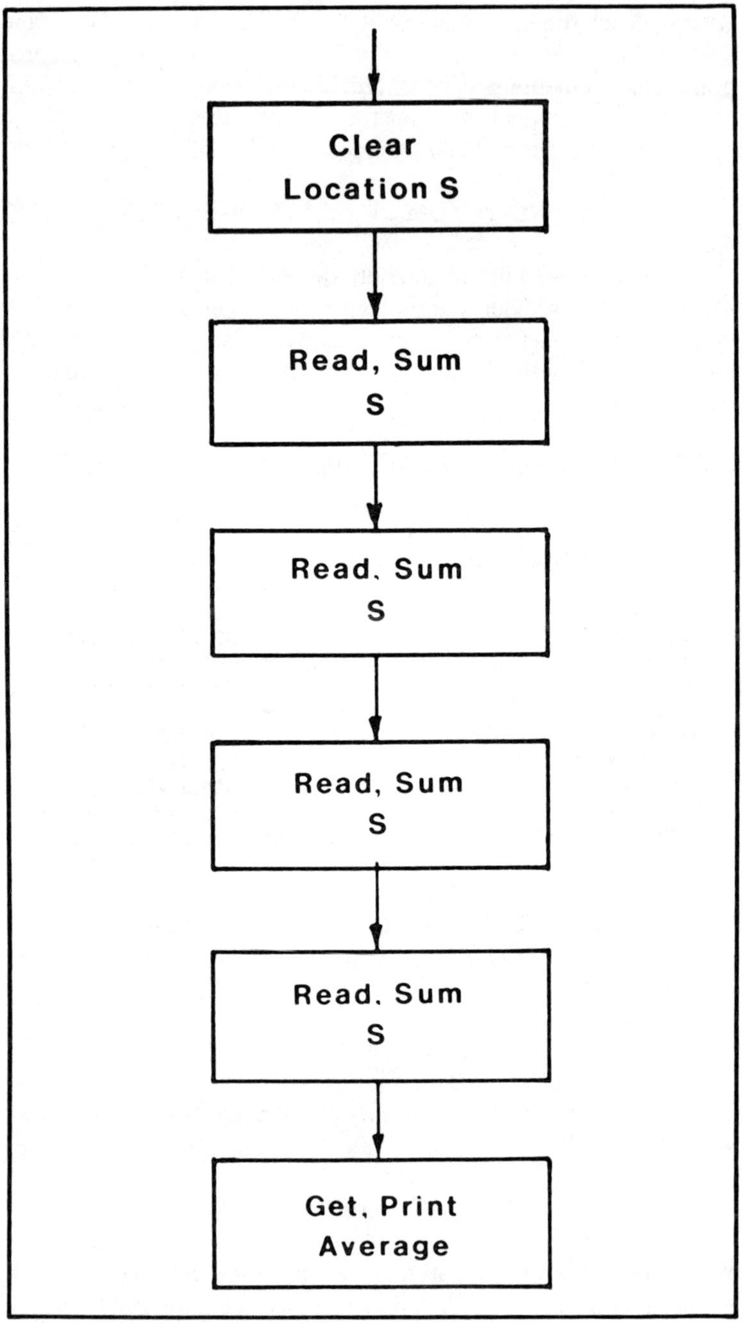

Fig. 2-4. Average with read flowchart.

In short, READ X means "go to the DATA statement, get the next value, and put it in X." It's not really all that complicated. Thus, in lines 50 and 60, I repeat the procedure.

```
50  READ X
60  S=S+X
```

When line 50 is encountered, the computer will return to the DATA statement, get the next bit of data, and put it into X. That next bit of data happens to be 120. Then, in line 60, 120 will be added to S, just as in the last program. The value of S will jump from 0 to 110 to 230 to 380 to 560 as we go through the S=S+X statements..The elements of the DATA statement are separated from each other by commas, by the way. The commas are usually ignored by the computer. This means, of course, that you shouldn't put commas in your numbers. Don't type 1,000,000 into the computer to indicate a million. The computer won't understand you. Type 1000000 and all will be well.

Lines 110 and 120 calculate and print the Average, just as in the last program.

By this point, you may be very discouraged. The programs get longer and longer, and it seems as though the Average is getting harder and harder to calculate. They'll get shorter soon.

Program 2-3 is important because, even more so than the last program, it uses the same elements over and over again. The combination

```
READ X
S=S+X
```

occurs many times. We can simplify Program 2-3 a great deal and make it usable with any sized group by figuring out a way of repeating those two statements automatically. That's what will happen in the next program!

THE AVERAGE WITH AN IF STATEMENT

The problem with the last program is the necessity to repeat the instructions READ X and S=S+X as many times as there are numbers to be averaged. What is needed is a simple procedure that will let us use the same two statements over and over again. Such a procedure would be much more efficient. There are many ways to do this, but Program 2-4 shows one that I like and have often used.

```
10 REM PROGRAM 2.4 AVERAGE WITH "IF" LOOP
20 S=0                            :REM CLEAR LOCATION
30 READ X                         :REM GET DATA
40 IF X=-1 THEN 80                :REM TEST FOR END
50 S=S+X                          :REM CUMULATE SUM
60 N=N+1                          :REM CUMULATE COUNT
70 GOTO 30                        :REM CONTINUE
80 A=S/N                          :REM GET AVERAGE
90 PRINT "THE AVERAGE IS",A       :REM PRINT RESULT
100 DATA 110,120,150,180,-1       :REM DATA TO USE
```

Type in this program and run it. It will give you the answer of 140. Now, change line 100 so that it reads:

```
100   DATA 100,110,120,130,140,150,160,170,180,-1
```

Now, there are nine data points instead of four. If you run the program, you find that you again get the Average. However, now there is a great deal more data. That's okay, because the main body of the program hasn't changed at all. As a matter of fact, this system will handle almost any amount of data with no changes between lines 10 and 90. All that we have to change is what's in the DATA statement, adding more DATA statements if necessary. This is obviously the solution we've been looking for.

How does the program work? It's very much like the last one, but there are a couple of new concepts (Fig. 2-5). We'll go through it line by line.

Line 20 clears the location for the sum, as always. As I've mentioned before, we don't really need that statement because most BASICs clear all variables before program execution begins. Shortly, I'll begin to omit this statement.

Line 30 states "READ X." This statement gets a piece of data from the DATA statement. The first time through READ X takes 110 from the data statement and puts it into the variable X. We've had that before.

Line 40 is new. It says:

```
40   IF X=−1 THEN 80
```

The program uses the statements READ X and S=S+X over and over again. Consequently, it must know when to stop adding in new data. In this example, I've elected to signal to the computer that all the data is used when it encounters the value −1 (minus one). The IF-THEN statement in plain English means "whenever X is found to

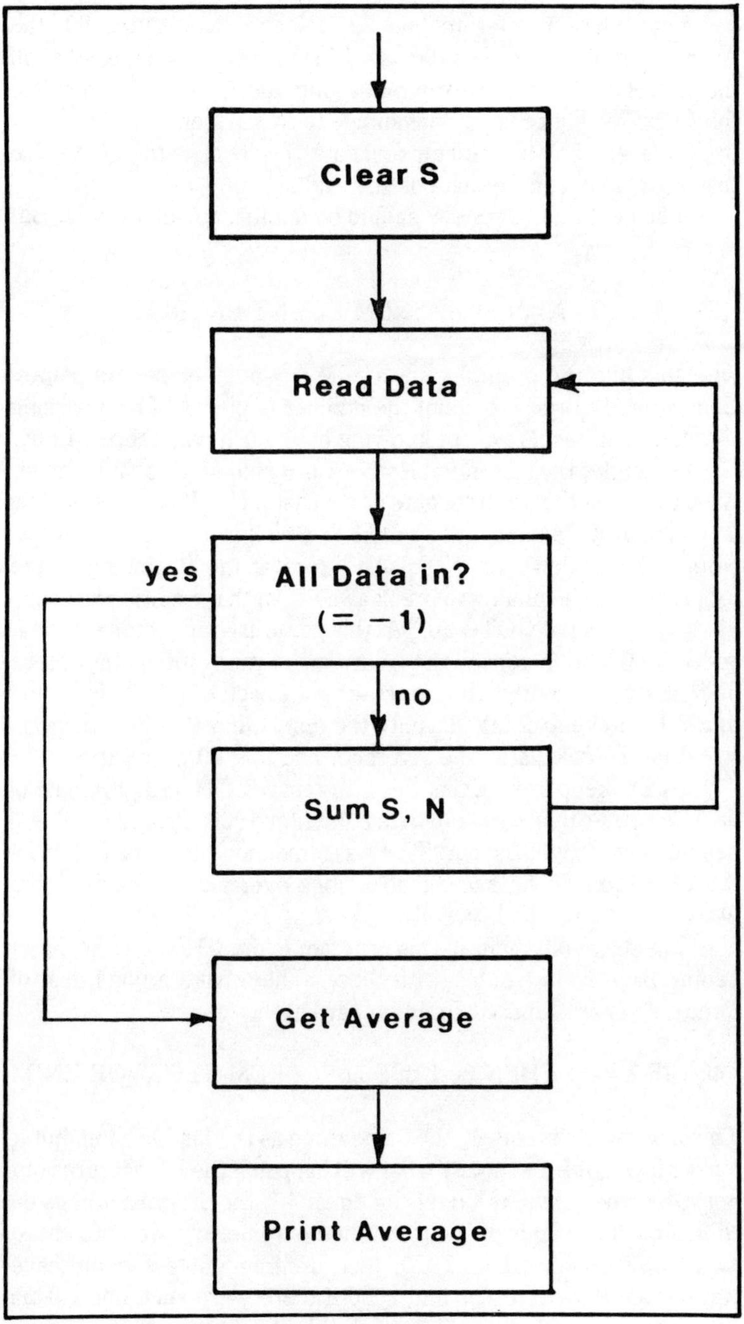

Fig. 2-5. Average with IF statement flowchart.

be minus one, transfer to line 80." Notice that in line 80, the Average is calculated. In short, a −1 in the data is the signal that all the data is in and that the program should then proceed to calculate the Average. There is a disadvantage to this system, of course, for the value −1 can't be used for real data. The trick is to pick a value that won't occur in the data as the end signal.

The next line, S=S+X, should be familiar to you by now. But line 60 is new.

```
60  N=N+1      :REM COUNT ENTRIES
```

with this line the program can handle any number of data points. Consequently, it has to count the number of pieces of data passing through it; it has no way of knowing how much data there will be. Consequently, the statement N=N+1 is included to count the data. Each time that the program gets a bit of data, N will be increased by 1. When all the data is in, N will be 4, which is just what we want. Notice that I didn't set N equal to zero at the beginning of the program; I am beginning to break away from that tradition already.

Line 70 says "GOTO 30." At this point, having gotten the data, added it up, and increased the count, we're ready for another piece of data. Consequently, the computer goes back to line 30 for some more. It will keep doing this until the test in line 40 tells it to stop.

Line 80 calculates the Average, and line 90 prints it out.

Again, keep in mind that this program will handle any amount of data. Whether there are 100 data points, or 1000, Program 2-4 will handle it just as well as four. That's a tremendous advantage. It is by the repetition of the same instructions over and over again that makes computers so efficient.

The new statement in this program is the IF statement. Most people have little trouble with these. They make a good deal of intuitive sense. Line 40 could be rewritten:

```
40  IF X=-1 THEN 80 ELSE 50      REM TEST FOR END
```

This line means precisely the same thing as the last version, but it makes it possible to specify what will happen if the IF test turns out not to be true. Thus, if X does not equal −1, the program will go on to line 50. Line 50 happens to be the next line, but we could have said ELSE 100 or ELSE 1000; then program control would have transferred to line 100 or line 1000 if there were such lines. This adds greatly to the versatility of the IF statement.

You can also test other relationships with the IF statement. For example, various other symbols can be placed in the line:

Symbol	Meaning
=	equal to
>	greater than
<	less than
< >	not equal to
<=	equal to or less than
=>	equal to or greater than

There are many other things that can be put in an IF statement. Now would be a good time to look up the IF statement in your computer's BASIC manual. It's one of the most useful statements in the language.

As mentioned above, this method of calculating the Average is my favorite. It's very efficient with respect to memory space, and is fairly trouble free. The principal disadvantage, of course, is the fact that one data value (−1) cannot be used for data since it is the one used to test for the end of the data. Usually, care in the choice of this value (it can be any number you like) solves the problem.

THE AVERAGE USING A FOR-NEXT LOOP

I don't like this method as much as the last one, but it demonstrates an important programming concept: the FOR-NEXT loop. You'll use these loops all the time in your programming. A program using this method looks like this:

```
10 REM PROGRAM 2.5. AVERAGE WITH FOR-NEXT LOOP
20 S=0                          :REM CLEAR LOCATION
30 READ N                       :REM GET GROUP SIZE
40 FOR I=1 TO N                 :REM START LOOP
50 READ X                       :REM GET DATA
60 S=S+X                        :REM CUMULATE SUM
70 NEXT I                       :REM END LOOP
80 A=S/N                        :REM GET AVERAGE
90 PRINT "THE AVERAGE IS",A     :REM PRINT RESULT
100 DATA 4,110,120,150,180      :REM DATA TO USE
```

This program looks a lot like Program 2-4. Run the program; it will provide the correct answer.

Lines 10 and 20 are the same (see Fig. 2-6), but in line 30, there is a difference.

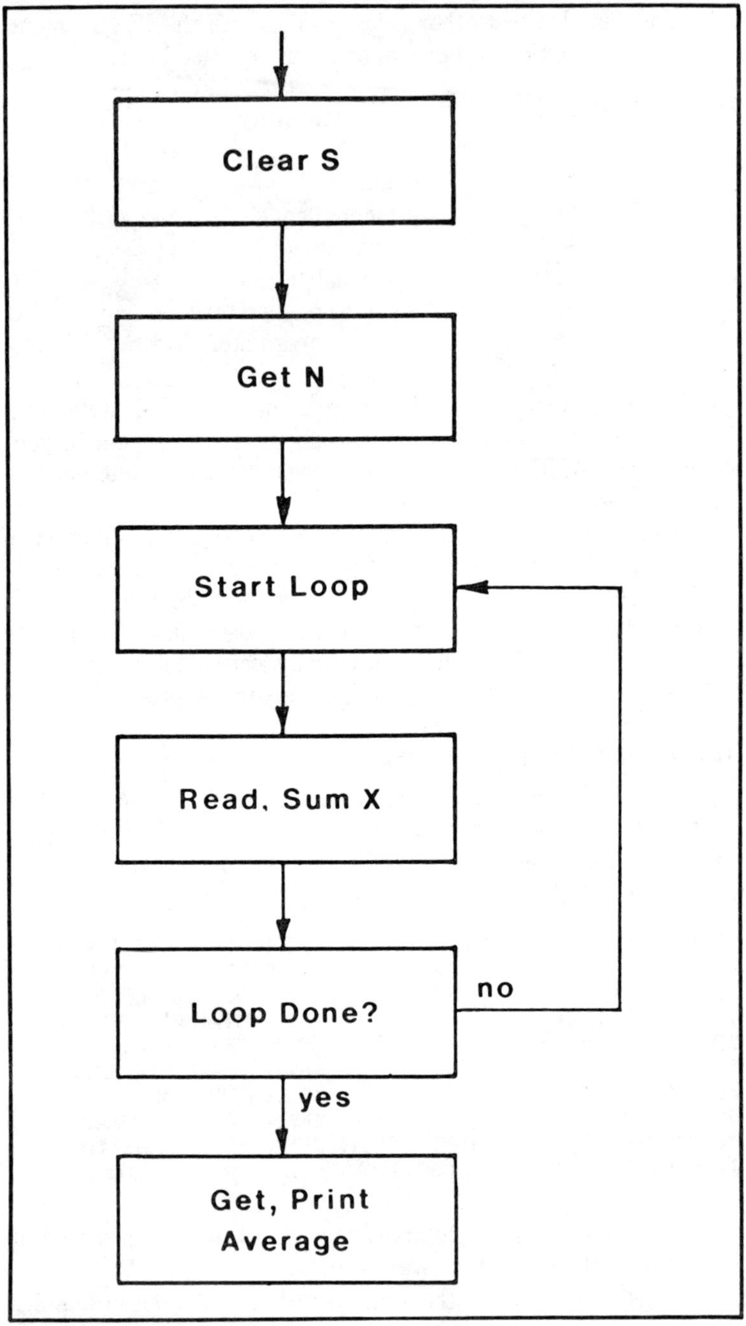

Fig. 2-6. Average with FOR-NEXT loop flowchart.

30 READ N

In Program 2-5, we must know the size of the group before doing any calculations. That's the disadvantage of this method. Consequently, N is read in from the DATA statement in line 30. Notice that in line 100, the first bit of data in the DATA statement is 4, the size of the group.

Lines 40 and 70 comprise the FOR-NEXT loop (detailed in Fig. 2-7).

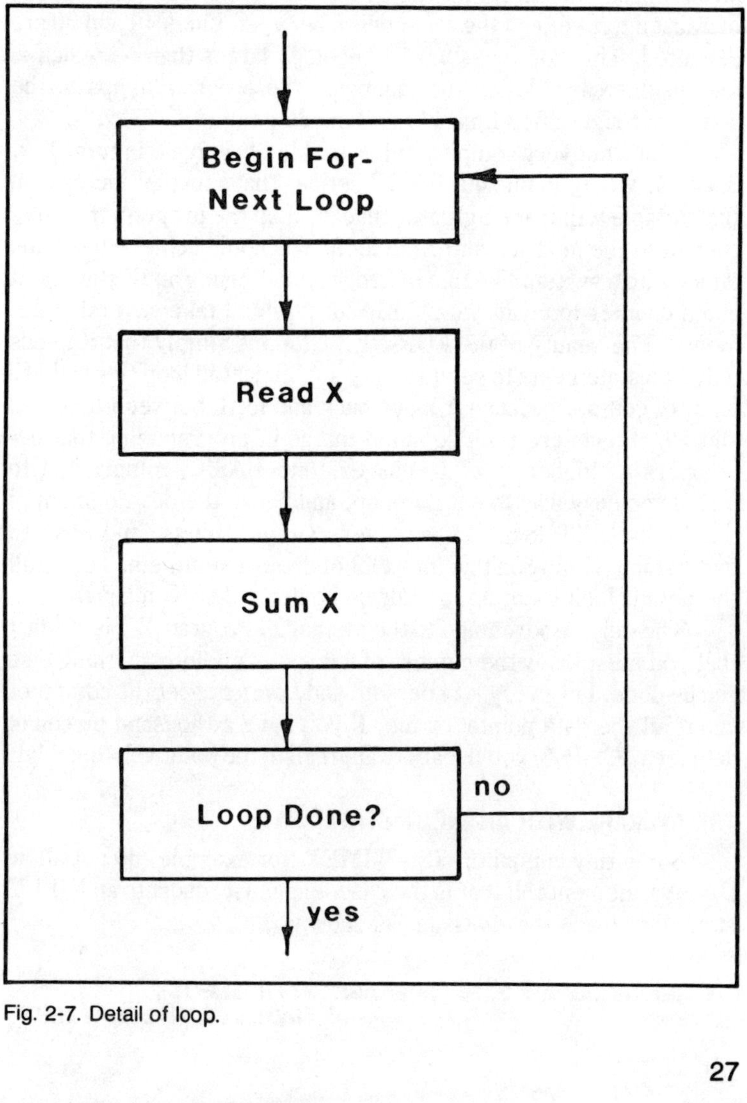

Fig. 2-7. Detail of loop.

```
40  FOR I=1 TO N        :REM START LOOP
    . . .
70  NEXT I              :REM END LOOP
```

FOR-NEXT loops are rather simple things. Lines 40 and 70 simply say, "Do everything between Lines 40 to 70 N times." The operation of the FOR-NEXT loop is fairly simple. The first time through the loop, variable I will be set equal to one, and the statements between lines 40 to 70 executed. When NEXT I in line 70 is encountered, the program jumps back to Line 40, variable I is increased to two, and the statements between Lines 40 and 70 are executed. This continues until I becomes larger than N, which is four in this case. When that happens, the program jumps to the statement right after Line 70, and executes it.

That's not very complicated. Variable I becomes, in turn, 1, 2, 3, and 4. When, in line 40, BASIC notices that a further increase in the variable will make I greater than N, then the program transfers control to the next line after the loop. You could achieve the same effect with a system like that of Program 2-4, using an IF statement and a counter to count variable I, but it would take several statements. The beauty of the FOR-NEXT loop is simply that it needs only two statements to set up a very sophisticated loop (it's called a loop, of course, because it loops back and forth between Lines 40 and 70). Loops are such common things in programming that developers of higher level languages, like BASIC, simply had to include an easy way to set them up, and hence the development of the FOR-NEXT loop. They're very simple things, and easy to understand, if you don't go into a lot of detail about them. To see all the detail, look them up in your computer's BASIC manual.

The only disadvantage to the method of Program 2-5 is the fact that you must know the number of data points before anything else can be done. I'm a very lazy person, and I prefer to let the computer count all the data points for me. If you have a thousand pieces of data, or 1,999,996, you'll learn to appreciate the point very quickly!

THE AVERAGE WITH AN INPUT STATEMENT

Some tiny computers (the TIMEX, for example) do not allow DATA statements. If that is the case, one must resort to an INPUT statement to do the Average, as shown:

```
10 REM PROGRAM 2.6. AVERAGE WITH INPUT
20 S=0                             :REM CLEAR LOCATION
```

```
30 INPUT X                        :REM GET DATA
40 IF X=-1 THEN 80                :REM TEST FOR END
50 S=S+X                          :REM CUMULATE SUM
60 N=N+1                          :REM CUMULATE COUNT
70 GOTO 30                        :REM CONTINUE
80 A=S/N                          :REM GET AVERAGE
90 PRINT "THE AVERAGE IS",A :REM PRINT RESULT
```

The operation of this program is exactly like that of Program 2-4, except for the fact that data is taken in from the keyboard, rather than from DATA statements in the program (see Flowchart 2-7). When the computer wants data, it displays ? on the screen. You type in the data and press Enter when it's all in. (Be sure to enter a −1 when all the data is in.) The disadvantage of this system lies in the fact that you can't edit the numbers as easily as in a DATA statement, but the INPUT statement is very handy in certain circumstances.

SOME OTHER PROPERTIES OF THE AVERAGE

Professional statisticians like to point out certain properties of various statistical procedures. For example, what happens to the Average or mean if you add a constant to all the data points before calculating the Average? That is, if we use the data points 120, 130, 160, and 190, rather than the original 110, 120, 150, and 180, what effect will this have on the Mean? It is possible to prove what will happen mathematically with some rigorous mathematical derivations, but it's much more fun to do it with the computer.

Assuming that we want to add 10 points to each value before calculating the Average, the following variant of Program 2-4 will do the job for you:

```
10 REM PROGRAM 2.7. AVERAGE WITH CONSTANT ADDED
20 S=0                            :REM CLEAR LOCATION
30 READ X                         :REM GET DATA
40 IF X=-1 THEN 80                :REM TEST FOR END
45 X=X+10                         :REM ADD 10
50 S=S+X                          :REM CUMULATE SUM
60 N=N+1                          :REM COUNT ENTRIES
70 GOTO 30                        :REM CONTINUE
80 A=S/N                          :REM GET AVERAGE
90 PRINT "THE AVERAGE IS",A :REM PRINT RESULT
100 DATA 110,120,150,180,-1 :REM DATA TO USE
```

This is precisely the same program as Program 2-4, except for the fact that Line 45 has been added (see Fig. 2-9). This line adds

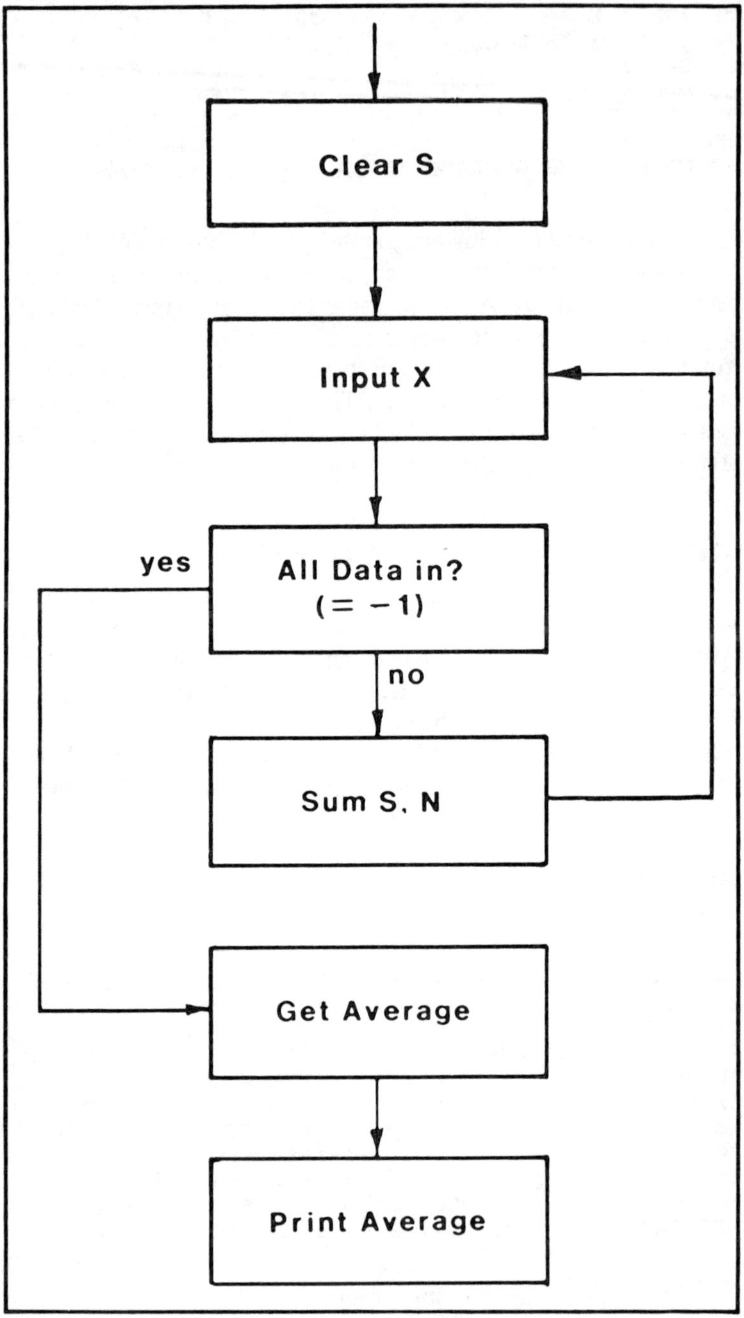

Fig. 2-8. Average with INPUT statement flowchart.

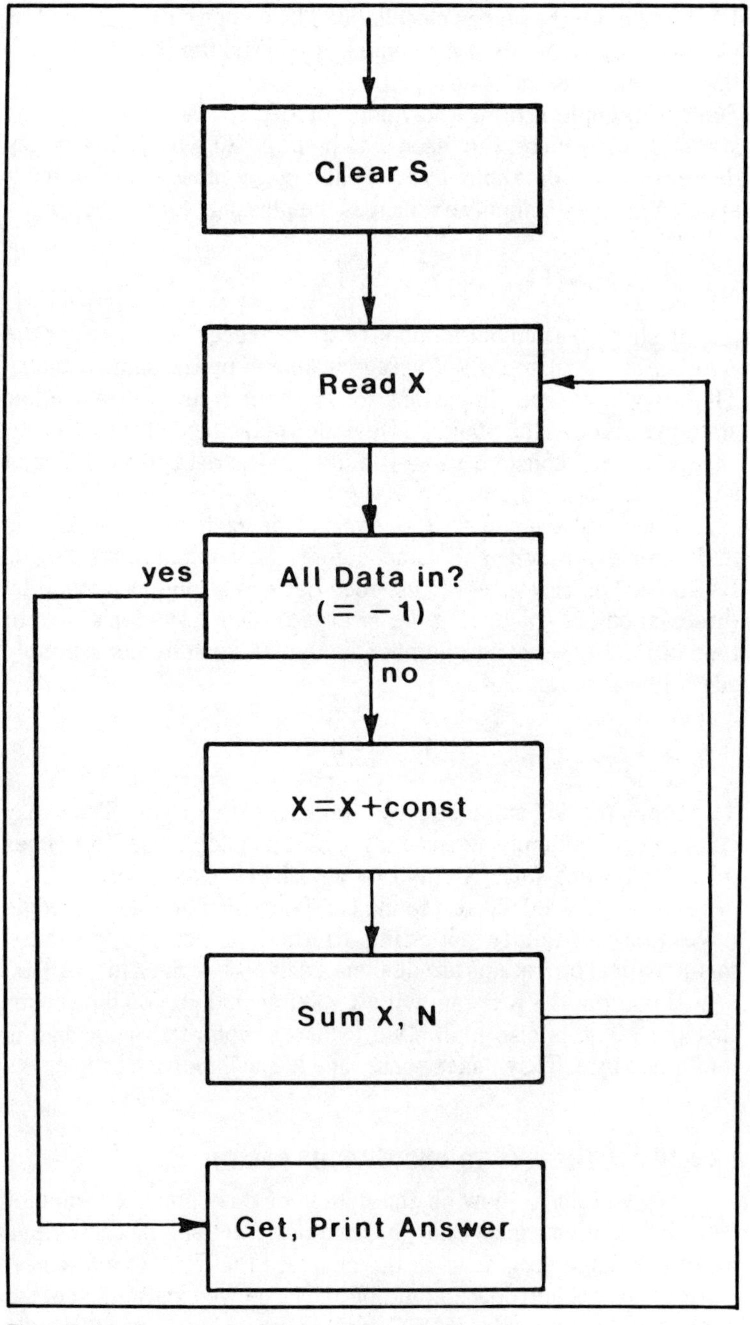

Fig. 2-9. Average with constant added flowchart.

10 to X, before the sum is calculated. Thus, each data point will be increased by 10 before it's summed. If you run the program, you'll find that the Average is now 150, rather than 140, as it was before. See how simple a little knowledge of BASIC can make learning statistics? Try increasing each data point by 20. Or, you could try decreasing each data point by 10. That might confuse you a bit, but it shouldn't. Line 45 simply exchanges the plus sign for a minus sign

```
45  X=X-10
```

In short, if a number is added to all the pieces of data before the Average is calculated, the Average is shifted by the same amount. This may not seem important to you, but it is. Certain other statistics, such as the Standard Deviation of the next chapter, do not change when a constant value is added to the data. Knowledge of such things is important.

What will happen to the Average if each piece of data is multiplied by a constant value before the computations begin? BASIC makes that easy for us, too. Let's multiply each value by three (3) before calculating the Average. Again, we don't have to type in a whole set of new numbers in the DATA statement. Simply add a line Program 2-4.

```
45  X=X*3
```

The computer will automatically alter that data for you. The * may confuse you. It simply means multiply. Most people use X for times or multiply, but since X could be a variable name in BASIC, its developers decided to use * to indicate multiplication. Most people have little difficulty adjusting to this. Thus, the principal mathematical operations are designated by + − * and / in BASIC.

If you run the program with X=X*3 included, you'll discover that the Mean is also multiplied by three when the input data is multiplied by 3. That makes sense, and it's an important property of the Mean.

I KNOW ALL THERE IS TO KNOW ABOUT BASIC?

No, you don't know all the details of this complex computer language. But you know enough, right now, so that you can write a lot of computer programs. In the chapters that follow, a few new concepts will be introduced, but, by and large, you know most of the important ones already. BASIC, as you can see, is a very friendly language. That's probably the reason for its enormous success.

We have used the following BASIC commands:

REM	PRINT	PRINT " "	READ
DATA	IF-THEN	ELSE	GOTO
FOR-NEXT	INPUT		

We know the following BASIC operators:

=	+	/	*	:	
<	>	< >	< =	= >	↑.

In the chapters to follow, I will introduce some new ones, but these few commands will form the framework of all the programs to follow.

This chapter has been a busy one! If you've followed along and worked the problems out on your computer, you now know all about the Average, and, believe it or not, you know most of the underlying concepts of BASIC. The rest of the chapters will be easy!

PROBLEMS

1. Find the Average, using the various programs in this chapter for the IQ scores 100, 115, 130, 145, and 170. (132)

2. Jones Motorcar company makes automobiles with top speeds as follows: 120, 180, 125, and 150. What is the average speed of these cars? (143.75)

3. Acme Detergent is used to wash dishes. A carefully controlled amount of detergent and dishwater washes the following numbers of dishes: 50, 33, 73, 84, 33, and 89. What's the average number of dishes that this product washes? (60.333)

4. Dismal Disk company finds the following number of bad tracks on its disks: 35, 38, 33, 21, and 5. What's the average number of bad tracks? (26.4)

5. Zippy Printer company makes printers which print the following number of characters per second: 80, 95, 150, 75, and 88. What's the Average? (97.6)

HOW DO YOU COMPARE THINGS?
+ − × ÷

3
Measures of Spread or Variability

The Average is the most important statistic used to describe data, but it ignores an important aspect of the information. Assume that there are two people, Person A and Person B. Both of them are given the same IQ test five times. The IQ test is a perfect one; it always gives exactly the right answer (there aren't any tests like that!). The two persons get scores, as follows:

Person A	Person B
102	120
101	110
100	100
99	90
98	80
------	------

Take both sets of numbers and plug them into Program 2-4 of the last chapter. What is the Average in each case?

Both Averages are 100, the same value. Since both people have the same Average, can it be inferred that they really are the same, even though they appear to be different? Obviously not, for while Person A is remarkably consistent in his performance, Person B is so variable that he might be considered to be mentally disturbed. If we plot data like that above, and assuming that the Normal

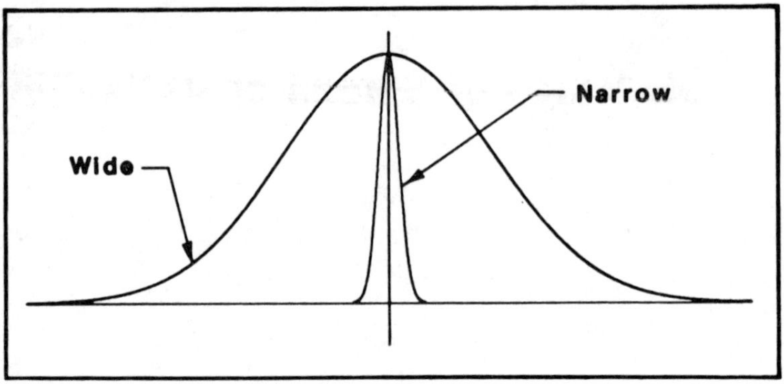

Fig. 3-1. Wide and narrow spreads.

Curve we'll discuss in Chapter 6 fits the data, we get curves like those shown in Fig. 3-1 (plotted to scale with my TRS-80 computer, and with my Radio Shack plotter). From the curves, notice that the scores of Person A don't spread out very far while those of Person B spread out a great deal. Person B is much more *variable* than Person A. It is this *variability* that we don't know how to describe, and it is the purpose of this chapter to detail the methods with which statisticians attack this problem.

THE RANGE

The simplest measure of spread is the *Range*, which is simply the highest score minus the lowest score. The Range is not a very valuable statistic, for it ignores all values except the top and bottom ones. Thus, the following two sets of data have the same Range, but look how different the numbers are!

120	120
80	120
80	120
80	120
80	80
-----	-----

Still, the Range is rather widely used, because it's so simple to calculate. It's even easy to do by computer. Program 3-1 shows how to get the largest number in a set of numbers:

```
10 REM PROGRAM 3.1. BIGGEST VALUE
20 B=0                          :REM INITIAL BIG #
30 READ X                       :REM GET DATA
```

```
40 IF X=-1 THEN 70                  :REM TEST FOR END
50 IF X>B THEN B=X                  :REM SAVE BIG VALUE
60 GOTO 30                          :REM CONTINUE
70 PRINT "BIGGEST VALUE:",B :REM PRINT RESULT
80 DATA 80,90,100,110,120,-1:REM DATA TO USE
```

Run this program. It will find the largest value in the set of numbers. Further, it works with a set of numbers of any size, assuming that they're all greater than zero, and that the series of numbers ends with −1. How does the program work? Look at Fig. 3-2.

Line 20, "B=0" sets variable B to zero. Variable B is eventually going to contain the largest value. Every other value will be compared with B to see whether the new value is larger than B. So I start out by setting it to the smallest possible value, zero. That system will work only if all the other values are greater than zero, obviously.

Line 30 gets the data, as in previous programs, and line 40 checks for the end of the data. See Program 2-4 of the previous chapter if you've forgotten how that works.

Line 50 finds the largest value:

```
50   IF X>B THEN B=X        :REM SAVE BIG VALUE
```

This line says: if the new value, X is larger than (>) the biggest value encountered so far, then take the new value, X, and put it in variable B. Recall that variable B is going to wind up with the largest value, when the program is all through. Thus, at the beginning of the program, B is 0. The first X, 80, is compared with B, and of course 80 is greater than 0. Consequently, according to line 50, B will be changed to 80. That's the largest value so far. (My Sharp PC-1500 needs LET B=X in line 50.)

The next time through the loop, the new value of X will be 90, and B will be 80. Line 50 says that if X is greater than B, that B should take on the value of X. So B now becomes 90. This process continues until all the data is in, and B winds up being 120. B is then the largest value. You can see the process by putting the line

```
55   PRINT B
```

into the program. That will print out all the intermediate values of B, as the program works its way to completion.

Now, that program doesn't yield the Range, only the largest

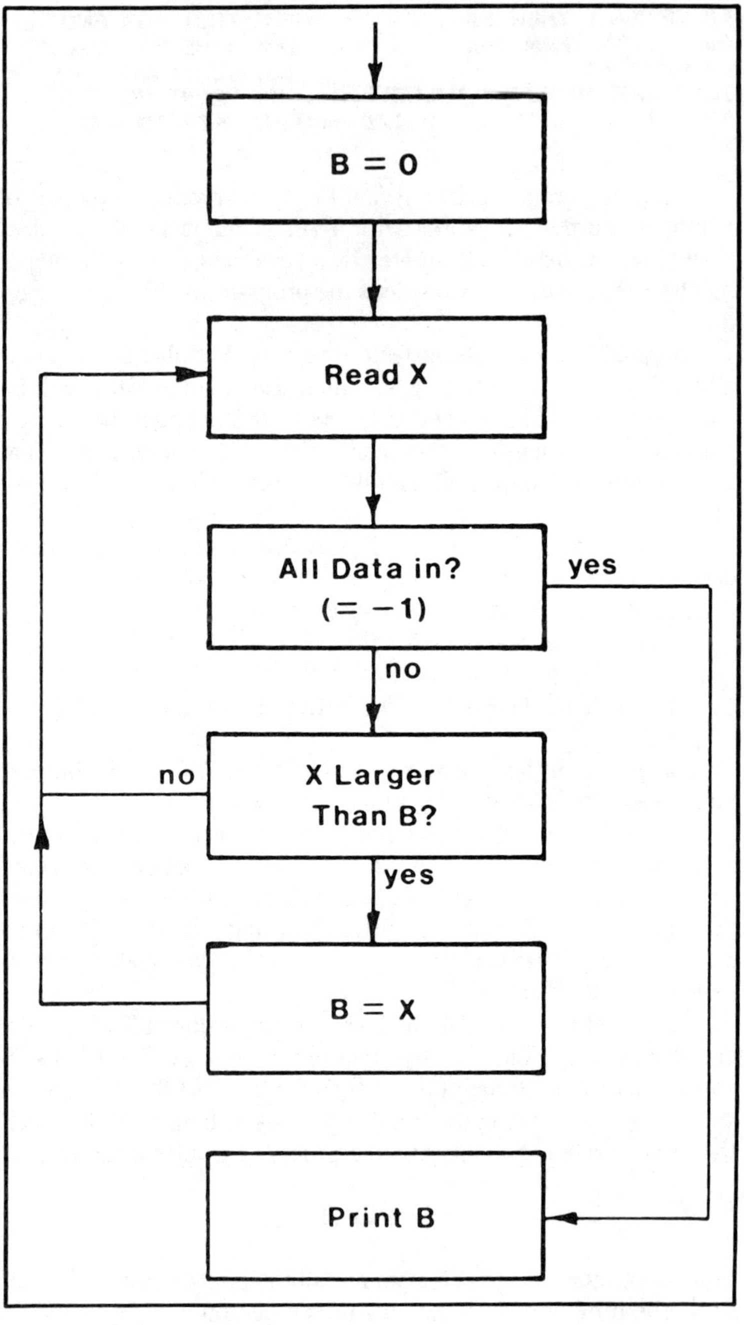

Fig. 3-2. Largest value program flowchart.

value. The following program gives the biggest value, the smallest value, and the Range:

```
1Ø REM PROGRAM 3.2. RANGE PROGRAM
2Ø B=Ø                             :REM INITIAL BIG #
3Ø L=1ØØØØØØ                       :REM INITIAL SMALL #
4Ø READ X                          :REM GET DATA
5Ø IF X=-1 THEN 9Ø                 :REM TEST FOR END
6Ø IF X>B THEN B=X                 :REM SAVE BIG VALUE
7Ø IF X<L THEN L=X                 :REM SAVE LOW VALUE
8Ø GOTO 4Ø                         :REM CONTINUE
9Ø PRINT "SMALLEST VALUE IS",L :REM PRINT VALUES
1ØØ PRINT "LARGEST VALUE IS",B
11Ø PRINT "RANGE IS",B-L
12Ø DATA 8Ø,9Ø,1ØØ,11Ø,12Ø,-1
```

The operation of this program is precisely like that of the previous one. Of course everything is backwards for the smallest value (see Fig. 3-3). In line 30, we must set up variable L, which will become the smallest value, with a very high number. In line 70, a test is made for the smallest value. It's just like the one for the biggest value, in line 60, but the sign, <, meaning less than is used, rather than >, or greater than.

Line 110 also contains something not seen before.

110 PRINT "RANGE IS",B-L

This might confuse you. The computer will print out "RANGE IS," and then it will subtract L from B and print out the difference. It's perfectly permissible to do arithmetic operations in a PRINT statement, although things are usually a bit clearer if the arithmetic is performed somewhere else. One could put parentheses around the variables to make it clearer, but it's not necessary on most machines.

110 PRINT "RANGE IS",(B-L)

Run the program. You'll find that the Range for the first set of data at the beginning of the chapter is four, while the range for the second set of data is 40.

Alas, the Range is not a very important, or very good statistic, so I won't spend more time on it. Indeed, it was included only because it's valuable to someone learning programming to know how to find the largest and smallest numbers in a set of numbers.

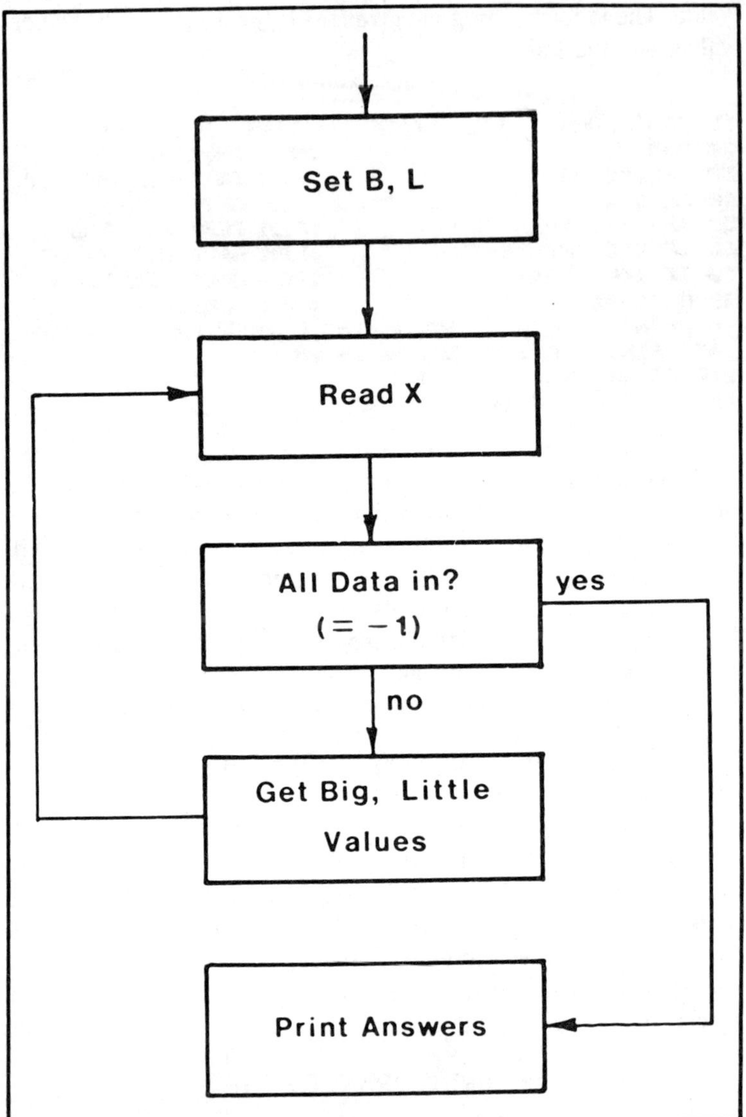

Fig. 3-3. Range program flowchart.

THE AVERAGE DEVIATION: A BETTER WAY

As mentioned, the difficulty with the Range is simply that it ignores so much data—only the top and bottom numbers are considered. What is needed is a procedure that uses every piece of information in some way. The simplest of these is called the Average Deviation, or the Mean Deviation. I'll work up to it gradually.

If we want to take into account every piece of information, one strategy is to compare each number with some other value. That is 80, 90, 100, 110, and 120 will all have to be compared with some other number. The Average, or Mean, is the best value to use for this purpose; this can be proven mathematically.

If you're not a mathematician, the idea of comparing values mathematically may be foreign to you. You may not have thought about this before. How do mathematicians compare numbers? Well, they compare numbers by subtracting. Say you have the numbers 80 and 90. To compare 80 and 90, you simply subtract them. The subtraction (80-90) yields three possible outcomes:

1. The result of the subtraction may be negative. A negative outcome means that the first number is smaller than the last. Thus, (80-90) is negative. That means that 80 is less than 90.
2. The result of the subtraction may be positive. A positive outcome means that the first number is larger than the last. Thus, (90-80) is positive. That means that 90 is greater than 80.
3. Finally, the result of the subtraction may be zero. If that's the case, a little thought will show that the two numbers must be the same.

This is all shown in Fig. 3-4.

Your little computer compares numbers in precisely that fashion. Most microcomputer chips have an instruction called "COMPARE." When that instruction is encountered, one number is subtracted from another. Action is then taken according to whether the result is positive, negative, or zero.

The result of this rather long digression is simply that to compare numbers in our little situation, we'll subtract them. Since we want to compare numbers with the Mean, we'll subtract the Mean from each one. Then, we'll add up all the differences. That will give the sum of the individual spreads. Then, finally, to make things legitimate, we'll have to divide the overall sum by the number of numbers, just as you do in the Average. The formula for this procedure looks like this:

$$AD = \frac{\Sigma(X - \overline{X})}{N} \qquad \textbf{3-1}$$

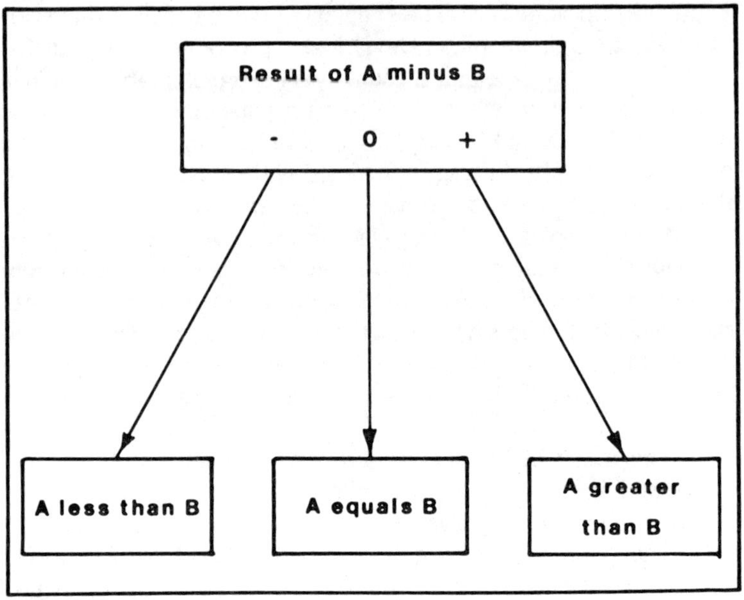

Fig. 3-4. The result of a computer comparison.

As you'll recall from the last chapter, sigma or Σ means "add up things." What is to be added up is the value $(X - \overline{X})$, or the difference between the piece of input data, X, and the Average. When all these differences have been added up, the sum will be divided by the number of data points. That's not so difficult.

Of course, we hate to do things like that by hand (although it wouldn't be bad practice for you to do it with your hand calculator, just to get the feel for it). Consequently, the following program will do all the work for you:

```
10 REM PROGRAM 3.3. 1ST MEAN DEVIATION TRY
20 A=100                          :REM SET AVERAGE
30 READ X                         :REM GET DATA
40 IF X=-1 THEN 90                :REM TEST FOR END
50 D=X-A                          :REM GET DIFFERENCE
60 S=S+D                          :REM SUM DIFFERENCE
70 N=N+1                          :REM CUMULATE COUNT
80 GOTO 30                        :REM CONTINUE
90 M=S/N                          :REM GET MEAN DEVIATION
95 PRINT S
100 PRINT "AVERAGE DEVIATION IS",M :REM PRINT IT
110 DATA 80,90,100,110,120,-1:REM DATA TO USE
```

This program is very, very similar to Program 2-4 (see Fig. 3-5). Notice, however, that for the first time, I've broken away from

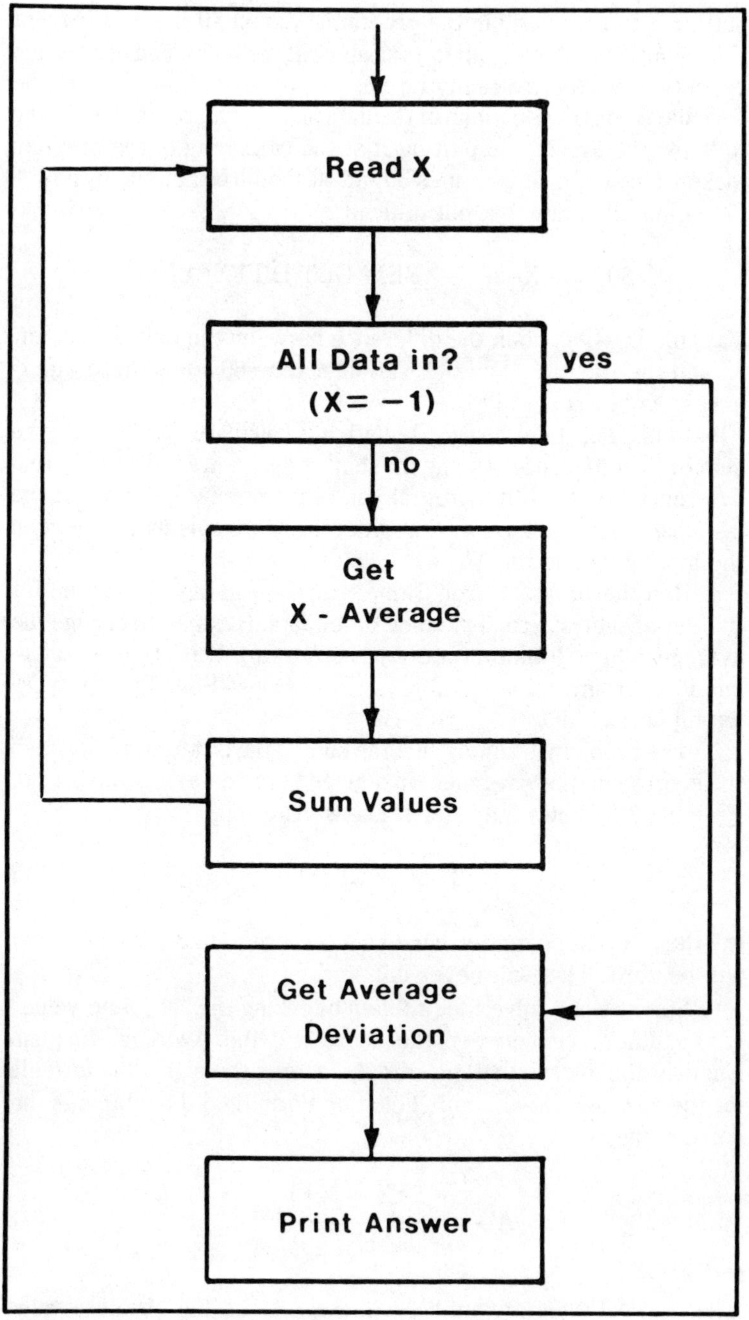

Fig. 3-5. Incorrect Average Deviation flowchart.

setting S to zero early in the program. Almost all BASICs (except that on my Sharp!) clear all the variables at the beginning of program execution, so there's really no need to do that. Line 20 sets the variable A to 100, the Mean of the data used. The program will have to know the Mean, and putting it at the beginning of the program makes it easy to change. Lines 30 and 40 should be familiar by now.

Line 50 is new but not difficult.

```
50  D=X–A      :REM GET DIFFERENCE
```

Variable D will contain the difference between the new data point, X, and the Average, for each variable. Line 60 sums these differences, as before.

In line 90, the Average Deviation is calculated by dividing the sum of the differences by the number of differences. The only real difference between this program and the Average program of the last chapter lies in the fact that differences are calculated between the input data and the Mean.

Run the program. You'll find that the Average Deviation for this set of data is zero. Try some other data. (Be sure to change the average in line 20 appropriately.) The Average Deviation, as calculated by Program 3-3, is always zero. Oops! Something is very wrong here.

The problem is simply that the sum of the differences between the scores and the Averages (in a given set of data) is always zero. Formula 3-2 shows how statisticians write this.

$$\Sigma (X - \overline{X}) = 0 \qquad \textbf{3-2}$$

In other words, no matter how we work Program 3-3, the Answer will be zero. This will never do!

Statisticians solve the problem by taking the "absolute value" of the difference between the score and the Average. In plain English, this means that one simply ignores the sign. The formula for the Average Deviation is given by Formula 3-3 (and this is the correct one):

$$AD = \frac{\Sigma | X - \overline{X} |}{N} \qquad \textbf{3-3}$$

The vertical lines around "$X - \overline{X}$" mean "take the absolute value of," or simply, "ignore the sign."

BASIC has a function which allows one to do just that. Replace line 50 of Program 3-3 with the following line:

```
50   D=ABS(X-A)
```

Now run the program. You'll find that the Average Deviation for the data in the program is now 12. (ABS means take the absolute value of.) Thus, the simple expedient of ignoring the sign works. We now have a meaningful way of calculating the Mean Deviation. The change is shown in Fig. 3-6.

The Average Deviation is still widely used, more in physics and the other physical sciences than in statistics. There is nothing really wrong with the Mean Deviation, although some people are made uneasy by the fact that the Average Deviation simply throws away the signs!

A MORE PRACTICAL WAY TO DO THE AVERAGE DEVIATION

Program 3-3, even with line 50 changed, is an inconvenient program. That's because you must know the Average before proceeding with the program. It's perfectly possible to write a program that calculates the Average Deviation all by itself, such as in the following program.

```
10 REM PROGRAM 3.4. AUTOMATIC MEAN DEVIATION
20 S=0                         :REM CLEAR LOCATION
30 READ X                      :REM GET DATA
40 IF X=-1 THEN 80             :REM TEST FOR END
50 S=S+X                       :REM CUMULATE SUM
60 N=N+1                       :REM CUMULATE COUNT
70 GOTO 30                     :REM CONTINUE
80 A=S/N                       :REM GET AVERAGE
90 RESTORE                     :REM RESET DATA
100 S=0                        :REM CLEAR LOCATION
110 FOR I=1 TO N               :REM BEGIN LOOP
120 READ X                     :REM GET DATA
130 D=ABS(X-A)                 :REM GET ABS DIFF
140 S=S+D                      :REM SUM DIFFERENCES
150 NEXT I                     :REM END LOOP
160 M=S/N                      :REM GET MEAN DEVIAT
170 PRINT "AVERAGE DEVIATION IS",M :REM PRINT IT
180 DATA 80,90,100,110,120,-1:REM DATA TO USE
```

That's a rather long and frightening program!—but if you look at it carefully, you'll notice that lines 10 through 80 are the old Average program, Program 2-4 (see the flowchart in Fig. 3-7). In short, lines 10 through 80 get the Average. Lines 100 through 180

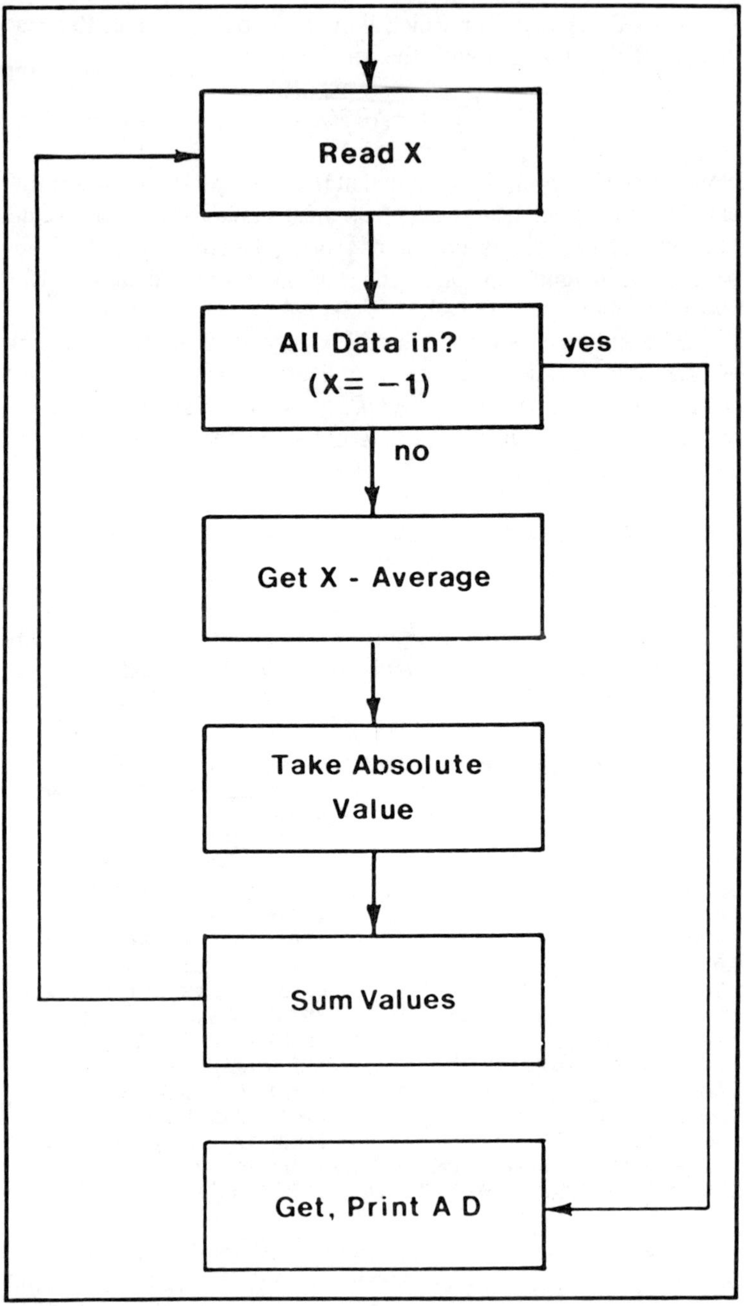

Fig. 3-6. Proper Average Deviation flowchart.

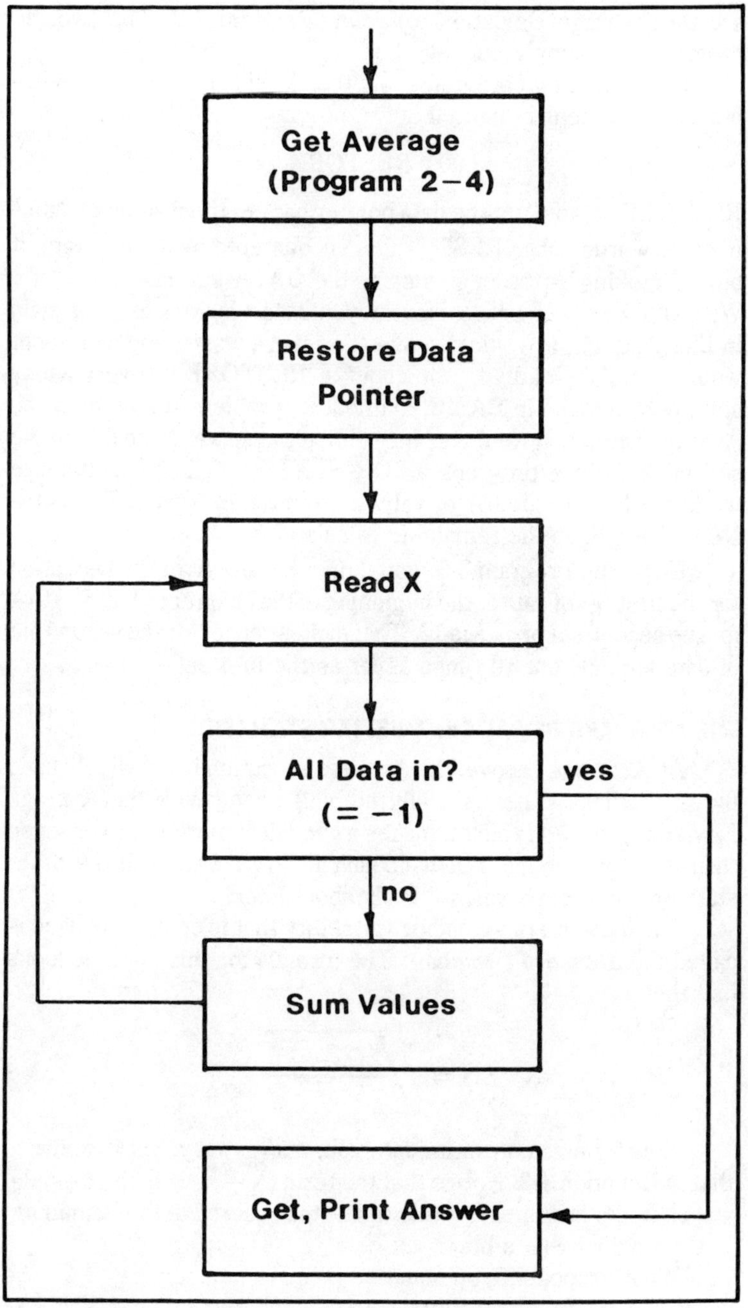

Fig. 3-7. Automatic Average Deviation flowchart.

are the Average Deviation program, Program 3-3. The two old programs are simply put together.

Putting the two programs together wouldn't be possible without the statement in line 90 of the program. Line 90 says:

```
90 RESTORE
```

RESTORE means "set the data pointer back to the first bit of data." In other words, when RESTORE is encountered by the program, it begins looking at the beginning of the DATA statement for data. Without RESTORE, the computer would start looking for data again in Line 120, but it would discover that there wasn't any more and, would complain loudly to you about it. RESTORE is a very handy statement to have in BASIC. Without it, problems like that of the Average Deviation would be very difficult to solve. Note that in the second half of the program, a FOR-NEXT loop can be used, since we know N, the number of values, from the first part. The FOR-NEXT loop does the remaining calculations.

If you run Program 3-5, you'll find that the Average Deviation for the first set of data at the beginning of the chapter is 1.2, and that for the second set of data is 12. That makes sense, for the second set of data spreads out 10 times as far as the first set.

THE STANDARD DEVIATION: A BETTER STATISTIC

As mentioned above, the Average Deviation is not widely used by statisticians. There's really nothing wrong with the Average Deviation; indeed I think it makes a good deal more intuitive sense than its replacement. But it doesn't fit in as well with the other statistical concepts which I'll talk about later.

The measure of spread or variability that is commonly used is called the *Standard Deviation*. The formula for this statistic looks like this:

$$SD = \sqrt{\frac{\Sigma(X - \overline{X})^2}{N}} \qquad \textbf{3-4}$$

This formula may intimidate you, but it's very, very similar to that of Equation 3-2. Notice that the term $(X - \overline{X})$ is in the formula, just as in the last one. But this time, it is squared before summing. Let's consider that a bit.

What happens if you square a positive number, say +4? The answer is 16. But what if you square a negative number, say −4? Minus four squared is again plus 16. You see, the principal difficulty with summing $(X - \overline{X})$, as done in the original Average Deviation formula (3-1) is that some of the values are negative, and some are positive. Therefore, they all add up to zero. This was avoided in Formula 3-3 by taking the absolute value of $(X - \overline{X})$ before summing.

Now, some mathematicians don't think that it is proper simply to ignore the signs of the differences, as is done in the case of the Average Deviation. In a way, the Standard Deviation in Formula 3-4 solves the problem, because, when $(X - \overline{X})$ is squared the negative values are eliminated. However, since squaring negative values makes them positive, perhaps the negative signs aren't lost quite so badly as in the case of the Average Deviation! I've never taken this argument very seriously, but it may help you to understand Formula 3-4.

In short, Formula 3-4 solves the problem of the negative values by squaring them. Then they're summed, just as with the Average Deviation, and the Average taken. Finally, since the values are squared in the beginning, it makes sense to end up by taking the square root. After all, we don't want things to remain squared forever, and consequently, the square root is taken of the result. However, keep in mind that Formula 3-4 is very nearly the same thing as Formula 3-3. The principal difference is in the squaring operation.

A program that will calculate the Standard Deviation follows:

```
10 REM PROGRAM 3.5. SLOW STANDARD DEVIATION
20 S=0                               :REM CLEAR LOCATION
30 READ X                            :REM GET DATA
40 IF X=-1 THEN 80                   :REM TEST FOR END
50 S=S+X                             :REM CUMULATE SUM
60 N=N+1                             :REM CUMULATE COUNT
70 GOTO 30                           :REM CONTINUE
80 A=S/N                             :REM GET AVERAGE
90 RESTORE                           :REM RESET DATA
100 S=0                              :REM CLEAR LOCATION
110 FOR I=1 TO N                     :REM BEGIN MAIN LOOP
120 READ X                           :REM GET DATA
130 D=(X-A)^2                        :REM GET SUM SQUARE
140 S=S+D                            :REM SUM SQRD DIFF
150 NEXT I                           :REM END LOOP
160 D=SQR(S/N)                       :REM GET STD DEVIAT
170 PRINT "STANDARD DEVIATON IS",D :REM PRINT IT
180 DATA 80,90,100,110,120,-1 :REM DATA TO USE
```

This program is almost identical with the last one for the Average Deviation (see Fig. 3-8). The variation occurs in line 130,

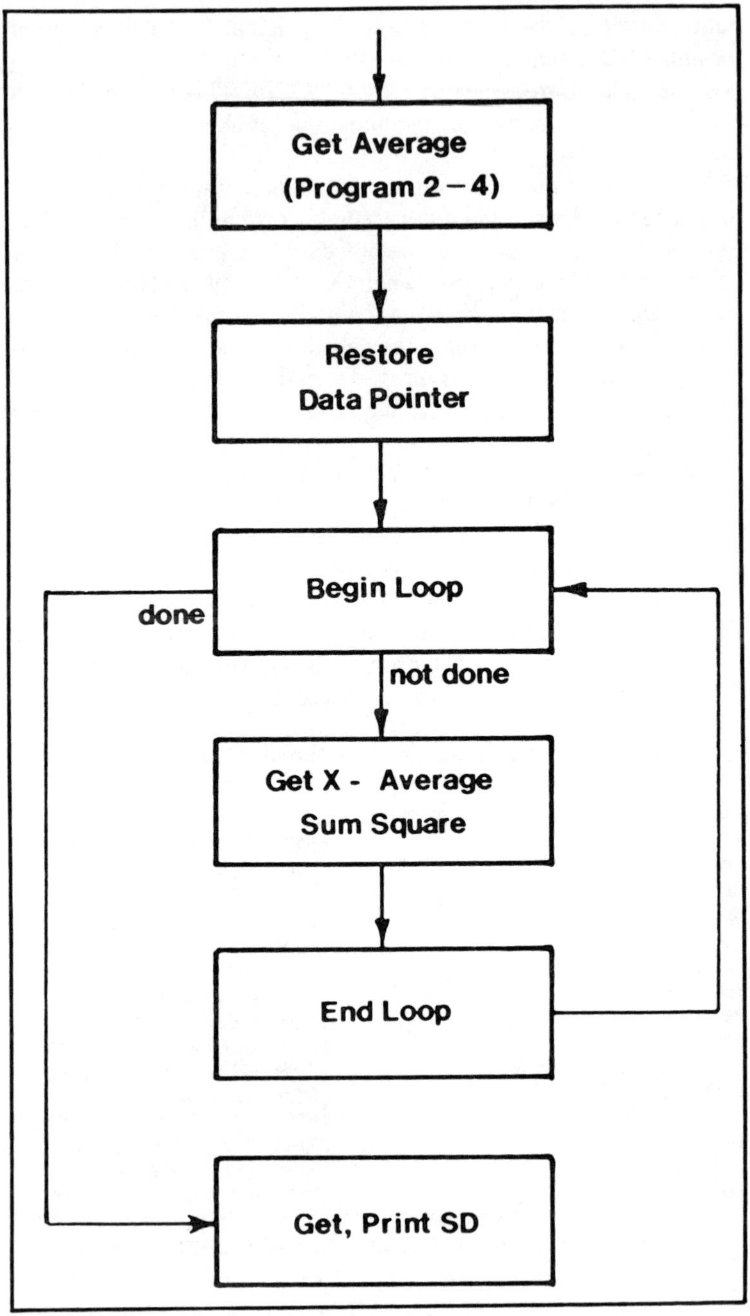

Fig. 3-8. Slow Standard Deviation program flowchart.

in which the difference between the score and the Average is squared, and in line 160, where the square root is taken. Lines 10 through 80 obtain the Average, just as in Program 2-4. Line 90 sets the computer back to the beginning of the data, and lines 100 through 170 actually calculate the Standard Deviation.

A new BASIC term in the program is SQR which means "take the square root of." BASIC has lots of built in functions of this sort. There's SIN for the sine of an angle, ABS for absolute value (which we used in Programs 3-4 and 3-5), ATN for arctangent, and on and on. The presence of these functions in BASIC means that you don't have to write programs to be able to use them. Some of the smaller computers, by the way, don't have these functions, but most do.

A second new BASIC term is the operation of exponentiation, shown by the up arrow (↑). In the printout from my Epson printer, and on many of your computer video screens, only the top part of the arrow is shown. The up arrow indicates that the power of a number is to be taken. Thus, 2 ↑ 3 means "two the third power", or 2^3. You should find the up arrow on the keyboard of your computer.

Type in and run the program. You'll find that the Standard Deviation is 14.1421, a value somewhat larger than the Average Deviation (12). The program works!

Notice just how much work we've gotten out of our small store of knowledge about BASIC. With just a few functions, one can do a great deal with this very accessible language.

THE STANDARD DEVIATION: A BETTER WAY

The method used for calculation of the Standard Deviation in Program 3-6 is very inefficient. You must go through the data once to obtain the Average, and then go through the data again to get the Standard Deviation. That's no problem when there are only five pieces of data, as in the last example. However, if we were doing the Census, with 215,000,000 numbers, this method would not be practical. What is needed is a method for calculating the Standard Deviation that requires only one pass through the data. Statisticians have developed a way of doing this, and it uses the following formula:

$$SD = \sqrt{\frac{\Sigma X^2 - \frac{(\Sigma X)^2}{N}}{N}} \qquad \textbf{3-5}$$

That's a rather frightening formula! Its origin is rather complex, but you can find a derivation for it in most college-level statistics textbooks. Let's look at it in some detail.

It uses two new terms: ΣX^2 and $(\Sigma X)^2$. The difference between them is subtle. ΣX^2 means "square all the numbers, and then add up the squared values." The term, $(\Sigma X)^2$, means, "add up all the numbers, and square the result of that sum." You have to think about that a bit.

Thus, with the data we've been using. ΣX^2 means, $6,400 + 8,100 + 10,000 + 12,100 + 14,400 = 51,000$. Notice that with ΣX^2, one squares the numbers before summing them. $(\Sigma X)^2$ means, $(80 + 90 + 100 + 110 + 120)^2$, which is $(500)^2$, or 250,000. The difference between these two is very hard for some people to understand. Make certain that you do before proceeding, for these terms will be used again and again and again in this book. It would be wise, at this point, for you to work out the Standard Deviation with Formula 3-5 using a hand calculator. It's good practice, and it will tell you whether you understand the formula.

Thus, to calculate the Standard Deviation by this method, we must get $\Sigma X^2$and $(\Sigma X)^2$. The nice thing about Formula 3-5 lies in the fact that you can get both of these terms in one pass through the data! You see, ΣX^2 can be cumulated by squaring each input data point and adding up the squared values. $(\Sigma X)^2$ only requires the sum of the unmodified data points. The squaring can be done after the summing is completed.

A BASIC program which will get the Standard Deviation through this method follows:

```
10 REM PROGRAM 3.6. BETTER STANDARD DEVIATION
20 READ X                           :REM GET DATA
30 IF X=-1 THEN 80                  :REM TEST FOR END
40 S=S+X                            :REM SUM NUMBERS
50 Q=Q+X*X                          :REM SUM SQUARED NOS
60 N=N+1                            :REM GET COUNT
70 GOTO 20                          :REM CONTINUE
80 V = Q -S*S/N                     :REM TOP OF FORMULA
90 D=SQR(V/N)                       :REM GET STD DEV
100 PRINT "STANDARD DEVIATION IS",D:REM PRINT IT
110 DATA 80,90,100,110,120,-1 :REM DATA TO USE
```

Type in and run this program. You'll find that it gives you the Standard Deviation, just as Program 3-5 did. But notice how much shorter it is! Program 3-6 is the way by which the Standard Deviation is most frequently calculated.

Operation of the program is fairly simple (see Fig. 3-9). Lines 10 through 40 are the same as in many previous programs. Line 50

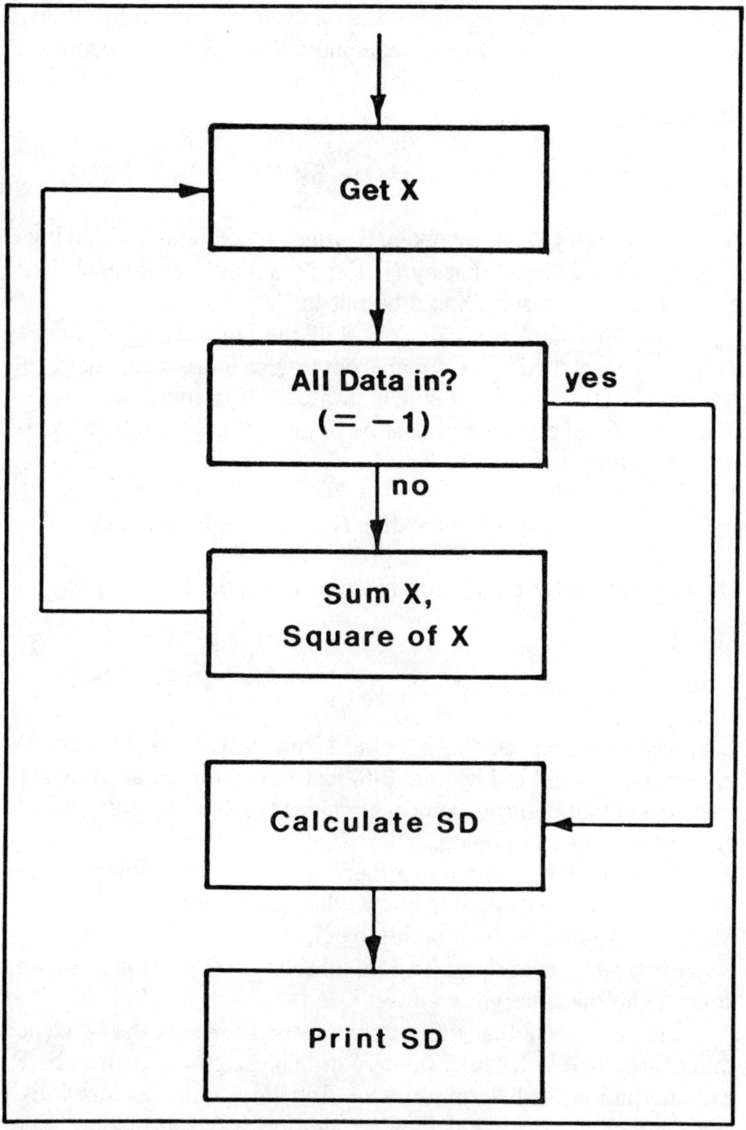

Fig. 3-9. Efficient Standard Deviation program flowchart.

```
50  Q=Q+X*X      :REM SUM SQUARED #S
```

says, take the data point, X, multiply it by itself, and add it together with Q. Q will then be ΣX^2 when we're all done. Instead of X*X you could of course use the up arrow for exponentiation to indicate the square, but most people prefer to use X*X because multiplication is a much faster process for the computer than taking exponents.

Lines 60 and 70 are familiar. Line 80 calculates the top of the Standard Deviation formula:

```
80   V = Q – S*S/N
```

It says, "first get $(\Sigma X)^2$ by taking S, which is ΣX, and multiplying it by itself. Then divide that by N. Finally subtract that result from ΣX^2, or Q. The result should be put in V."

Line 90 completes calculation of the Standard Deviation by dividing V by N, taking the square root (SQR), and putting the result in variable D. Then the result is printed out in line 100.

It would of course be possible to get lines 80 and 90 into one line, as follows:

```
80   D=SQR((Q–S*S/N)/N)      :REM GET SD
```

Or, you could save even more room by using the following for line 80:

```
80 PRINT SQR((Q–S*S/N)/N)      :REM PRINT ANSWER
```

You may not understand either one! I frequently break up formulas into parts, as I did in Program 3-6, just to make it clear to myself. Usually computer space is not as precious to me as my being able to understand what I'm doing.

You may have noticed that there is a potential problem in Line 80. How does the computer know which operation to do first? We want it to square S, then divide by N, and then subtract from Q. However, this order doesn't seem inherent in the formula, so why doesn't the machine get confused?

The answer is that there is an *inherent order* in the way such things are done in BASIC. In any formula, functions such as SQR, exponentiation, and so on are done first. Multiplications and divisions are done next, and finally, addition and subtraction. Consequently S*S/N will be first, since it involves multiplication and division, and the subtraction operation last (see Fig. 3-10).

S*S/SN may also confuse you. It seems that there may be a potential problem here, too. Well, there really isn't. S is 500, and N is 5. Notice that one way to do it is to multiply 500 by 500, getting 250,000, and then divide by 5, yielding 50,000. Or, you could divide

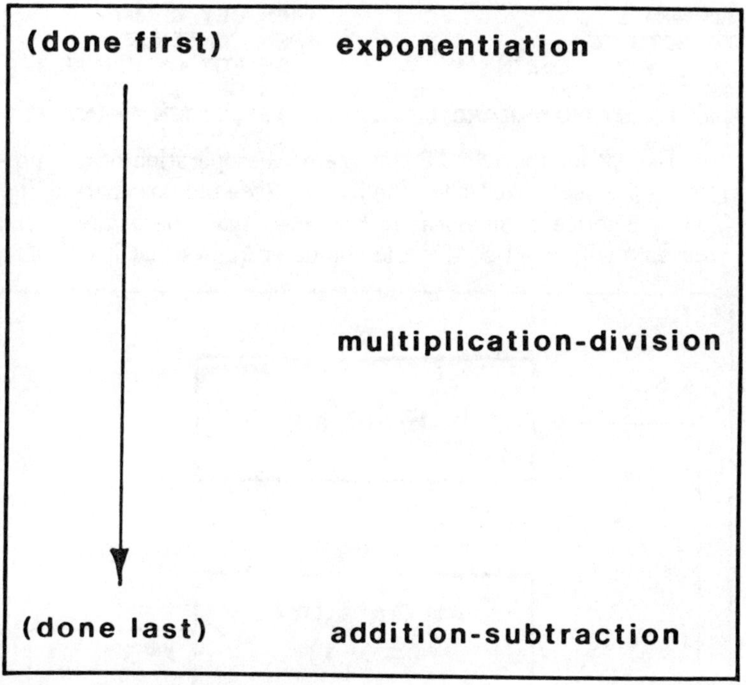

Fig. 3-10. The order of mathematical operations.

500 by 5, yielding 100, and multiply by 500, leading to 50,000. The result is the same in either case. If this troubles you, you can make it clearer by putting parentheses around things that go together, as follows:

```
80  V = Q – ((S*S)/N)
```

The parentheses make it clear which things are to be done first. The parentheses override the inherent order in which BASIC does things.

STANDARD DEVIATION WITH INPUT STATEMENT

It is of course perfectly possible to calculate the Standard Deviation with an INPUT statement. Program 3-7 shows this.

```
10 REM PROGRAM 3.7. INPUT STANDARD DEVIATION
20 INPUT "X=";X                    :REM GET DATA
30 IF X=-1 THEN 80                 :REM TEST FOR END
40 S=S+X                           :REM SUM NUMBERS
50 Q=Q+X*X                         :REM SUM SQUARED NOS
```

```
60 N=N+1                              :REM GET COUNT
70 GOTO 20                            :REM CONTINUE
80 V = Q - S*S/N                      :REM TOP OF FORMULA
90 D=SQR(V/N)                         :REM GET STD DEV
100 PRINT "STANDARD DEVIATION IS",D:REM PRINT IT
```

Except for the INPUT statement, the operation of this program is precisely like that of the last one (see the flowchart in Fig. 3-11, and notice its similarity to Fig. 3-9). Again, be certain to end your data with a −1 so that the computer knows that it's all in.

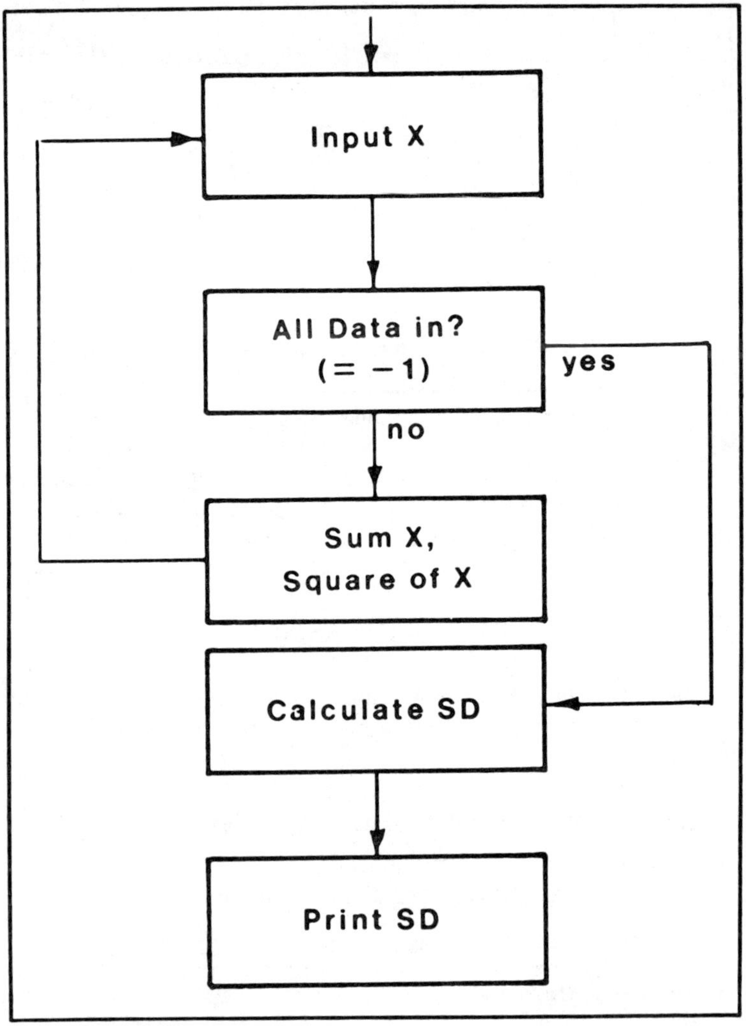

Fig. 3-11. Standard Deviation with INPUT program flowchart.

SOME EXAMPLES

It is useful to use some fictitious data to illustrate this complicated concept of the Standard Deviation. Let's start with the following data:

	IQ
	140
	120
	100
	80
	60

Average	100
SD	28.28

Let's assume that these are IQ scores on a group of psychotic or crazy people. Such people are very inconsistent in their behavior, and you'd expect their scores to vary a great deal. Notice that the data spreads out rather widely (see Fig. 3-12); this wide spread is mirrored in the rather large SD of 28.28, calculated by Program 3-6.

Fictitious data that might be collected on a set of normal people is shown next:

	IQ
	120
	110
	100
	90
	80

Average	100
SD	14.14

Figure 3-13 shows that while these data spread out a fair amount, it's not nearly so far as that in the last example. The Standard Deviation is just half as much as in the last example. Of course this is to be expected, since the numbers are just half as large.

The next example completes the progression. Here, the numbers are again halved:

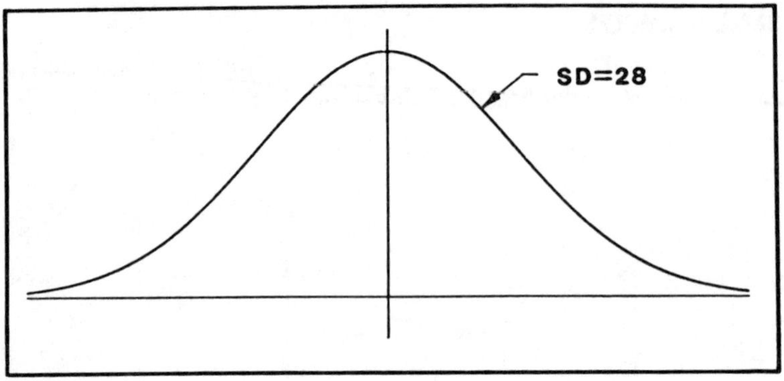

Fig. 3-12. Plot of example with SD of 28.

	IQ
	110
	105
	100
	95
	90

Average	100
SD	7.071

Figure 3-14 shows, again, that the numbers spread out far less in this example. These scores spread out a good deal less than one would expect from random sampling from the population. A Standard Deviation of this size would indicate that some systematic selection of people had occurred in the collection of data.

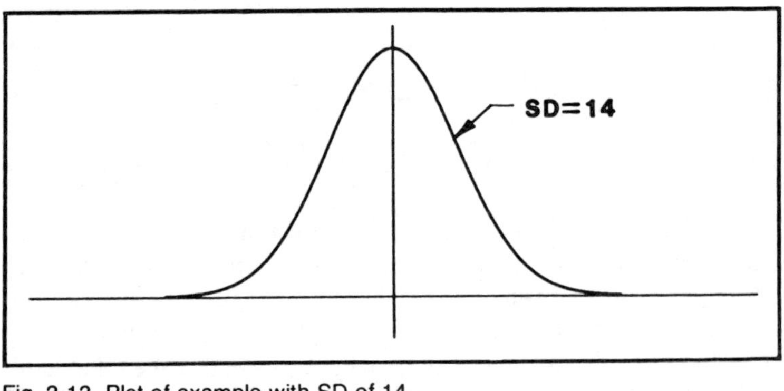

Fig. 3-13. Plot of example with SD of 14.

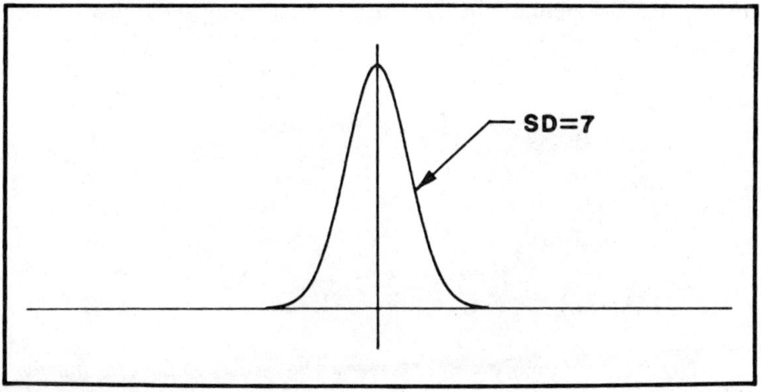

Fig. 3-14. Plot of example with SD of 7.

In Figs. 3-12 through 3-14, note that the Standard Deviation has steadily decreased while the Average has remained the same. Figure 3-15 plots all three curves on the same graph. The decrease is spread is very clear indeed.

EXPERIMENTING WITH THE STANDARD DEVIATION

Now that you have Program 3-6 running, you may calculate the Standard Deviation again and again, in order to get a feel for it. It is a statistic that, quite frankly, most people don't understand well because of the complexity of the formula. Students of statistics in precomputer days didn't have the luxury of being able to do the Standard Deviation as easily as you can, and hence couldn't get a feel for it. By changing the values in the DATA statement in

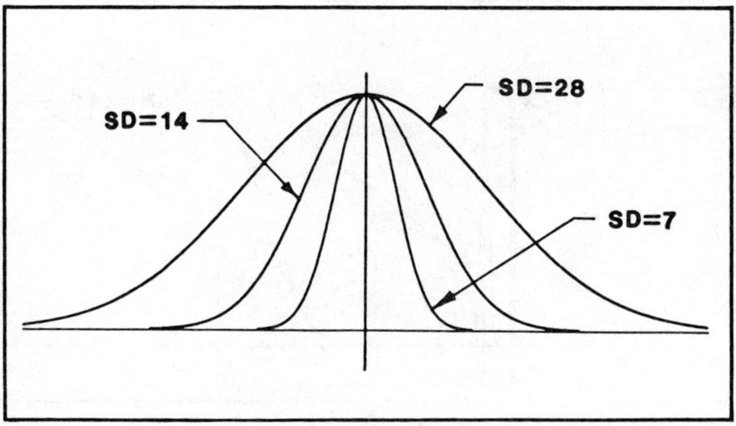

Fig. 3-15. All three plotted together.

Program 3-6, calculate the Standard Deviation for the following sets of numbers:

1200	120	12	1.2
1100	110	11	1.1
1000	100	10	1.0
900	90	9	.9
800	80	8	.8
-------	-------	------	-----
(141.42)	(14.14)	(1.414)	(.1414)

Notice how easily you can work the problems! (The answers are in parentheses, if you want to check them.) The computer makes working out the Standard Deviation child's play.

In the examples above, notice that each set of numbers is simply the last set, divided by 10. Similarly, notice that each Standard Deviation is the last one, divided by 10. That's the way the statistic works. Multiplying and dividing the data by a constant value changes the Standard Deviation in the same way. Remember, that was also true for the Average. You could also prove this by adding a line to Program 3-6:

```
35  X=X/10
```

or

```
35  X=X*10
```

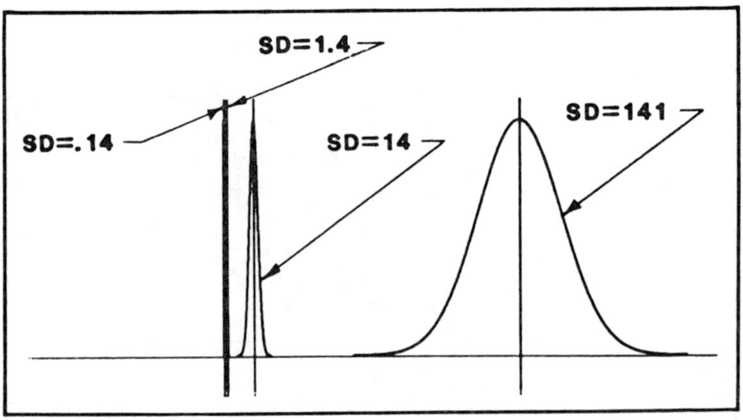

Fig. 3-16. The effect of multiplication on the SD.

Such added lines would change the data values for you, so that you wouldn't have to retype the data each time. Again, see how valuable knowledge of a little BASIC can be? Figure 3-16 is a plot of the previous data, assuming that the Normal Curve to be discussed in Chapter 6 applies. Notice that the Standard Deviations vary so widely that the plotter really isn't able to capture them, but the spread clearly increases from left to right, and so does the Average, shown by the center line.

Try these numbers in Program 3-6:

120	130	140	150
110	120	130	140
100	110	120	130
90	100	110	120
80	90	100	110
------	------	------	------
(14.14)	(14.14)	(14.14)	(14.14)

This is interesting. Each set is simply 10 points higher than the last set. Notice that adding a constant value to each of the numbers before calculating the Standard Deviation does not change the value of the Standard Deviation. Remember, the Average did change under similar circumstances. One may also prove this through the addition of a line to Program 3-6:

```
35 X=X+10
```

Again, that line will increase the data automatically for you, without changing the DATA statement. I have always hated work, and so I like such time-saving devices. Figure 3-17 plots the data just above. Notice that the spread of the numbers remains constant, while the Averages shift.

Try some more values in Program 3-6. If you do 20 or so little problems, you'll understand the Standard Deviation very well indeed.

You now know two of the most important concepts in statistics. All the remaining topics in statistics are based on the Average and the Standard Deviation, and you have the means at your disposal to know them thoroughly.

Even better, you now know enough BASIC to do almost any simple statistical programming job. Indeed, we have covered the following BASIC statements to this point in the book:

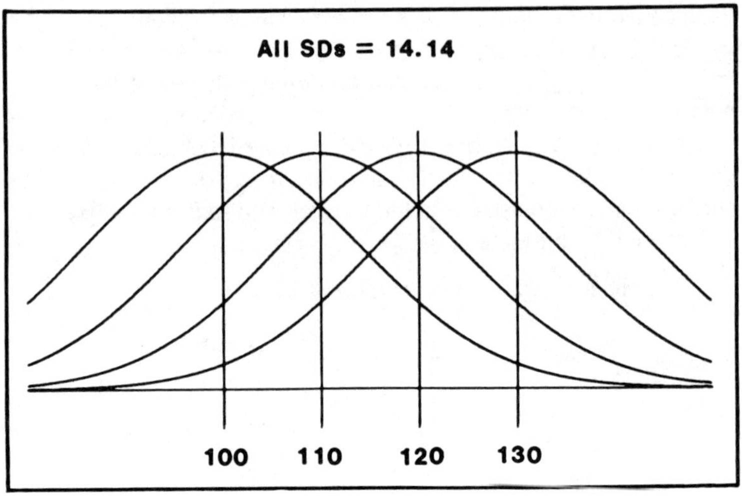

Fig. 3-17. The effect of addition on the SD.

ABS	IF-THEN	READ
DATA	PRINT	REM
FOR-NEXT	PRINT " "	RESTORE
GOTO	READ	SQR
INPUT		

Also, you have learned the following operators:

```
=    +    -    *    /
<    >    <>   <=   =>    (    )
```

It's hard to believe that such a limited vocabulary of terms can let you do complex statistical calculations, but in the succeeding chapters, we'll do just such elegant calculations, adding only a very few new statements to those listed above.

These few concepts go a long way. For example, what if you wanted to have Program 3-6 also print the Average, in addition to the Standard Deviation? To do that, simply add the line

```
95 PRINT "THE AVERAGE IS",S/N
```

and the computer will print the Average as well! It wouldn't be very much more difficult to have the computer print out the various sums, N, and so on. Or for that matter, you now have the ability to write a program that will calculate the Standard Deviation and

Average Deviation at the same time. Such a program would probably have to be based on Program 3-5, rather than 3-6, because of the way the Average Deviation must be calculated.

In short, you know the underlying concepts of computer programming, and of statistics—and only in three chapters!

PROBLEMS

1. Find the Average Deviation and the Standard Deviation for the IQ scores 100, 115, 130, 145, and 170 (20.4 and 24.2074) (notice that these problems are the same as those in Chapter 2).

2. Jones Motorcar company makes automobiles with top speeds as follows: 120, 180, 125, and 150. What's the Average Deviation and Standard Deviation? (21.25 and 23.8157)

3. Acme Detergent is used to wash dishes. A carefully controlled amount of detergent and dishwater washes the following numbers of dishes: 50, 33, 73, 84, 33, and 89. What are the Average Deviation and Standard Deviation? (21.6667 and 22.8886)

4. Dismal Disk company finds the following number of bad tracks on its disks: 33, 38, 33, 21, and 5. Find the Average Deviation and Standard Deviation. (10.4 and 11.8996)

5. Zippy Printer company makes printers that print the following number of characters per second: 80, 95, 150, 75, 88. What are the Average Deviation and Standard Deviation? (20.96 and 27.0747)

MOM, I GOT A ZERO ON MY TEST.
YOU GOT A ZERO!!

4
Standard Scores

Assume that there are two people, Sam and John. Both of them are given intelligence tests, Sam taking one kind of test, and John taking another kind. The tests that both of them take are perfect tests; they measure intelligence exactly, with no mistakes. (Needless to say, there are no such tests!) However, there is no guarantee that the measurement scales from both tests are the same. Sam and John get scores as follows (see Fig. 4-1):

Sam	**John**
600	115

Now, which person is the smarter? If you're like me, you'll jump to the conclusion that Sam is brighter than John.

"Why not?" you query. "Sam's score is 600, and John's is only 115. Obviously Sam is much brighter than John. Where's the problem?"

The problem lies, simply, in the fact that we know nothing about the sorts of scores that the tests provide. It's like knowing one student who has a grade point average of 3.5 at Harvard, and another student who has a grade point average of 3.5 at Podunk college. One suspects that the standards at Harvard may be a bit tougher than those at Podunk. A 3.5 at one school may mean something very

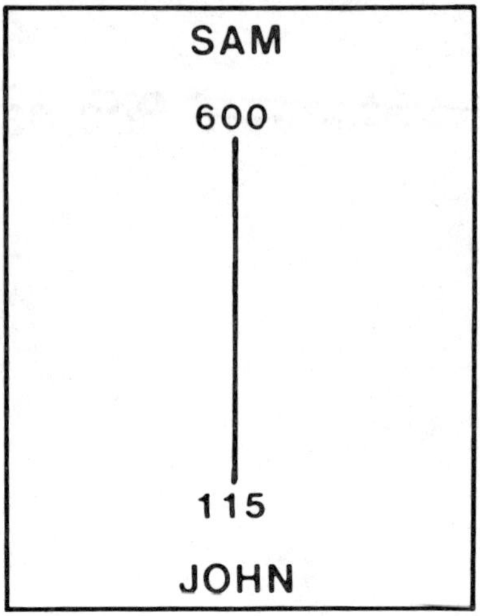

Fig. 4-1. Sam's and John's scores.

different from a 3.5 at another school. In short, the average grade may be a bit lower at Harvard than at Podunk.

Thus, it seems reasonable that we should control for the effect of the Average or Mean in comparing two scores. Obviously, the Average might be very different on two completely different tests. So we'll list the Average for each person's test:

	Sam	John
Score	600	115
Average	500	100
	------	------
Difference	+100	+ 15

That helps a good deal. We now find that Sam is 100 points above average, while John is only 15 points above average (see Fig. 4-2).

"Sam is, then, much smarter than John," you say. "Obviously, Sam is 100 points above average, while John is only 15 above average. Our problem is solved! All we need do is calculate the score minus the Average, or $(X - \overline{X})$, and we've got it!"

Alas, you're wrong. There is another complication. You see, we don't know the value of the points for Sam and John. Each one of John's points might well be worth a great deal more than Sam's, or vice versa. Indeed, as an illustration, our little example was set up so that Sam and John would be exactly equal in intelligence. In other words, Sam's score of 600 indicates precisely the same level of

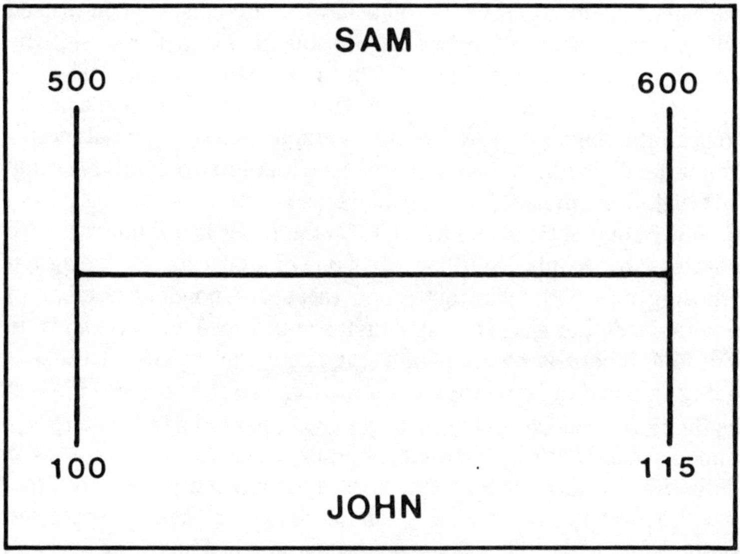

Fig. 4-2. Sam's and John's scores with Averages.

intelligence as John's score of 115. A diagram of the situation is shown in Fig. 4-3.

Under the assumption that Sam and John are equally bright, and that a score of 600 on Sam's test is equal to a score of 115 on John's,

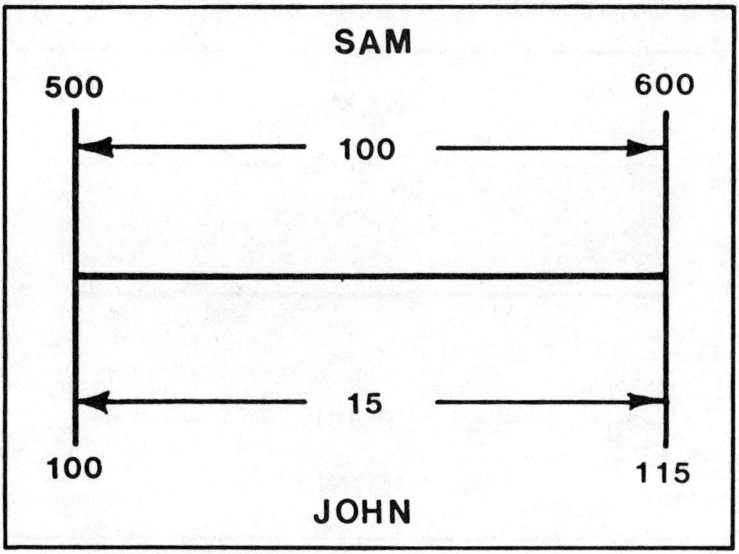

Fig. 4-3. Effect of differing spread.

we can get an idea of the difference in worth of the sort of units used with each person. It takes a bit of thought, but ask yourself the question, "Whose points are worth more, Sam's or John's?"

John's points are worth more. It takes only 15 of John's points to span the distance between the Average and their mutual intelligence level, while it takes 100 of Sam's points to span the same interval. This means that one of John's points is worth 100/15 = 6.6667 of one of Sam's points. Or, on the other hand, one of Sam's points is worth only 15/100 = .15 of one of John's points. A diagram showing one of John's points and the corresponding number of points needed to span the same distance in Sam's units is shown in Fig. 4-4. There is nothing confusing about this, really. All that is being said is that some things are worth more than others. It's just as though one were saying that there are ten dimes in a dollar, or a dime is one-tenth of a dollar. The point is that unless a clear definition is made, one cannot know how much a point is worth.

Look at Fig. 4-3 again. It should be clear that the more the scores spread out, the less the points are worth. Thus, Sam's scores spread out 100 points, but each one is worth far less than one of John's. This means that if a method of comparing scores is wanted, it must take into account this backward relationship between spread and worth of points. Thus, we shall have to divide by whatever measure of spread we use. You have to think about this a bit. Sam's scores spread out much more than John's, so each of Sam's points is worth less. Consequently, the measure of how much

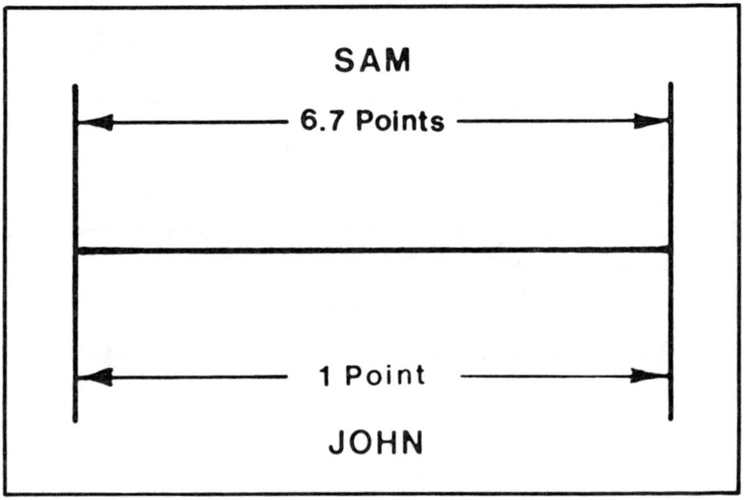

Fig. 4-4. Sam's points compared with John's.

the points spread out should be on the bottom of the equation. It makes sense to use the Standard Deviation as the measure of spread, since it is the proper statistic for this. The top of the equation should, of course, contain the original score minus the Average or $(X - \overline{X})$. A formula which will allow us to compare scores properly is:

$$z = \frac{X - \overline{X}}{SD}$$

This formula does the whole job: it removes the effect of the Average, and it also corrects for the worth of the points, by dividing by the Standard Deviation. This formula is called the *z-score* formula. It is very, very important in statistics.

Let's calculate Sam's and John's z-scores. For Sam,

$X = 600$
$\overline{X} = 500$
$SD = 100$

$$z = \frac{X - \overline{X}}{SD} = \frac{600 - 500}{100} = +1.00$$

For John,

$X = 115$
$\overline{X} = 100$
$SD = 15$

$$z = \frac{X - \overline{X}}{SD} = \frac{115 - 100}{15} = +1.00$$

Consequently, as calculated by Formula 4-1, Sam and John have equal z-scores, both being equal to +1.00. We said that Sam and John were defined as being equally bright, and the z-scores bear that out.

Sam's scores have a Mean of 500 and a Standard Deviation of 100. These are the sort of scores used by Educational Testing Service, which constructs the various college entrance examinations, such as the Scholastic Aptitude Test (SAT). John's scores have an Average of 100, and a Standard Deviation of 15, and they're the sort of scores used on what are perhaps the best-known individual intelligence tests, the Wechsler tests.

A PROGRAM TO CALCULATE z-SCORES

It is somewhat difficult to grasp the concept of z-scores just by reading about them, but we have an immensely valuable aid to understanding available to us—our microcomputer. Let's use it!

A program to calculate z-scores is a direct derivation of Program 3-6, for the Standard Deviation, in the last chapter. It looks like this:

```
10 REM PROGRAM 4.1. Z-SCORE PROGRAM
20 READ X                           :REM GET DATA
30 IF X=-1 THEN 80                  :REM TEST FOR END
40 S=S+X                            :REM SUM NUMBERS
50 Q=Q+X*X                          :REM SUM SQUARED NOS
60 N=N+1                            :REM GET COUNT
70 GOTO 20                          :REM CONTINUE
80 V=Q -S*S/N                       :REM TOP OF SD EQUAT
90 D=SQR(V/N)                       :REM GET STD DEV
100 A=S/N                           :REM GET AVERAGE
110 RESTORE                         :REM RESET DATA
120 PRINT "Z-SCORES FOLLOW"
130 FOR I=1 TO N                    :REM LOOP TO GET Z'S
140 READ X                          :REM GET DATA AGAIN
150 Z=(X-A)/D                       :REM CALCULATE Z
160 PRINT I,Z                       :REM PRINT Z-SCORE
170 NEXT I                          :REM END LOOP
180 DATA 80,90,100,110,120,-1 :REM DATA TO USE
```

Following the flowchart in Fig. 4-5, we see that lines 10 through 90 are identical with those in Program 3-6. If you didn't understand them in Chapter 3, you'd better go back and look them over again. They calculate the Standard Deviation.

Line 100 should be no stranger. It's from Program 2-4, and calculates the Average. Formula 4-1 needs the Average, and so it is gotten in line 100. Line 110 resets the data pointer, so that the data can be read in again.

The loop from lines 130 to 170 calculates the z-scores. We've determined the number of data points from the first part of the program; it's found in variable, N. Consequently, we can use a FOR-NEXT loop to calculate and print the z-scores. The loop begins in line 130.

Line 140 gets a piece of data, and puts into X. Line 150 subtracts the Average from X, and divides the result by the Standard Deviation. This is precisely what is called for in Formula 4-1.

Line 160 prints out the results for each piece of data, and line 170 finishes up the loop.

As you can see, the calculation of z-scores is very simple

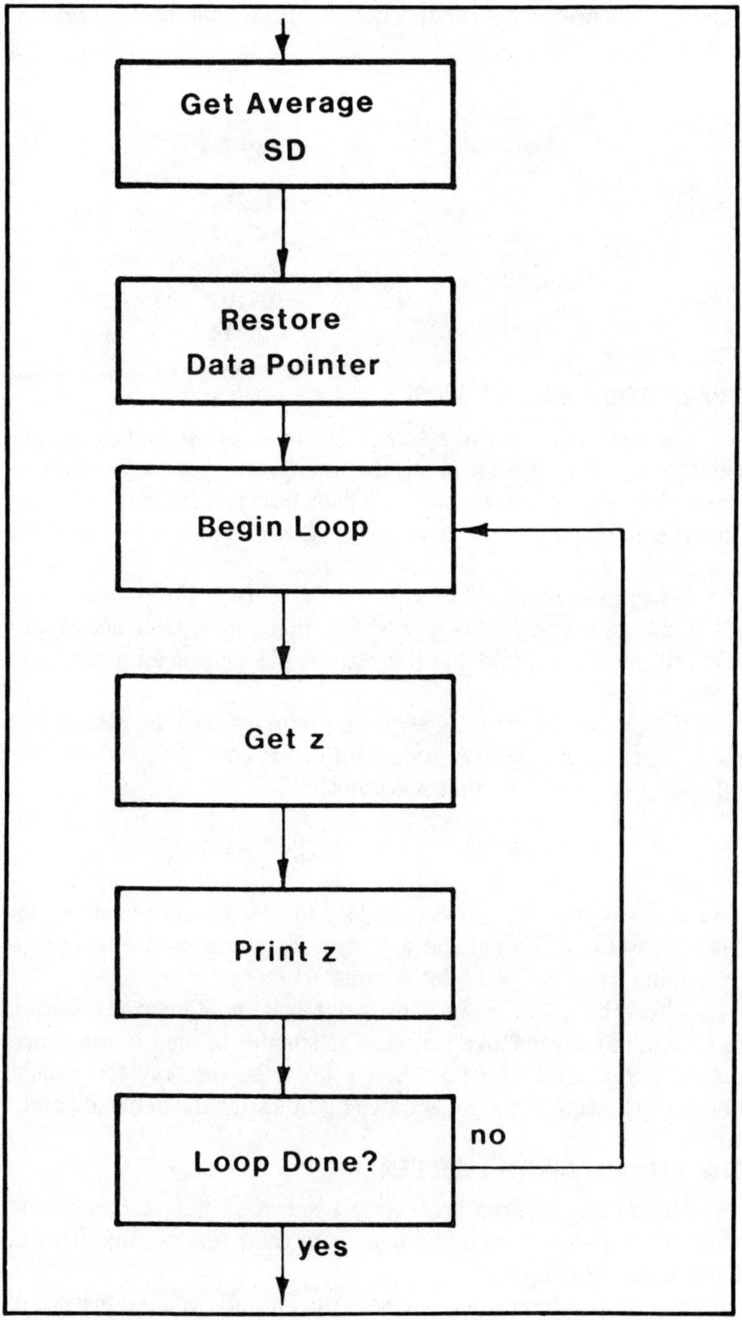

Fig. 4-5. Flowchart for the z-score program.

indeed. If the program is run (try it!), you'll find that the z-scores are as follows:

Original Score	z-score
80	−1.414214
90	− .707107
100	.000000
110	.707107
120	1.414214

CHARACTERISTICS OF z-SCORES

Certain properties of z-scores are interesting, and can easily be obtained. Notice, first of all, that the Average of the z-scores in the little table just above is zero. That's obvious, but if it isn't, just plug the numbers into the Average program, Program 2-4, and you can prove it.

What's the Standard Deviation of z-scores? Program 3-6 will tell you that! If you put the numbers from the little table above into that program, you'll find that the Standard Deviation for z-scores is 1.00.

(If you don't want to type those programs back in, simply put the z-scores in the DATA statement of Program 4-1, and add the following statement to that program:

```
115  PRINT A,D      :PRINT AVG, SD
```

You see, Program 4-1 must calculate the Mean and Standard Deviation before it can get the z-scores. Knowledge of a little programming can save you a lot of time!)

Thus, the Average of z-scores is 0, and the Standard Deviation is 1. Mathematicians love numbers as simple as 0 and 1, and hence they are very fond of z-scores. z-scores, by the way are almost always written with a sign, and two figures after the decimal point.

THE DIFFICULTY WITH z-SCORES

However, z-scores are rarely used in situations where one must tell nonsophisticated people about their test results. This is fairly easily explained.

Say, for example, that you're a student, and you got a score of 100 on the test shown in the data just above. A score of 100 is precisely average, as you know. It's not a good score, but at the same time, it's no disgrace, either. In college, such a score would probably get a student a grade of C, and many people are happy with such scores.

What's the z-score associated with the real score of 100? It's 0.00, zero. The fact that z-scores have an Average of zero is troublesome for ordinary purposes. People are so used to thinking of zero as a terrible score that one simply couldn't think of using z-scores for any ordinary purpose.

Even worse is the fact that z-scores can have *negative* values. If you think that parents would get mad with a z-score of 0, think what would happen if you had a score on the test of 90. Then your z-score would be −.7071. That would seem to be worse than zero.

You see, most non-technically-oriented people don't understand negative numbers. Worse, they don't understand decimals. As someone with some technical interests (you do have them, you know, or you wouldn't have bought a microcomputer), it is difficult to understand that some people simply don't understand decimals, *unless they are associated with money.*

As an example of this, my mother, rest her soul, near the end of her life had pretty much forgotten how to do arithmetic. Consequently, I bought her a nice desk calculator, so that she could balance her checkbook. When she first tried to use it, the calculator attempted to tell her that the sum of her numbers was 70 cents. However, it printed it as

0.7

Mom was totally confused by that.

She said, "This calculator is no good. It's broken. Thirty cents and 40 cents is 70 cents, not "0.7."

I said, "But 0.7 means 70 cents! That's obvious."

Mom said, "0.7 means 0.7. It should say 0.70. It's no good. Why do you buy me such useless junk?"

Thus, the problem with z-scores is simply that ordinary people can't understand them, although they're wonderful from a mathematical standpoint.

OTHER TYPES OF STANDARD SCORES: A SOLUTION

Obviously, some other sort of Standard Scores will have to be used if people are to understand them. Fortunately, by doing a little algebraic manipulation with Formula 4-1, one can arrive at a formula which allows us to make up Standard Scores with any Mean and Standard Deviation. The result looks like this:

$$\text{Desired Score} = (z)*(\text{desired SD})+\text{desired Average} \quad \textbf{4-2}$$

For clarity, this formula is written out in plain English, rather than algebraic symbols. It says, simply, that to go from one set of scores to another, one first must get a z-score from the first set of scores. Then one multiplies the z-score by the new desired Standard Deviation, finally adding in the desired Average. In other words, the Standard Deviation is changed first (see Fig. 4-6), and then the Average is shifted (see Fig. 4-7).

Let's assume that we want to transform the scores 80, 90, 100, 110, and 120 to Standard Scores with a Mean of 500 and a Standard Deviation of 100. Let's transform the score of 80 to the new scores. We know from our previous computer example that the z-score

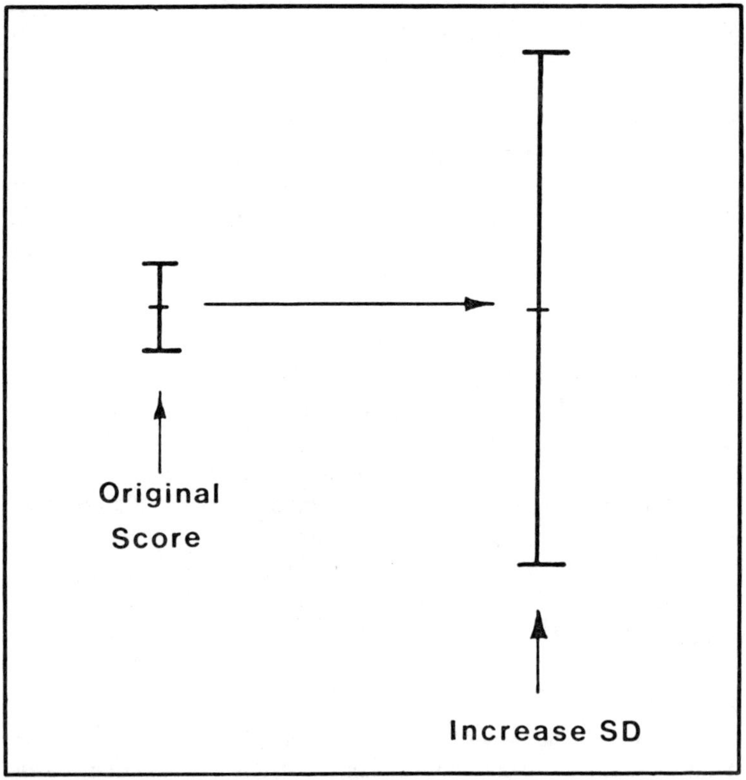

Fig. 4-6. The effect of changing the SD.

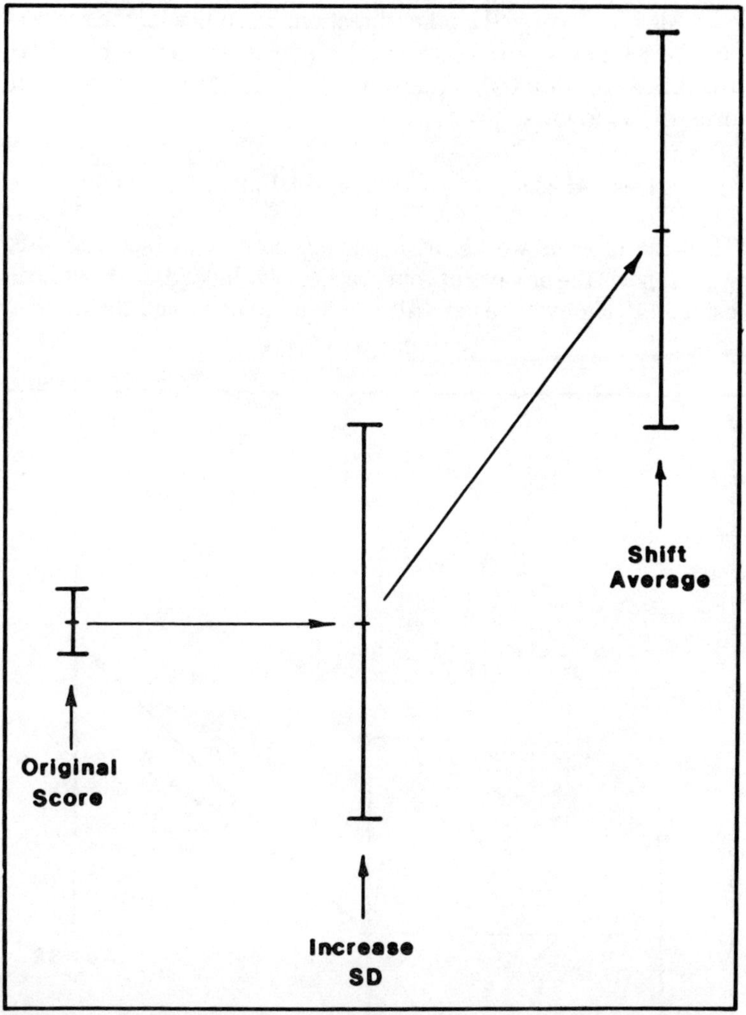

Fig. 4-7. The effect of changing the SD plus that of changing the Average.

corresponding to the score of 80 is −1.414214. (z = −1.414214 = (80 − 100)/14.1421, if you've forgotten). We want scores with a Mean of 500 and a Standard Deviation of 100. Putting all these values in Formula 4-2, we get:

$$\text{Desired score} = (-1.414214)*100 + 500 = 358.57864$$

You see, that wasn't very difficult.

Also, we may go the other direction. Starting with the score of 358.57864, we calculate the z-score to be −1.414214 ((358.57864−500)/100). Then that score is changed back to the original, as follows:

Desired score =(−1.414214)*14.1421 + 100 = 80

Thus the process works, as it should, both ways (see Fig. 4-8, which shows the process of going back to 80). Indeed, with our little computer, it's even easier. All you need to do is add the line

```
155  Z=Z*100+500
```

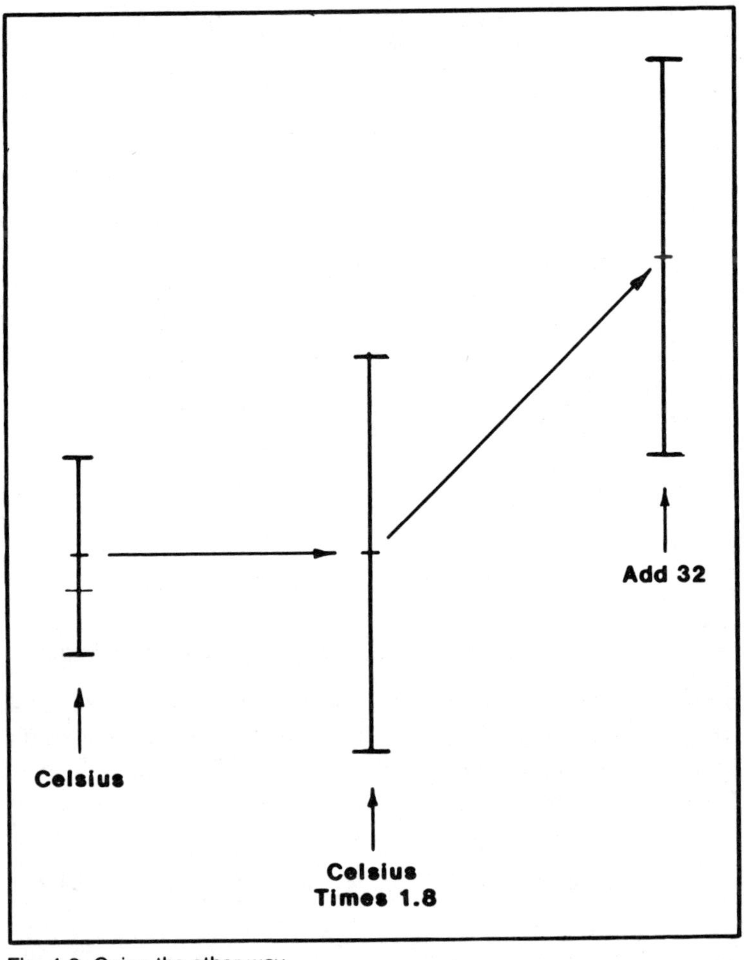

Fig. 4-8. Going the other way.

to Program 4-1 to do the calculations automatically. This may be a bit confusing, so the whole program is listed here:

```
10 REM PROGRAM 4.2. TRANSFORMED Z-SCORES
20 READ X                          :REM GET DATA
30 IF X=-1 THEN 80                 :REM TEST FOR END
40 S=S+X                           :REM SUM NUMBERS
50 Q=Q+X*X                         :REM SUM SQUARED NOS
60 N=N+1                           :REM GET COUNT
70 GOTO 20                         :REM CONTINUE
80 V=Q -S*S/N                      :REM TOP OF SD EQUAT
90 D=SQR(V/N)                      :REM GET STD DEV
100 A=S/N                          :REM GET AVERAGE
110 RESTORE                        :REM RESET DATA
120 PRINT "SCORES FOLLOW"
130 FOR I=1 TO N                   :REM LOOP TO GET Z'S
140 READ X                         :REM GET DATA AGAIN
150 Z=(X-A)/D                      :REM CALCULATE Z
155 Z=Z*100+500                    :REM TRANSFORM Z
160 PRINT I,Z                      :REM PRINT SCORE
170 NEXT I                         :REM END LOOP
180 DATA 80,90,100,110,120,-1 :REM DATA TO USE
```

You don't have to retype the whole program! It's only necessary to add the single new line (see Fig. 4-9). The full listing is only for your convenience.

If this program is run, the Standard Scores printed out will have the desired Mean and Standard Deviation, and will look like this:

Original Score	New Score
80	358.5788
90	429.2893
100	500.0000
110	570.7106
120	641.4214

Notice that your little computer does this new sort of score as rapidly as it did the z-scores! Line 155 may confuse you, however. This sort of statement confuses many people. Z=Z*100+500 may not strike you as a proper statement, but it is. Line 155 simply says, "take variable Z, multiply it by 100, then add 500 to the result, and put it all back in variable Z." This procedure will, of course destroy the original z-score in variable z. If you'd like to keep the z-score,

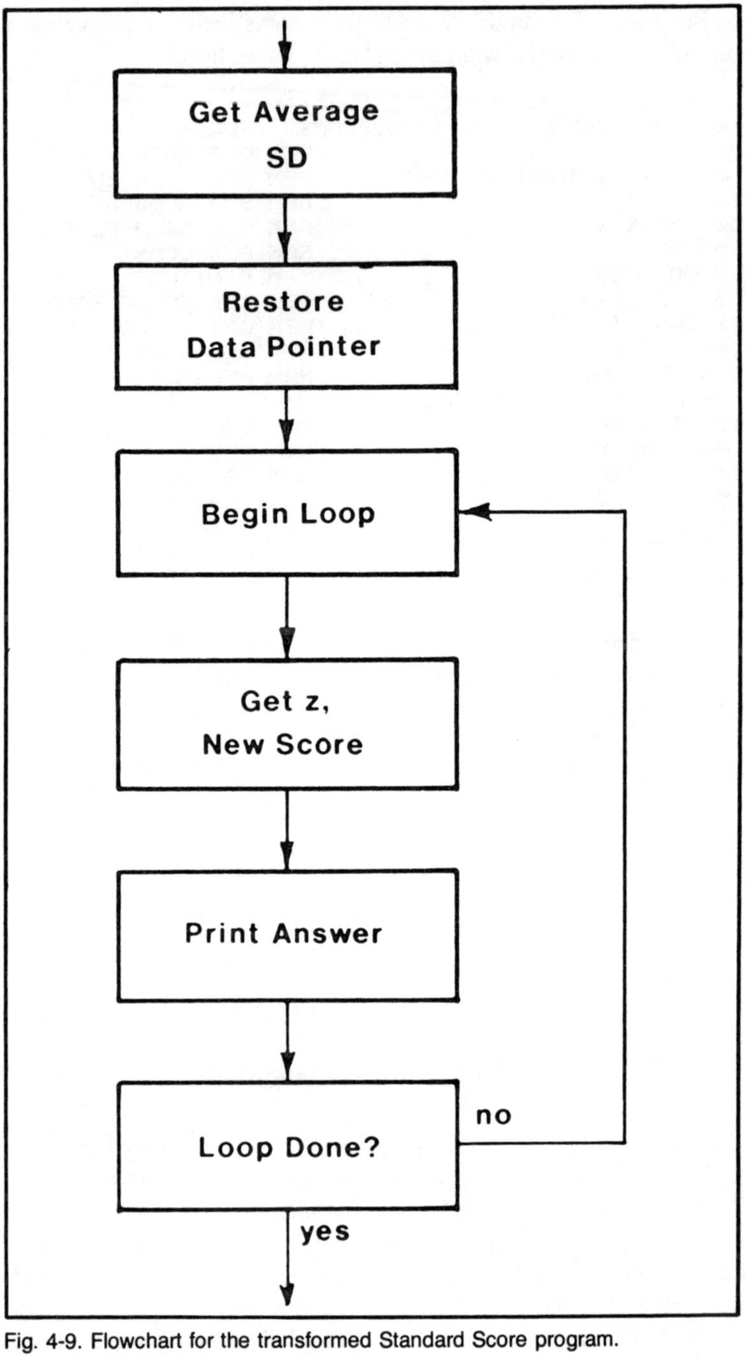

Fig. 4-9. Flowchart for the transformed Standard Score program.

and print out both scores, you could change the program as follows:

```
155 P=Z*100+500
160 PRINT I,Z,P
```

Such a procedure saves both scores in the program, and prints out both.

A PLETHORA OF TYPES OF STANDARD SCORES

Because of the ease with which transformations of this sort may be made, scientists have developed many types of Standard Scores. This is particularly so in my field, psychology, where there are so many different types of Standard Scores that people just can't keep up with them. For example, a z-score of +3.00 is precisely equivalent to each of the following Standard Scores:

Type of Score	Value Corresponding to z=+3.00
Wechsler IQ	145
Stanford-Binet IQ	148
Army Tests	160
Scholastic Aptitude Test (SAT)	800
T-Score	80

And there are many others! All of these scores are intended to indicate precisely the same amount of intelligence, and yet notice how widely they vary. Someone with a Wechsler IQ test score of 145 would obtain scores of 148, 160, 800, or 80 on other tests, assuming, of course, that all of the tests measured intelligence perfectly. It's no wonder that people become confused about test scores, or that schools are reluctant to release IQ test results. An illustration of this madhouse of scores is shown as Fig. 4-10.

If you're curious about the Averages and Standard Deviations of these various scores, the following list.

Type of Score	Average	SD
Wechsler IQ	100	15
Stanford-Binet IQ	100	16
Army Tests	100	20
SAT	500	100
T-Score	50	10

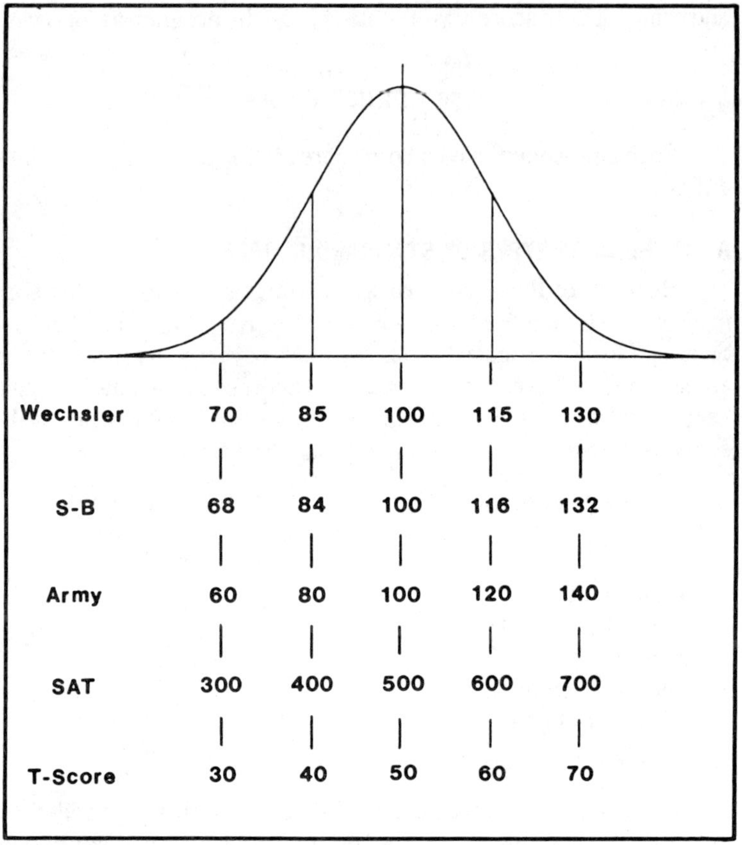

Fig. 4-10. Various types of Standard Scores.

You can have Program 4-2 print out any of these scores, simply by changing the Average and Standard Deviation in line 155. Or, if you like you have the programming skills by this time to have the program print out all of those scores, so that you can compare them. To do this, you'd need a lot of lines like line 155, one for each of the transformations you'd like to have, and a separate PRINT statement for each.

Until statisticians get their house in better order, keep the following fact in mind: A SCORE FOR WHICH THE MEAN AND STANDARD DEVIATION ARE NOT KNOWN IS MEANINGLESS. You simply can't compare scores until they have been reduced to scores with the same Average and Standard Deviation. There is almost no agreement among the various Standard Score systems, making comparison most difficult.

A VERY SIMPLE SCORE TRANSFORMATION PROGRAM

Programs 4-1 and 4-2 are fine, but sometimes you want to move from one Standard Score system to another when you already know the Averages and Standard Deviations. In such a case, calculation of the Averages and Standard Deviation becomes unnecessary. A very simple program will make the transformation for you.

You have the programming skills to write such a program without any assistance, but at the risk of insulting your intelligence, I'll give one. It's short:

```
10 REM PROGRAM 4.3. SIMPLE SCORE PROGRAM
20 INPUT "SCORE:";X              :REM GET SCORE
30 Z=(X-100)/15                  :REM GET Z-SCORE
40 P=Z*100+500                   :REM TRANSFORM
50 PRINT "NEW=";P                :REM PRINT IT
60 GOTO 20                       :REM CONTINUE
```

As set up, this program will convert Wechsler Intelligence Test scores, with an Average of 100 and a Standard Deviation of 15, to SAT scores, with an Average of 500 and a Standard Deviation of 100. The Mean and Standard Deviation for the original scores are found in line 30, and the same values for the desired, transformed scores are found in line 40. The very, very simple flowchart is shown in Fig. 4-11.

Type in this simple program, and run it. Try a few values. See how simple your little microcomputer and the BASIC language makes the understanding of statistics?

Notice that this program, and this chapter, uses no new programming concepts! We've almost completed a new chapter with the same old concepts, and you could easily have written the program yourself!

SOME PERSONAL COMMENTS ABOUT STANDARD SCORES

Some of my experiences about various Standard Scores systems might be helpful to you. I've used most of the various types of Standard Scores over the years, and there are certain things that can be learned only by long usage:

People seem to have particular difficulty in understanding T-scores, which have a Mean of 50 and a Standard Deviation of 10 (see table above). At the university where I teach, the results on classroom exams are routinely returned from the computer as T-scores. Students simply do not understand them. They feel that they must somehow be related to the 70 percent to 100 percent that

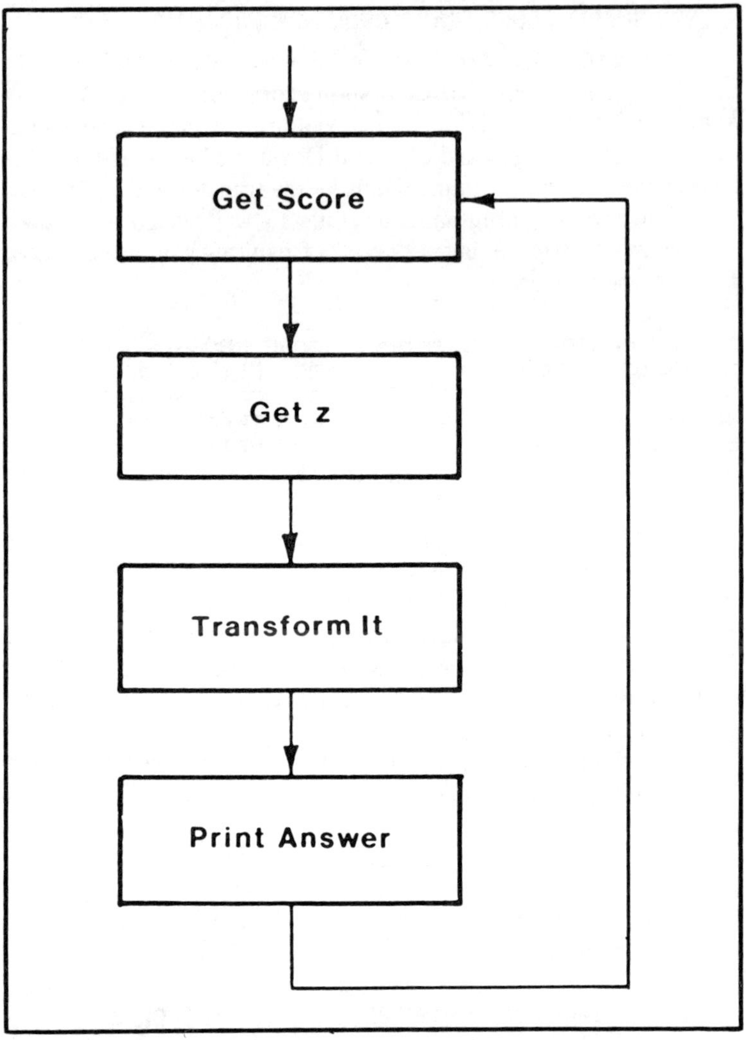

Fig. 4-11. Flowchart for the simple score transformation program.

they have become used to as the scoring scale in elementary and high school. They cannot understand that a score of 50 is average.

IQ-type scores might be more satisfactory, one might think, for people have become accustomed to IQs over the years. Acting on this idea, I once tried reporting class grades in the form of Standard Scores with an Average of 100 and a Standard Deviation of 20, calling these things “TIQs, for “Test IQ. The problem that developed with this system was simply that a few students got the idea that somehow I was giving them their intelligence test scores publicly, and objected strenuously.

Among college students, scores of the type used in the Scholastic Aptitude Test (SAT) (Average about 500, SD of 100) are more or less understood, but they tend to be a bit too fine in their scale. That is, on scores of this type it is possible for one student to obtain a score of 601, and another to obtain a score of 600. In this situation, it is possible that a college would accept the person with a score of 601, and not accept a person with a score of 600. Yet, of course, it is totally impossible for any existing test to measure with that sort of accuracy. The company that gives the SAT has for some time been reporting these scores to only two significant figures (that is, scores like 610, 620, but no 621s) to get around this problem. But the potential for unnecessary reliance on the scale's fineness remains, especially when people average scores (that is, 620 and 630 averaged yield 625, etc.)

LINEAR TRANSFORMATIONS

Standard Scores often seem obscure to students when they first encounter them, but you've been using things like them for years. Suppose you want to convert from degrees Celsius to degrees Fahrenheit. The formula that you would use has this form:

$$\frac{\text{Fahrenheit}}{\text{Degrees}} = \frac{9}{5}\ (\text{Celsius Degrees}) + 32$$

Thus, 20 degrees Celsius corresponds to 68 degrees Fahrenheit. Notice that this formula is of precisely the same form as Formula 4-2. The Celsius Degrees corresponds to z, 9/5 is like the Desired Standard Deviation, and 32 is very similar to the Average. The diagram of the Celsius to Fahrenheit conversion in Fig. 4-12 is remarkably similar to those earlier in the chapter. The Celsius to Fahrenheit and the Standard Score equations are very similar in form to this one:

$$Y = mX + b$$

If you've had Algebra, you should recognize that as the formula for a straight line. All of the conversions from one scale to another that have been discussed in this chapter are known as *linear transformations*, because of their similarity to the straight line just noted. Most

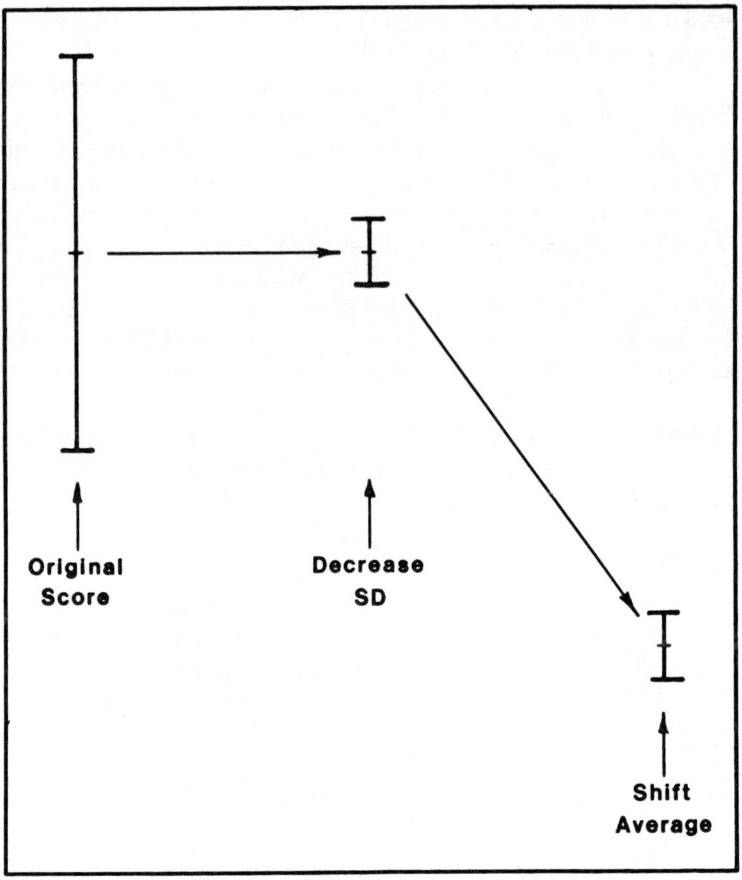

Fig. 4-12. Fahrenheit to Celsius conversion.

Standard Scores are simple linear transformations of precisely this sort.

PROBLEMS

1. Calculate z-scores for the IQ scores 100, 115, 130, 145 and 170. (First one is −1.3219.)

2. Calculate scores with an Average of 500 and a Standard Deviation of 100 for the same scores (First one is 367.8092.)

3. On one of the original IQ tests (Average = 100, SD = 16), it used to be said that people with IQs above 140 were geniuses. What

would that value change to in scores with an Average of 100 and a Standard Deviation of 20? (150)

4. Andy got an IQ of 100 on an IQ test (Average is 100, SD=15). What's his score on a test with an Average of 500 and a Standard Deviation of 100? (500—notice that the Average yields the Average).

5. Albert got a score of 70 on one IQ test (Average =100, SD =15). What should his score be on a test with an Average of 100, SD of 16? (68)

POKE IN 15360?

5
Plotting

The ability to make plots is frequently of great value to someone studying statistics. Your microcomputer is particularly well adapted to making fast, if crude, plots. With such plots, you can peruse the data and get a general idea of the form that it takes. In this brief chapter, I shall develop a procedure for making simple plots; the program which results will be of great value in later chapters. This will be a rather difficult chapter for some of you. It may be omitted without loss of continuity if it looks too intimidating, but it will be invaluable to you for what follows.

Unfortunately, the procedure for making plots is heavily dependent on the particular way in which your computer is designed and built. First, it must have a video screen of some kind. This means that my little pocket computer, the Sharp PC-1500, which can display only one line at a time, would not be a suitable machine for the purpose. Second, the various microcomputers vary greatly as to the number and width of line they display. My TRS-80, Model III has 16 lines of 64 characters each, and so on. This must be taken into account in the program.

Third, the precise way in which a dot is placed on the screen varies from machine to machine. The TRS-80 and the VIC-20 use the POKE command to put data into the screen, while the ATARI, Apple, and Texas Instruments computers use rather different systems. We'll emphasize the POKE method, but a little reading on your part should make you familiar with the other systems—they're all very similar.

HOW THE SCREEN IS ARRANGED

Most microcomputer screens are *memory mapped.* That fancy term means that the various spaces on the computer screen are addressed as part of the computer memory. I'll talk about the TRS-80 computer intially, and generalize to other computers later

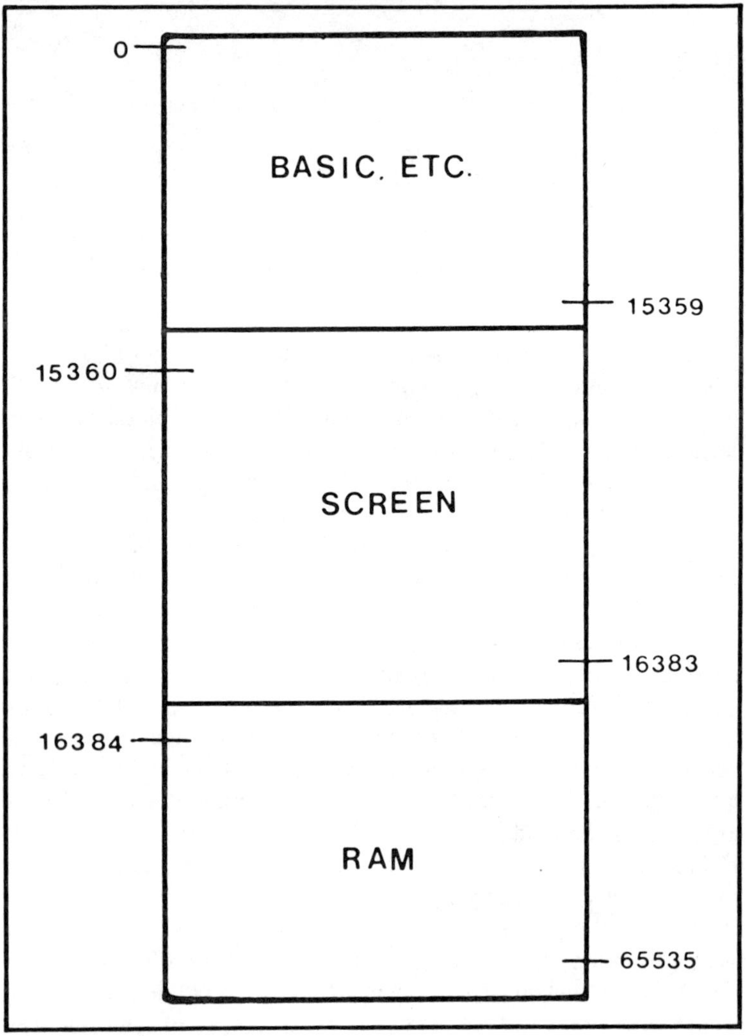

Fig. 5-1. TRS-80 memory map.

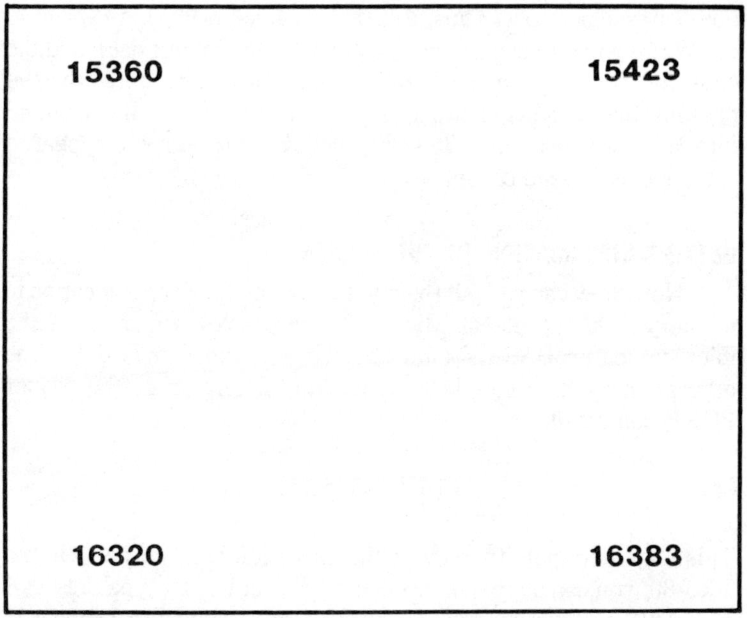

Fig. 5-2. TRS-80 screen.

in the chapter. Figure 5-1 shows the way in which the memory of my TRS-80 is laid out.

The lower 15359 memory locations (bytes) are used by the computer for its own purposes—BASIC is located there, the keyboard, and other things. The area from 16384 up is called random access memory, and that's where BASIC programs are located. It's generally available for your use in storing programs.

The video screen takes up the area from 15360 to 16383. This area corresponds directly to the screen itself, as shown in Fig. 5-2.

Thus, each place on your video screen corresponds directly to a location in memory (in the VIC-20, two locations in memory, but we'll get to that later). Now, if we can just put the proper number in each of the screen memory locations (POKE it), we can draw pictures on the screen.

Almost any of the characters found on your keyboard can be placed on the screen, for they all have different number representations. Virtually all computers use a coding scheme called ASCII, which stands for American Standard Code for Information Interchange, which is no doubt shown in the back of your computer's manual. In this code, each letter, number, and symbol has a specific number assigned to it. Thus, A is 65, B is 66, Z is 90, 1 is 49, and so on. We're going to use an asterisk, or star, to plot our data, and the code for * is 42. We won't talk about ASCII any more, for the present, but keep in mind that 42 is the code for a star. If you put 42 into Memory Location 15360 on a TRS-80., the star will appear in the upper left hand corner.

GETTING INFORMATION TO THE SCREEN

Now, how can you put the number 42 into the proper location in memory? The people at Microsoft, who developed most of the microcomputer BASICs (and set the *de facto* standard for the others), made this simple for you. You simply enter the proper POKE command.

```
POKE 15360,42
```

This statement puts the star in the upper left hand corner. On the TRS-80, you can do this in direct mode. That is, just type in POKE 15360,42, with no line number needed, and the star will appear in the upper left hand corner (see Fig. 5-3):

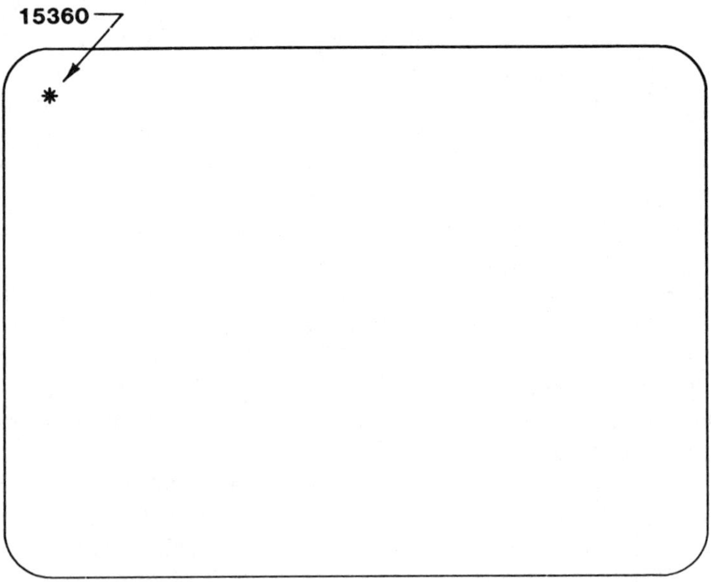

Fig. 5-3. A star in the upper left-hand corner.

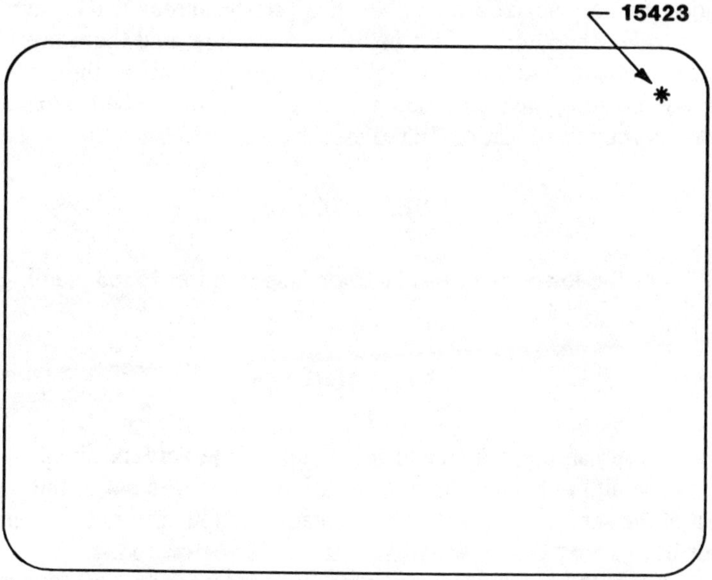

Fig. 5-4. A star in the upper right-hand corner.

WARNING!!! Be very careful in using POKEs. The POKE command can change any changeable part of memory, and if you type in the wrong number, and accidentally change any of the parts of memory that are important for the computer's operation, you will get the most unbelievable system crash that you've ever seen—the computer will go wild. (I usually get odd little symbols appearing all over the screen.) You can't hurt the computer this way, but you will certainly lose any program that you're working on, so be sure that you have a copy of the program on cassette or disk before you begin doing many POKEs. If you have a disk system, it's wise to take the disk out, too, for the computer becomes completely unpredictable at such times, and may decide to wipe your disk out too, for good measure.

I've probably scared you away from POKEs forever. They're really not that dangerous, and they're terribly useful.

Now, to put the star in the upper right hand corner of the TRS-80 screen, enter the following POKE (see Fig. 5-4).

```
POKE 15423,42
```

Notice that 15423 is 63 more than 15360, which is the start of the video memory space. Remember, the TRS-80 screen is 64 characters wide. 15360 is the first position in the line, and there are 63 more, making 15423 for the right hand position (this arithmetic is tricky; be sure you understand it). To put the star in the lower left hand corner use location 16320 (see Fig. 5-5):

POKE 16320,42

For the lower right hand corner, use location 16383 (see Fig. 5-6):

POKE 1683,42

Thus, you can place a star at any point on the screen, simply by referencing the proper memory location. If a star is put at the top left of the screen, that is, memory location 15360, a star at the very left in the next line down will be at 15360+64 or 15424.

If you think about it a bit, you could put a star anywhere you wanted, just by keeping track of where the stars are. Indeed, the

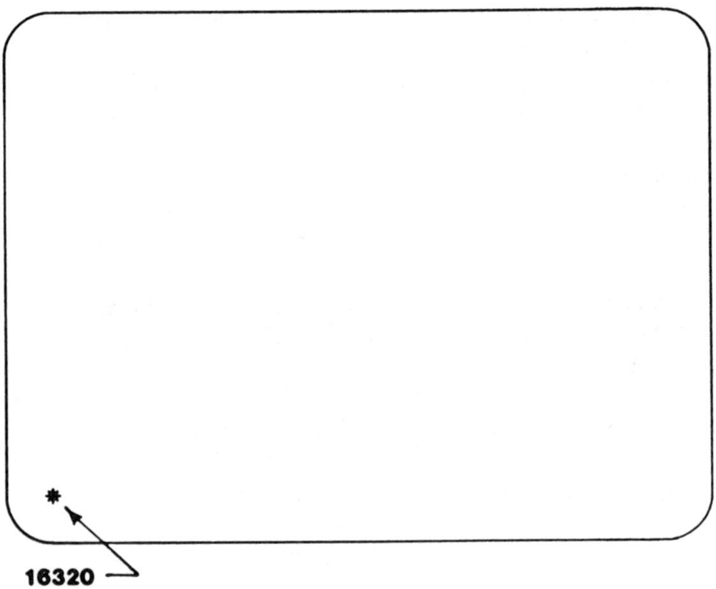

Fig. 5-5. A star in the lower left-hand corner.

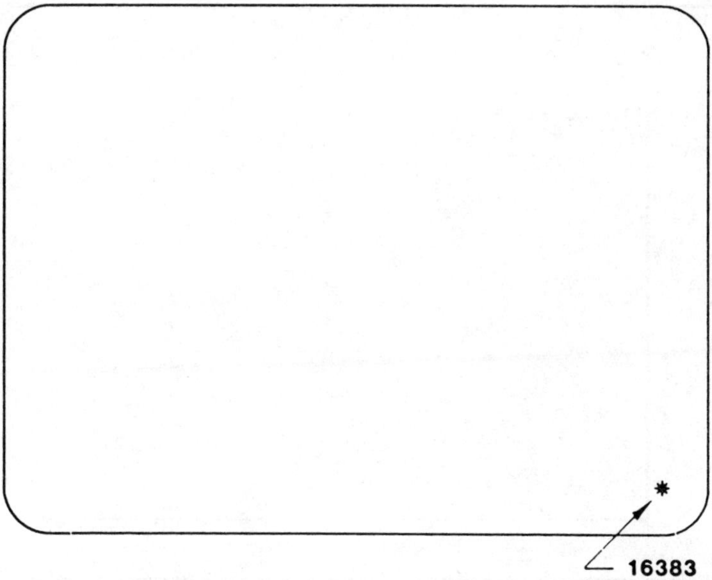

Fig. 5-6. A star in the lower right-hand corner.

following little program will fill the whole screen full of stars on the TRS-80:

```
10 FOR I=15360 TO 16383:POKE I,42:NEXT I
```

Other computers have different screen sizes, and different locations for the video memory, that is, the location in memory of the screen, but the principle is the same. We'll get to them later.

MATHEMATICAL AXES

Thus, the screen memory locations on your computer probably increase from left to right, and also increase from top to bottom. Alas, mathematicians have set up their graphs so that values increase from bottom to top. Figure 5-7 shows this arrangement.

Notice that the horizontal axis is called X, and the vertical axis is called Y. X increases to the right, just like our microcomputer, but Y increases from bottom to top, unlike our machine. This will, as you can imagine, make for certain complications in writing a simple plot program, but these are not insurmountable. To prove this, we'll now take up the plot program itself!

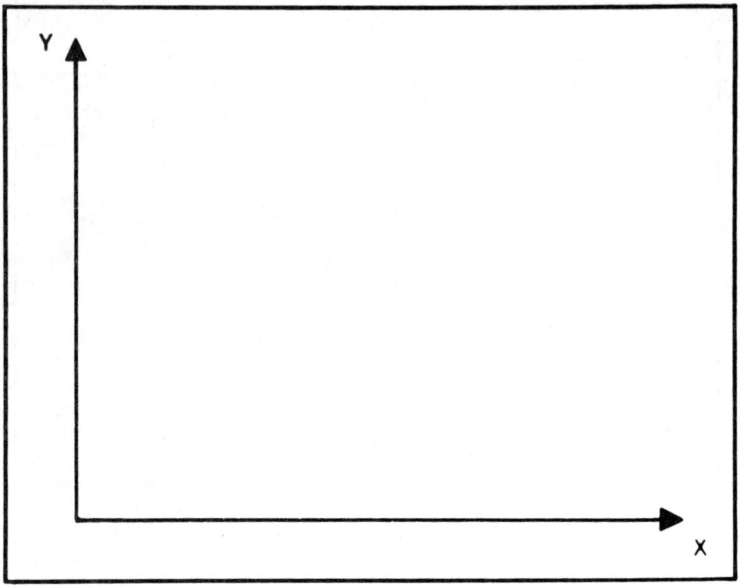

Fig. 5-7. The mathematical axes.

A SIMPLE PLOTTER

The following program, which looks a lot more complex than it really is, will do the plotting for us. It's about the simplest program that will do the job:

```
10 REM PROGRAM 5.1. GENERAL PLOTTER
40 T=16                              :REM SCREEN HEIGHT
50 L=64                              :REM SCREEN WIDTH
60 S=15360                           :REM SCRN MEM START
70 A=0:B=1000000                     :REM HI, LO X VALUES
80 C=0:D=1000000                     :REM HI, LO Y VALUES
200 FOR I=1 TO 10000                 :REM START LOOP
210 READ X,Y                         :REM GET DATA
220 IF X=-1 THEN 300                 :REM CHECK FOR END
230 IF X>A THEN A=X                  :REM HI  X?
240 IF Y>C THEN C=Y                  :REM HI  Y?
250 IF X<B THEN B=X                  :REM LOW X?
260 IF Y<D THEN D=Y                  :REM LOW Y?
270 N=N+1                            :REM KEEP COUNT
280 NEXT I                           :REM END LOOP
300 W=A-B:W=(L-1)/W                  :REM WIDTH  FACTOR
310 H=C-D:H=(T-1)/H                  :REM HEIGHT FACTOR
320 RESTORE                          :REM RESET DATA
330 CLS                              :REM CLEAR SCREEN
340 FOR I=1 TO N                     :REM LOOP FOR PLOT
350 READ X,Y                         :REM GET DATA
```

```
360 P=INT((X-B)*W)                   :REM WIDTH  CALCULAT
370 Q=INT((Y-D)*H)                   :REM HEIGHT CALCULAT
380 F=L*(T-1-Q)+P                    :REM SCREEN LOC CALC
390 POKE (S+F),42                    :REM PUT * ON SCREEN
410 NEXT I                           :REM PLOT LOOP
420 GOTO 420                         :REM WAIT TO VIEW
500 DATA 1,1                         :REM DATA TO USE
510 DATA 2,2
520 DATA 3,3
530 DATA 4,4
540 DATA 5,5
550 DATA -1,0
```

Don't panic. It isn't as bad as it looks. Let's take it line by line. Notice that some line numbers are missing. Don't worry about it now; it's for a purpose. The flowchart in Fig. 5-8 outlines the program, although in very abbreviated form.

Lines 10 through 60 set up certain things that the plotter has to know. Specifically, the screen height (16 on my TRS-80, 23 on my VIC-20), the screen width (64 on my TRS-80, 22 on the VIC-20), and the starting point for video memory locations (15360 and 7680, respectively). The program needs these to make its calculations.

Lines 70 through 280 find the high and low values for the X and Y variables. You see, we want the plot to fill the whole screen, and so we must scale the plot to fit the screen. That is, if the X numbers go from .01 to .11, we want it to fill the screen just as well as the numbers 100 to 1100, and to do that we must know the maximum and minimum values for each variable. This is done just exactly as we did it in Program 3-2 for the range. For example, variable A, which will contain the high value for X, is compared again and again with the new X. If the new X is higher than A (the largest value to date), then A is changed to the new value. The same procedure is followed for C, the high value for Y. Similarly, B and D are made to the low values for X and Y, respectively. As usual, we know the data is all in when a minus 1 (−1) is encountered. Finally, the program is cycled through this section by a FOR-NEXT loop; this method is used to simplify a later program in this chapter.

Now things get tough! Line 300 calculates the scaling factor for the width, the X variable. The first part of line 300 states: W = A − B. This is the *range* of the X values, since A is the largest X value, and B is the smallest. Assuming we use the data in the data statements at the end of the program, A-B will be 4. Now, we must scale this to the width of the screen. The second part of line 300 says, W=(L−1)/W. L is the length of the screen, or 64 for the TRS-80. To make things work properly, we need one less than this

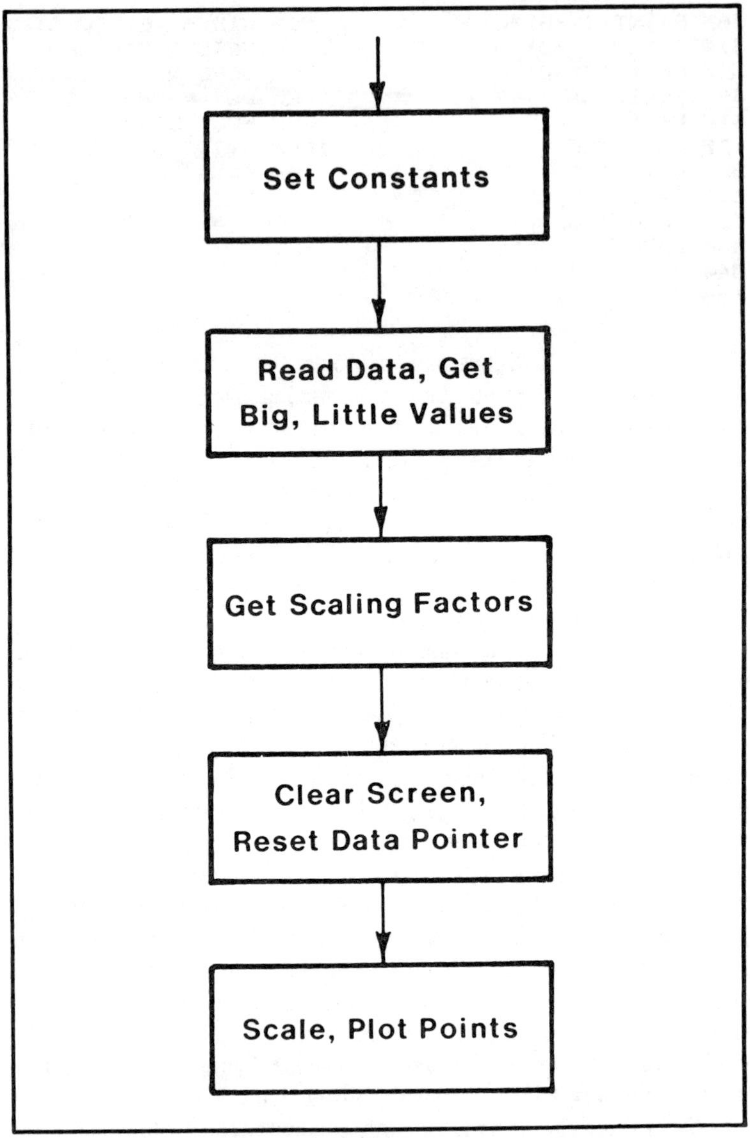

Fig. 5-8. Flowchart for the general plotter program.

value, or 63. Thus, 63 is divided by 4, making the width factor 15.75. That's complex isn't it? However, it's necessary to do it that way to make things work! 15.75 will be the number by which each value is multiplied to make things fit best on the screen. Let's see if it works. An X value for the data point, 5, should be at the right edge of the screen. Assuming it's on the top line, the left edge of the screen would be at memory location 15360. Now, 15.75* (5–lowest X) or 15.75* (5–1)=63. 15360+63=15423, the right edge of the top line (see Fig. 5-2).

Line 310 does the same thing for the values on the Y axis. You need a scaling factor for those values, also.

Line 320 resets the data pointer so that you can reread the data when the actual plot is made. Line 330 clears the screen so that the plot won't be cluttered up with a lot of old symbols.

Lines 340 through 410 actually plot the points. Line 340 sets up the loop that plots each point. Line 350 gets the data back again. Line 360 calculates the distance the point should be plotted from the left edge, just as shown above. That is, the lowest X value (B) is subracted from each new X value, and then that's multiplied by the width scaling factor, W.

Then a new BASIC function is used, one we haven't encountered before. INT means take the integer value of. INT simply drops the decimal value from a number. Thus, if we have the number 22.25, the function INT (22.25) yields 22. We want the integer value, since there are only 64 possible points on the X axis. If the computer tries to plot a point at memory location 15382.25, it's going to run into trouble, for there are only locations 15382 and 15383. It wouldn't know where to put 15382.25. INT takes care of this problem for us.

Line 370 figures out the vertical distance needed in precisely the same fashion as line 380 did for the horizontal distance.

Line 380 is the grand finale, and is the most complex. It does two things: first it calculates the point on the screen at which the asterisk is to be placed, and then it takes care of the problem with the vertical axis. Remember that mathematicians want the axis to increase as we go down. For the TRS-80, Line 380 can be written as follows:

```
380  F= 64  *  (15–Q) + P
```

You see, each line is 64 characters long. Thus, for each Q, or vertical value, we need to include 64 characters. So you might think that we need to multiply 64 by Q. Aha, but the screen is upside down! Therefore, we must subtract Q from 15, which is one less than 16. The difference of one is necessary to make things work properly. Finally, we add the horizontal distance, P, back in again at the end.

Line 390 plots the point by POKEing an asterisk, 42, at the proper memory location, gotten by adding the value for F, obtained in the last line, to the starting address for the video memory, which for the TRS-80 is 15360.

Good heavens, that's complicated! And so it is. There is nothing more difficult, or tricky to write, or understand, than plotting programs. They can drive any programmer berserk without any difficuly at all. It took me a week to get Program 5-1 working properly. If you don't understand the program, don't worry about it; you can always come back to it later. It works, and it will allow you to examine your data.

TRYING THE PROGRAM

If you run the previous program, the screen will have the plot of Fig. 5-9. **Be sure to save the program on cassette or disk before running it.** One false POKE and all will be lost. Notice that this is a nice, straight line. If the data in Program 5-1 were as follows the plot will look like the one in Fig. 5-10.

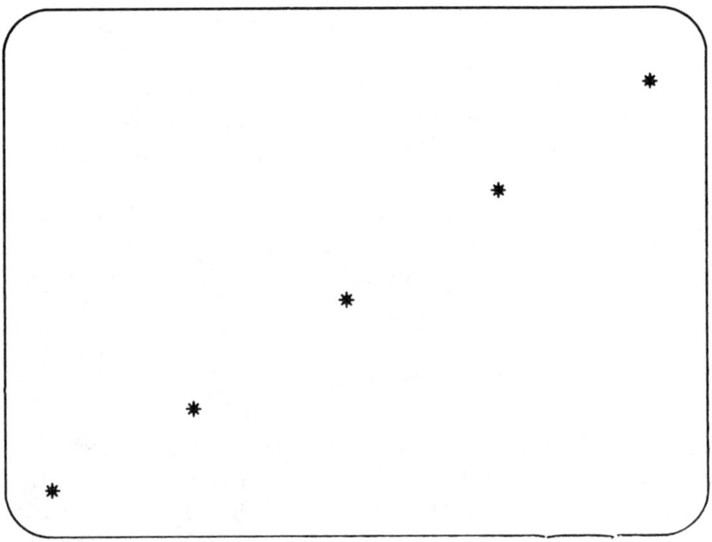

Fig. 5-9. A straight line plot.

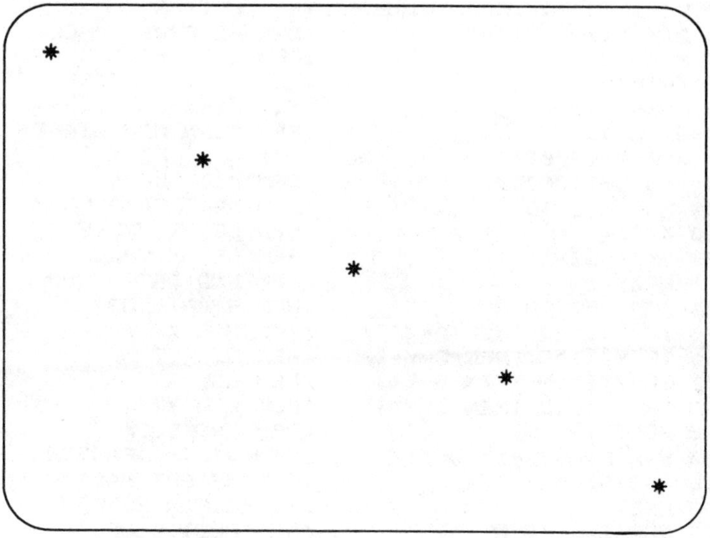

Fig. 5-10. A straight line plot, sloping down.

```
500  DATA   1,5
510  DATA   2,4
520  DATA   3,3
530  DATA   4,2
540  DATA   5,1
550  DATA  -1,0
```

The plots are not very elegant, but they give an excellent idea of the arrangement of the data. The illustrations, by the way, are screen printouts of the screen of my TRS-80. (Actually, they were done on my Radio Shack pen plotter; real screen printouts are oddly compressed. Therefore, the vertical spacing has been increased somewhat over that of the internal TRS-80 screen printout routine.)

A SECOND PLOTTING PROGRAM

Program 5-1 is fine, but it has one defect: you must supply all the data. It is fun to plot mathematical functions, many of which are provided directly in your microcomputer. However, if you try to do that with Program 5-1, you have to type in all the data. Program 5-2 allows you to have the computer do all the work. After all, that's what computers are for!

```
10 REM PROGRAM 5.2. DIMENSIONED PLOTTER
20 DIM X(70),Y(70)             :REM RESERVE SPACE
30 N=30                        :REM # OF POINTS
40 T=16                        :REM SCREEN HEIGHT
50 L=64                        :REM SCREEN WIDTH
60 S=15360                     :REM SCRN MEM START
70 A=0:B=1000000               :REM HI,  LO X
80 C=0:D=1000000               :REM HI,  LO Y
100 FOR I=1 TO N               :REM MAKE DATA LOOP
110 X(I)=I                     :REM SET X VALUE
120 Y(I)=SIN(I/4)              :REM SET Y VALUE
130 NEXT I                     :REM END DATA LOOP
200 FOR I=1 TO N               :REM START LOOP
230 IF X(I)>A THEN A=X(I)      :REM HI  X?
240 IF Y(I)>C THEN C=Y(I)      :REM HI  Y?
250 IF X(I)<B THEN B=X(I)      :REM LOW X?
260 IF Y(I)<D THEN D=Y(I)      :REM LOW Y?
280 NEXT I                     :REM END LOOP
300 W=A-B:W=(L-1)/W            :REM WIDTH  FACTOR
310 H=C-D:H=(T-1)/H            :REM HEIGHT FACTOR
330 CLS                        :REM CLEAR SCREEN
340 FOR I=1 TO N               :REM PLOT LOOP
360 P=INT((X(I)-B)*W)          :REM WIDTH  CALC
370 Q=INT((Y(I)-D)*H)          :REM HEIGHT CALC
380 F=L*(T-1-Q)+P              :REM SCREEN CALC
390 POKE (S+F),42              :REM PLOT POINT
410 NEXT I                     :REM END LOOP
420 GOTO 420                   :REM WAIT FOR BREAK
```

Now you know the reason for the missing line numbers in Program 5-1. The numbers are used only in Program 5-2. By keeping the numbering systems consistent, you can compare the programs easily. They're almost identical.

Line 20, however, contains a new concept—that of a dimensioned statement. In mathematics, variables are frequently identified with subscripts, such as X_1, X_2, X_3, and so on. It would be nice to use subscripts with BASIC, but of course there's no way to write little numbers below the line on your microcomputer display. Consequently, to indicate a subscript in BASIC, one simply uses parentheses, ( ) to indicate the subscript. Thus, X_1 becomes X(1), X_2 becomes X(2), and so on. You can see an example of a subscripted variable in line 110 of Program 5-2.

```
110   X(I)=I            :REM SET X VALUE
```

X(1) indicates a subscript.

The nice thing about subscripts in BASIC is that the subscript itself may be a variable. If you think about it a bit, that means that the program itself can select which bit of data it will use by changing the subscript. Subscripted variables allow an enormous amount of flexibility to programmers; without them many, many computations would be impossible, or at best very difficult. Subscripted variables, such as X(I), above, are called *arrays* or sometimes *vectors* by computer experts.

If a subscripted variable is used in a program, it makes sense that the program has to know the range of the subscripts before it can proceed. Thus, there may be 100 possible Xs—X(1) through X(100). Or there may be a thousand of them—X(1) through X(1000). Since the program has to divide the memory of the computer up between storage for programs and storage for numbers, it has to know how large the arrays are before it can proceed. That's the purpose of the dimension statement.

```
20    DIM X(70),Y(70)
```

in line 20 of Program 5-2, above. It says, in plain English, reserve 70 spaces for X and 70 spaces for Y. DIM stands for dimension. When BASIC encounters line 20, it will set aside 70 spaces for each subscripted variable.

It follows that the DIM statement must come before any of the subscripted variables are used in the program. If it doesn't, your microcomputer will object loudly! Also, it follows that a DIM statement can rapidly use up all the memory in your computer. Most BASICs use four or more bytes to store a number, and arrays take a bit more than that. Consequently, DIM X(70) uses 280 bytes, minimum.

Thus, the two new concepts in Program 5-2 are both related to the need for subscripts. DIM sets aside the storage needed for the arrays (subscripted variables), and the subscripting itself is indicated by parentheses, as in A(18). That means the eighteenth value of A.

PROGRAM 5-2 EXPLAINED

Program 5-2 is almost exactly like Program 5-1, with the exception of the subscripted variables (see Fig. 5-11). Adding the subscripts necessitates the addition of the DIM statement in line 20. In addition, wherever X or Y appeared in Program 5-1, X() or Y() (the ( ) indicating subscripts) appears in Program 5-2. Let's look at the similarities; we'll take up the new elements later.

Lines 10 through 80 are very similar to those in the previous

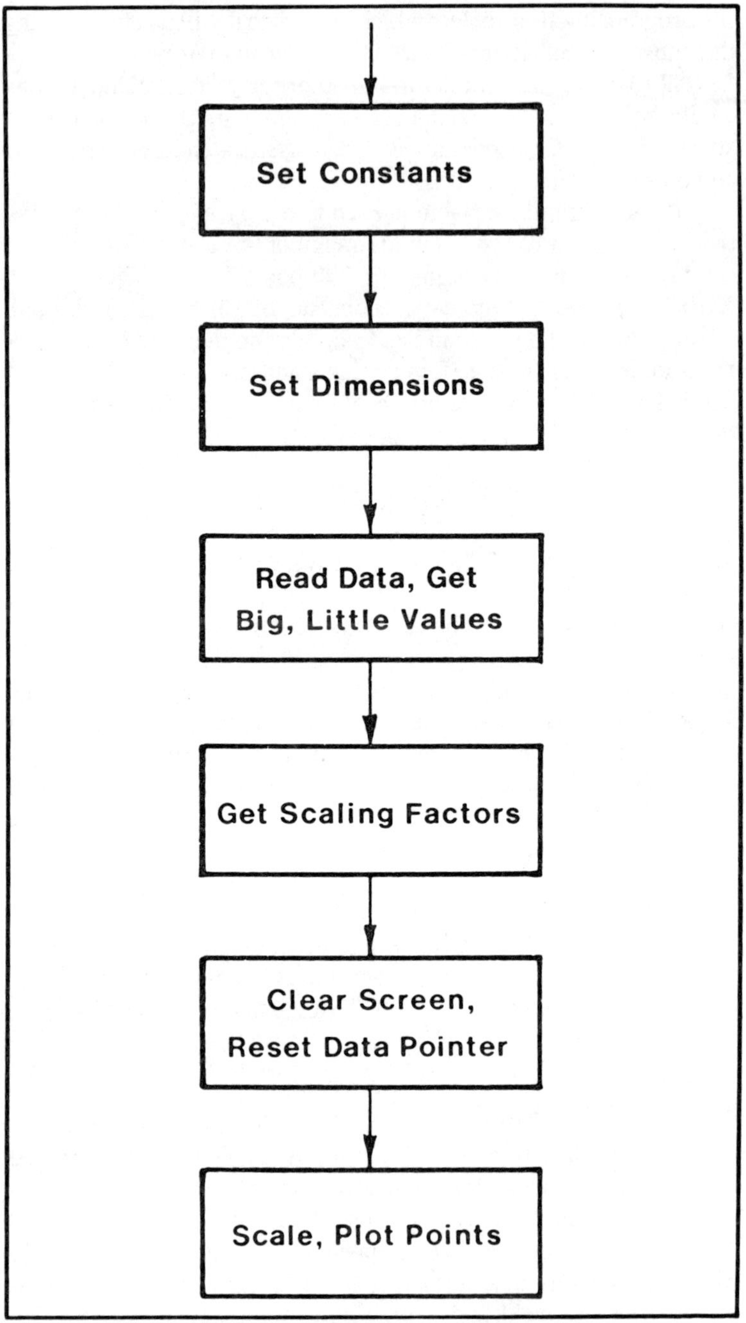

Fig. 5-11. Flowchart for the dimensioned plotter program.

program. Of course we need the DIM statement, and, in Line 20, we have to specify the number of data points we want, since they won't be counted as they're read in from the DATA statements.

Lines 200-410 are almost identical in both programs, except for the fact that subscripted variables are used in the newer program. Whenever X appears in the old program, X( ) appears in the new program, and the same is true for Y. We don't have to read in any data, since it will be generated earlier in the program, and consequently no READ or RESTORE statements are needed. Also, we know how many data points there are (in Variable N), so there's no need to count them as in line 270. If you understood lines 200-410 of Program 5-1, you should understand the lines in the newer program. Of course, the DATA statements in Lines 500 and above are not needed.

Lines 100-130 generate the data for the plot. Let's reproduce them here, so that we can examine them separately:

```
100   FOR I=1 TO N        :REM MAKE DATA LOOP
110   X(I)=I              :REM SET X VALUE
120   Y(I)=SIN(I/4)       :REM SET Y VALUE
130   NEXT I              :REM END DATA LOOP
```

In this loop, each of the subscripted variables to be used is given a value. First of all, the various X values are set equal to the I variable of the FOR-NEXT loop. Thus, X will take on the values 1, 2, 3, . . ., 30 in this program.

In line 120, the Y values are set equal to the sine (SIN) of the I values. Actually, the plot looks better if they're set equal to the sine of a quarter of the I values, and thus the /4 in line 120. BASIC has a great number of built-in mathematical functions: sines, cosines, tangents, logarithms, exponentials, and a great many more. Notice, however, that the trig functions, such as sine and cosine, are in radians, not degrees. Most people don't understand radians very well; one radian corresponds to 57.2958 degrees. It's easier for your computer to evaluate angles in radians, and so it does them that way. (My little Sharp PC-1500 pocket computer lets you use degrees; that's a nice feature).

Thus, in Lines 100 to 130, the X and Y arrays will be set up as follows:

Subscript	**X/4**	**Y**
1	.25	.2474
2	.50	.4794

Subscript	X/4	Y
3	.75	.6816
4	1.00	.8415
	(and so on)	

Thus, Program 5-2 generates all the data for you, automatically. That's neat. Even better, it will generate all the data that you want. If you want 70 data points, just change the number in line 30 to 70. If you want 200 data points, change the numbers in lines 20 and 30 to 200. Of course, it's necessary to make certain that you have enough storage saved for the data in the DIM statement (you might run out of computer memory in this way, if you have a small computer).

TRYING THE DIMENSIONED PROGRAM

If you run Program 5-2, you'll get a sine curve on the screen of your computer, as shown in Fig. 5-12.

Figure 5-12 is a direct copy, on my plotter, of the screen of my TRS-80. That's not the most elegant plot of a sine curve that you'll find, but it's certainly recognizable. Figure 5-13 shows a much finer-grained plot, a sine curve plotted without the screen size limitations of the TRS-80. Obviously, the screen plot is severely

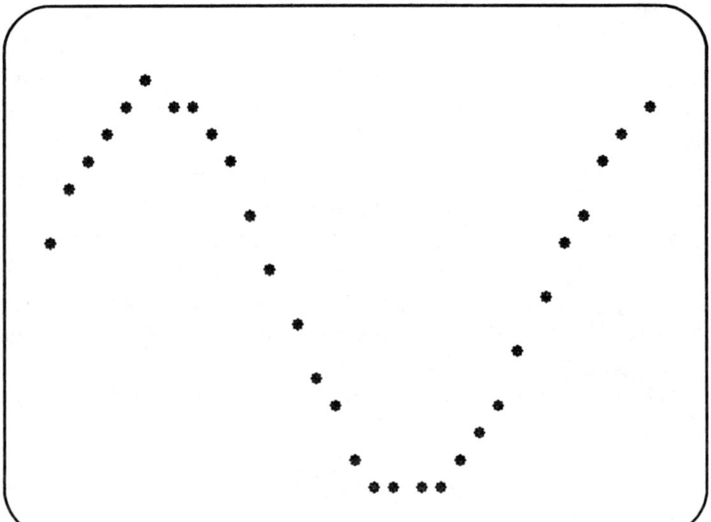

Fig. 5-12. A sine curve.

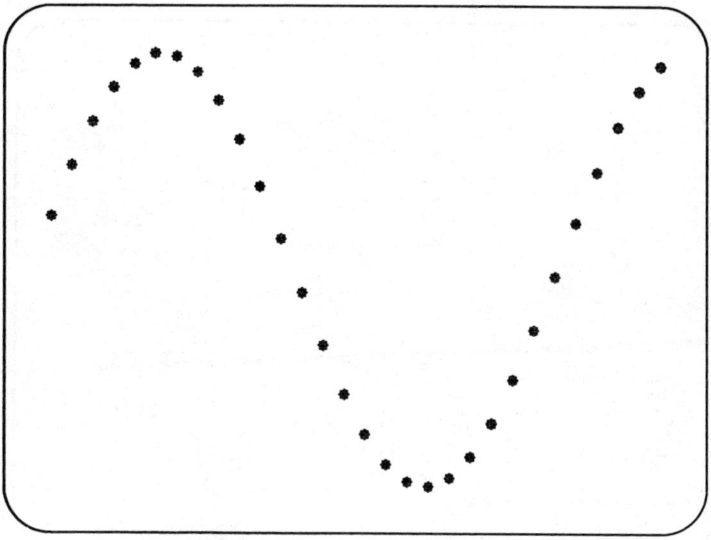

Fig. 5-13. A finer-grained sine curve.

limited by the lack of detail in the 16 × 64 screen. Figure 5-14 is a line plot of a sine wave by the plotter, and shows what it really looks like.

Change line 120, above, so that it reads:

```
120  Y(I)=I ↑ 3
```

That is, the Y values will be set to the cube (third power) of I. If you

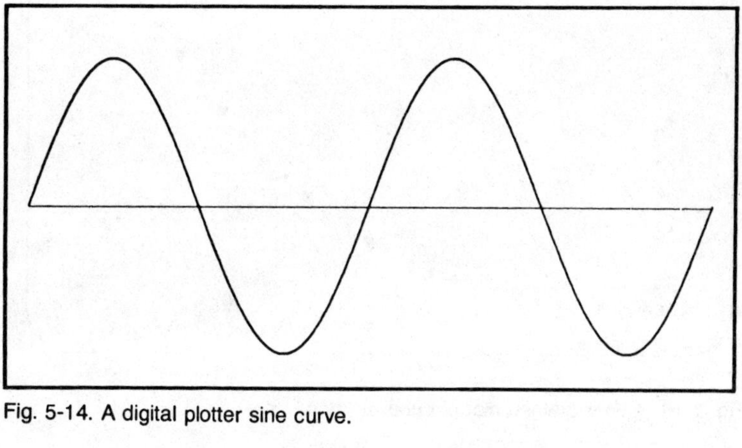

Fig. 5-14. A digital plotter sine curve.

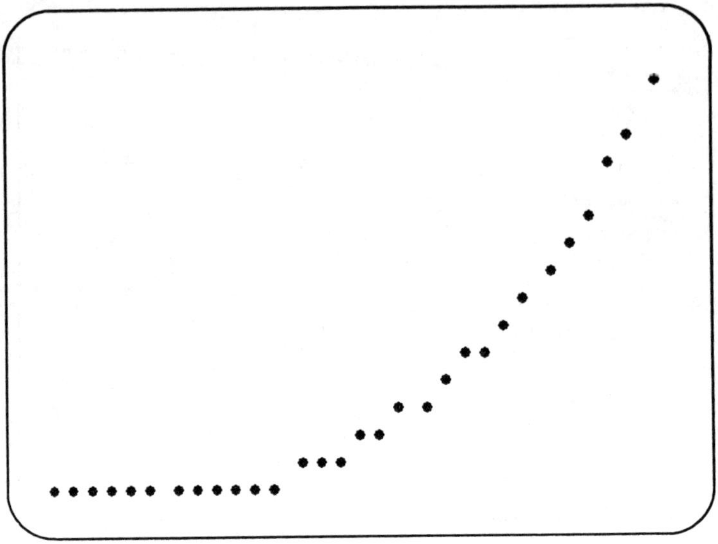

Fig. 5-15. A TRS-80 plot of cube function.

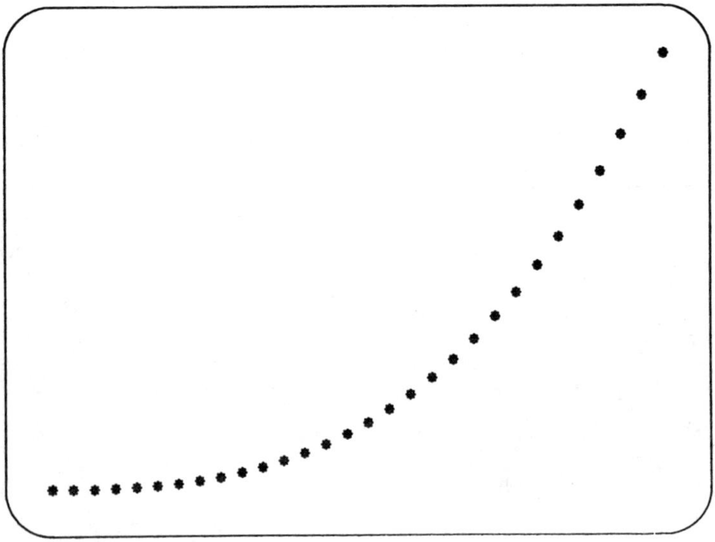

Fig. 5-16. A finer-grained plot of cube function.

run the program, you'll get a plot like that of Fig. 5-15. This plot is, again, rather crude, but you can see the rapidly increasing values obtained by taking the cube of a number. Figure 5-16 is the same plot, without the screen limitations.

Now, try the line

```
120  Y(I)=TAN(I/8)
```

This will plot a curve of the tangent function, and the screen plot from my computer looks like that in Fig. 5-17. Oops! That's not so clear. The limited screen resolution rather fails us in this case. Even the finer-grained plotting program has trouble with this one (See Fig. 5-18). The difficulty with the tangent function lies in the fact that as an angle approaches 90 degrees (1.57 radians), the tangent becomes infinity. You can't plot infinity on any normal plotter! And so things look very strange. Indeed your computer may object strenuously if the program tries to take the tangent of precisely 90 degrees, although that value is unlikely. This may be seen clearly in Fig. 5-19, a true line plot of the tangent function.

To plot a parabola (Y = X squared), change line 120 to:

```
120  Y(I) = (1-N/2) ↑ 2
```

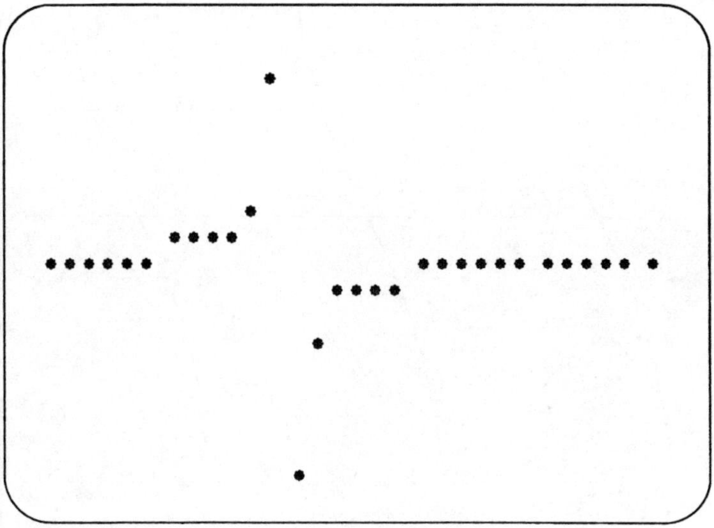

Fig. 5-17. A TRS-80 tangent plot.

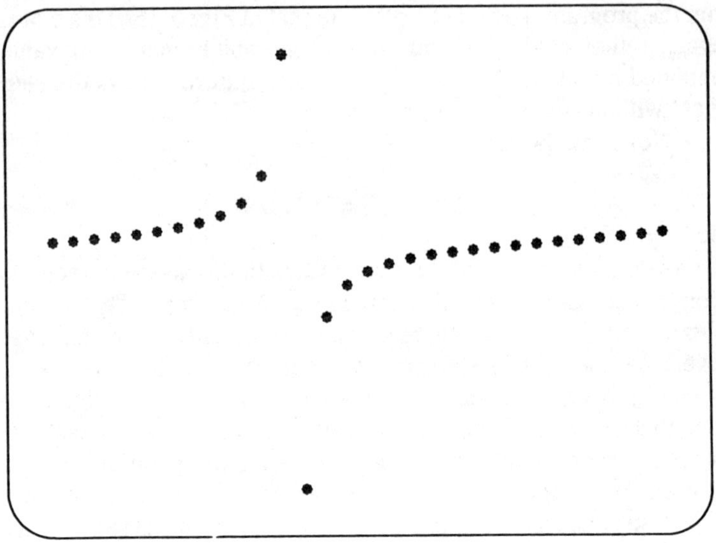

Fig. 5-18. A finer-grained tangent plot.

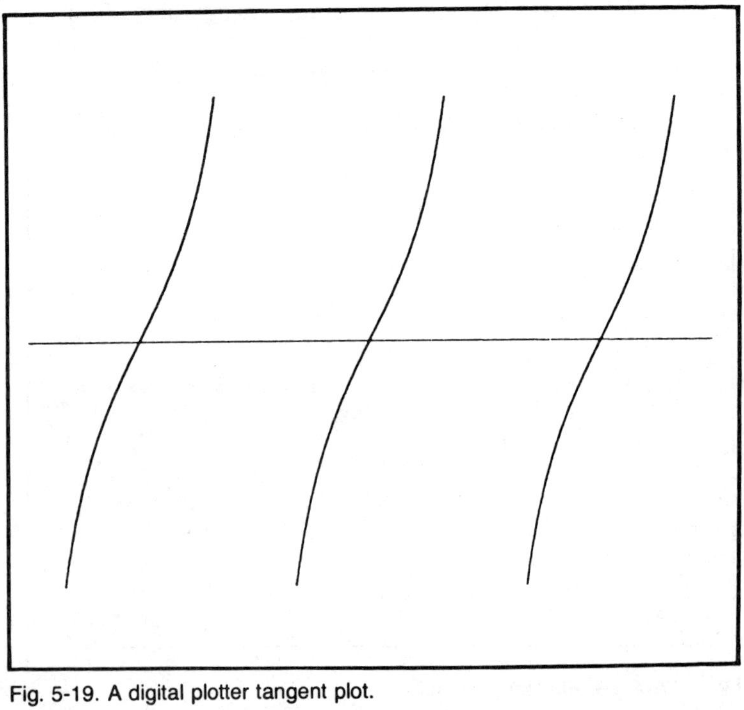

Fig. 5-19. A digital plotter tangent plot.

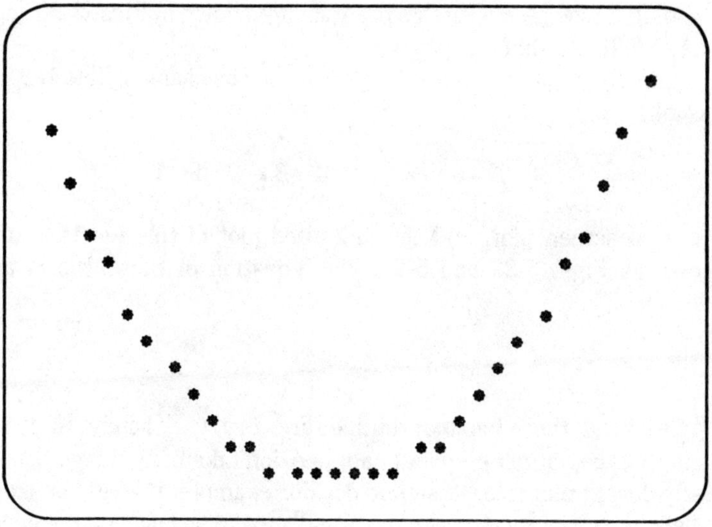

Fig. 5-20. A TRS-80 parabola plot.

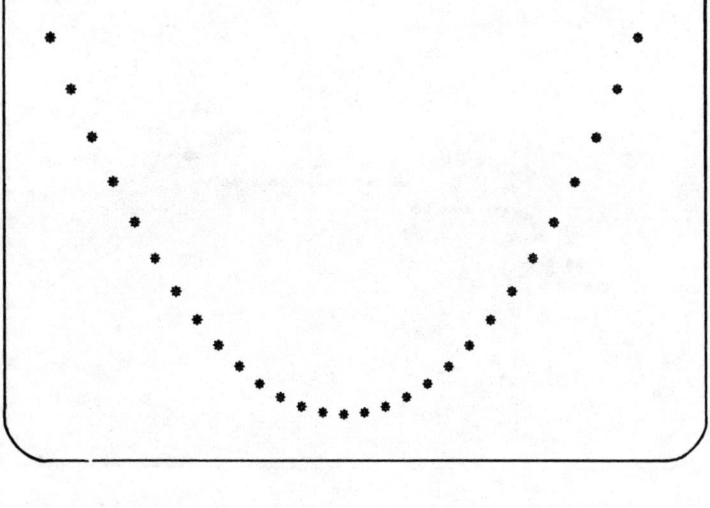

Fig. 5-21. A finer-grained parabola plot.

The plots associated with this program are shown in Figs. 5-20 and 5-21. We'll use them in Chapter 7.

Finally, a cubic equation may be plotted by changing line 120 to the following:

```
120 II=I-N/2:Y(I)=II ↑ 3 - 3*II ↑ 2
```

The real screen plot, and the unlimited plot of this function are shown as Figs. 5-22 and 5-23. The equation of the curve is as follows:

$$Y = X^3 - 3X^2$$

All in all, these two plot routines are very, very handy. Keep in mind that they aren't elegant at all, and don't do many things that a really decent plot routine should do. For example, it would be nice to have the program also plot the X and Y axes, and show the values on these axes, or show how many points there are at each screen location. However, if those features were incorporated into the plot programs, things would become much, much more complicated, and the programs couldn't be understood very easily by a novice. Consequently, we'll stick with the crude little plot routines that we have

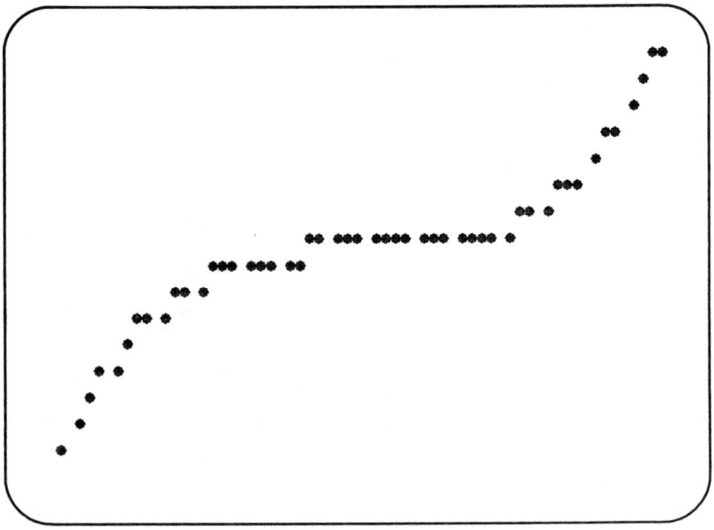

Fig. 5-22. A TRS-80 cubic equation plot.

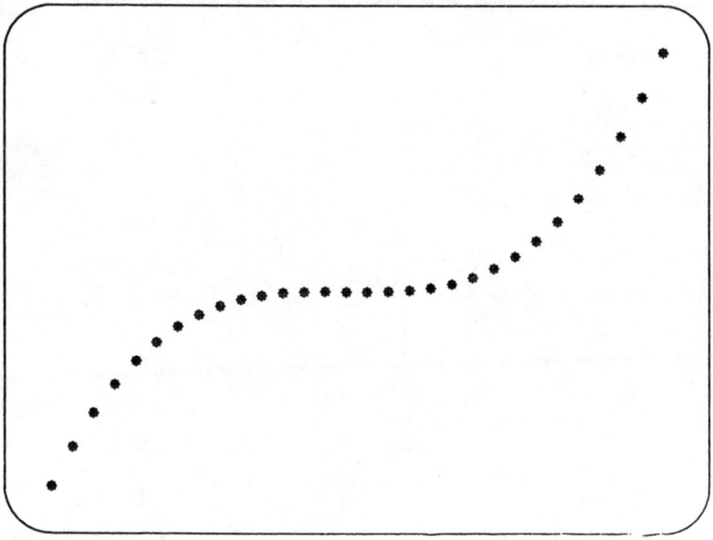

Fig. 5-23. A finer-grained cubic equation plot.

here. If you want more elegant ones, there are no doubt books available for your computer that show very sophisticated plot routines—and of course you could always use a pen plotter to draw really elegant plots.

USE OF THE PROGRAMS WITH OTHER COMPUTERS

As mentioned above, there are great variations in the ease with which these programs may be adapted to other computers. Basically, it is necessary to change the screen width, screen height, and location of video memory, to suit the computer.

For my VIC-20, the length of a line in the screen is 22 columns, the number of columns on the page is 23, and the video memory starts at memory location 7680. Consequently, the values in lines 40 to 60 should be changed as follows:

```
40 T=23      :REM SCREEN HEIGHT
50 L=22      :REM SCREEN WIDTH
60 S=7680    :REM SCRN MEM START
```

In addition, since the VIC-20 is a color machine, it is necessary to tell it which color to use to plot the asterisk. This is done by adding line 400.

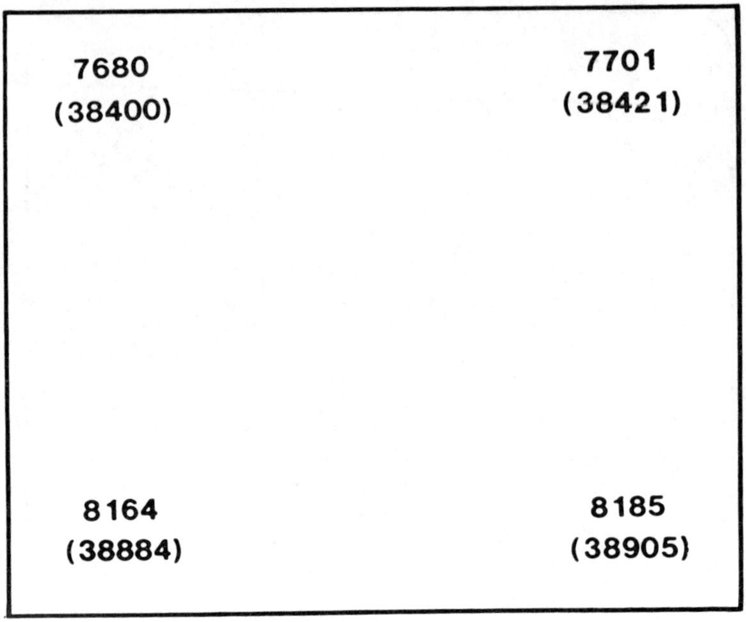

Fig. 5-24. The VIC-20 screen.

```
400    POKE (S+F+30720),0          :REM SET COLOR
```

This places the code for black in the color memory of the VIC-20. The color memory is set up just like the screen memory, but it starts at memroy location 38400, or 30720 units away. A diagram of the VIC-20 screen is shown as Fig. 5-24. Indeed, there are so many changes in the VIC-20 program, that Program 5-1 is reproduced here, with the changes:

```
1Ø REM PROGRAM 5.1A. VIC-2Ø GENERAL PLOTTER.
4Ø T=23                            :REM SCREEN HEIGHT
5Ø L=22                            :REM SCREEN WIDTH
6Ø S=768Ø                          :REM SCRN MEM START
7Ø A=Ø:B=1ØØØØØØ                   :REM HI, LO X VALUES
8Ø C=Ø:D=1ØØØØØØ                   :REM HI, LO Y VALUES
2ØØ FOR I=1 TO 1ØØØØ               :REM START LOOP
21Ø READ X,Y                       :REM GET DATA
22Ø IF X=-1 THEN 3ØØ               :REM CHECK FOR END
23Ø IF X>A THEN A=X                :REM HI  X?
24Ø IF Y>C THEN C=Y                :REM HI  Y?
25Ø IF X<B THEN B=X                :REM LOW X?
26Ø IF Y<D THEN D=Y                :REM LOW Y?
27Ø N=N+1                          :REM KEEP COUNT
28Ø NEXT I                         :REM END LOOP
```

```
300 W=A-B:W=(L-1)/W                :REM WIDTH  FACTOR
310 H=C-D:H=(T-1)/H                :REM HEIGHT FACTOR
320 RESTORE                        :REM RESET DATA
330 PRINT "CLR"                    :REM CLEAR SCREEN
340 FOR I=1 TO N                   :REM LOOP FOR PLOT
350 READ X,Y                       :REM GET DATA
360 P=INT((X-B)*W)                 :REM WIDTH  CALCULAT
370 Q=INT((Y-D)*H)                 :REM HEIGHT CALCULAT
380 F=L*(T-1-Q)+P                  :REM SCREEN LOC CALC
390 POKE (S+F),42                  :REM PUT * ON SCREEN
400 POKE (S+F+30720),0             :REM SET COLOR
410 NEXT I                         :REM PLOT LOOP
420 GOTO 420                       :REM WAIT TO VIEW
500 DATA 1,1                       :REM DATA TO USE
510 DATA 2,2
520 DATA 3,3
530 DATA 4,4
540 DATA 5,5
550 DATA -1,0
```

In line 330 you should use the odd symbol that the VIC uses for clear screen. A plot of the sine curve on the VIC-20 is shown as Fig. 5-25.

Other computers use different plotting schemes, but our little program will figure out the horizontal and vertical plotting distances for you in lines 360 and 370, and you can put them into whatever plotting functions that your computer provides. It may be necessary

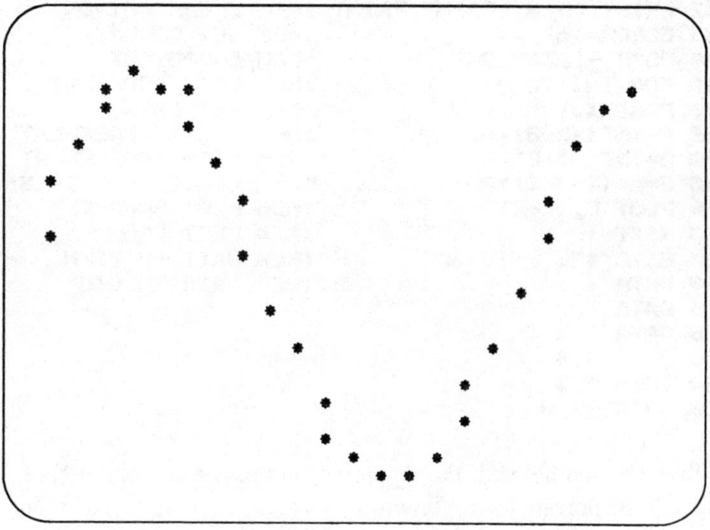

Fig. 5-25. A VIC-20 sine wave plot.

to add or subtract 1 to the values with some computers, or perhaps even invert the values, but it should work.

The Applc computer uses a very different screen arrangement. It can POKE values to the screen, but the stars don't appear in the proper places. Since many of you have Apples, a separate listing is provided here. This works, although the four protected lines at the bottom of the screen remain. Perhaps an Apple genius can tell me how to eliminate them! In any case, the plots are thoroughly usable.

```
10 REM PROGRAM 5.1B. APPLE GENERAL PLOTTER
40 T=40                          :REM SCREEN HEIGHT
50 L=40                          :REM SCREEN WIDTH
60 S=15360                       :REM SCRN MEM START
70 Z=0:B=1000000                 :REM HI, LO X VALUES
80 C=0:D=1000000                 :REM HI, LO Y VALUES
200 FOR I=1 TO 10000             :REM START LOOP
210 READ X,Y                     :REM GET DATA
220 IF X=-1 THEN 300             :REM CHECK FOR END
230 IF X>Z THEN Z=X              :REM HI  X?
240 IF Y>C THEN C=Y              :REM HI  Y?
250 IF X<B THEN B=X              :REM LOW X?
260 IF Y<D THEN D=Y              :REM LOW Y?
270 N=N+1                        :REM KEEP COUNT
280 NEXT I                       :REM END LOOP
300 W=Z-B:W=(L-1)/W              :REM WIDTH  FACTOR
310 H=C-D:H=(T-1)/H              :REM HEIGHT FACTOR
320 RESTORE                      :REM RESET DATA
330 GR                           :REM SHIFT MODE
332 PRINT:PRINT:PRINT:PRINT      :REM CLEAR BOTTOM
333 COLOR=15                     :REM SET COLOR
334 POKE -16302,0                :REM UNPROTECT
340 FOR I=1 TO N                 :REM LOOP FOR PLOT
350 READ X,Y                     :REM GET DATA
360 P=INT((X-B)*W)               :REM WIDTH  CALCULAT
370 Q=INT((Y-D)*H)               :REM HEIGHT CALCULAT
380 F=L*(T-1-Q)+P                :REM SCREEN LOC CALC
390 PLOT P,(L-Q)                 :REM PLOT SQUARE
410 NEXT I                       :REM PLOT LOOP
420 GOTO 420                     :REM WAIT TO VIEW
500 DATA 1,1                     :REM DATA TO USE
510 DATA 2,2
520 DATA 3,3
530 DATA 4,4
540 DATA 5,5
550 DATA -1,0
```

In line 230 and beyond, the Apple objects to the combination of A and T, interpreting it as AT, which is a legitimate Apple command. Thus, A was changed to Z. The other differences were necessary to adapt to the Apple low resolution graphics plotting mode. Some of the lines in the program are not needed with the Apple, but have been left in to maintain compatibility with the other programs. See the Applesoft manuals for details.

Program 5-2 can be adapted for other computers in precisely the same fashion.

Now we are ready to use our plotting ability in the next chapters. It will be of immense use.

In this chapter we have added a few new functions to our knowledge of BASIC: INT, SIN, and TAN. We probably won't have much use for these in future chapters. But the idea of dimensioned variables (DIM), and subscripting, will appear many times.

YOU'RE NOT A NORMAL COMPUTER.

6
The Normal Curve

The curve shown in Fig. 6-1 is the most important in all of statistics, and is called the *Normal Curve*. This curve is important because almost all human characteristics, and a great many physical phenomena, fit the Normal Curve: things such as height, weight, intelligence, and so on are said to be *distributed normally*. The Normal Curve is the curve on which students in schools are graded. The few things that are not arranged according to the Normal Curve generally can be shown, in the long run, to behave very much in accordance with it. The Normal Curve, plotted to scale (see Fig. 6-2) using Formula 6-1, below, is a rather squat-looking thing. It is generally exaggerated in height, as shown in Fig. 6-1, for clarity. We shall follow that practice here.

The formula for the Normal Curve is rather frightening to the student who has not had advanced mathematics. It looks like this:

$$y = \frac{1}{\sqrt{2\pi}} e^{-z^2/2} \qquad \textbf{6-1}$$

In Formula 6-1, y refers to the height of the curve at any given point, and z is our old friend, the z-score, the distance from the center line to any given point on the curve. The remainder of the frightening-looking symbols in the formula are constants. You probably already know that π is pi, or 3.1415927, the number that relates the diameter of a circle to its circumference.

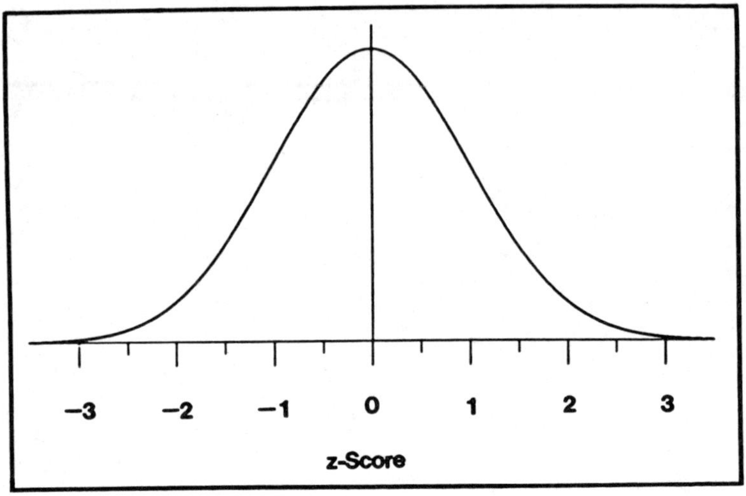

Fig. 6-1. The normal curve.

Another mathematical constant that you may not know about if you havén't had calculus is e. The value of e is 2.7182818, and it is widely used in mathematics. Indeed, it is so widely used that your computer probably has it in its BASIC. On your computer, in direct mode (that is, without using line numbers), type

PRINT EXP(1)

and the screen of your computer will have the value of e on it. To prevent confusion with variables that might be called E, and by mathematical convention, e is called EXP on computers.

Now, let's experiment with Formula 6-1 a bit. The following little program will allow you to get values of the height of the curve for various z-score values:

```
10 REM PROGRAM 6.1. NORMAL CURVE HEIGHT
20 INPUT Z                      :REM GET DATA
30 C=1/SQR(2*3.1415927)         :REM GET CONSTANT
40 Y=C*EXP(-Z*Z/2)              :REM GET HEIGHT
50 PRINT Y                      :REM PRINT ANS
60 GOTO 20                      :REM CONTINUE
```

Line 20 gets the data, as usual. Lines 30 and 40 calculate the height of the curve by Formula 6-1. You could put it all on one line, but it's easier to read if split over two lines. Line 30 calculates the constant, and line 40 calculates the height of the curve. Line 50 prints the result.

Type in the program, and let's find out some things about the Normal Curve. First of all, where is it the highest? Run the program, and enter several numbers as data. You'll get values like the following:

z-score	**Height**
−3.0	.0044
−2.0	.0540
−1.0	.2420
0.0	.3989
+1.0	.2420
+2.0	.0540
+3.0	.0044

Try several numbers with your computer; if you're ambitious, you could even plot the curve. Notice that the curve is highest, with a height of .3989, at a z-score of zero. Remember that the z-score of 0 corresponds to the Average. Consequently, the Normal Curve is highest at the Average! Chalk up another important characteristic of the Average.

Mathematicians almost always examine a curve to see whether it is symmetrical or not, that is, whether the height is the same at equal distances on both sides of the middle. Try z-scores of +2 and −2. You'll find that the height is .0540 in both cases. Similarly, for z-scores of eight −1 or +1, the height will be .2420. Thus, the curve is symmetrical.

Mathematicians also concern themselves with the question of whether curves will ever touch the X-axis. If a curve gets very close

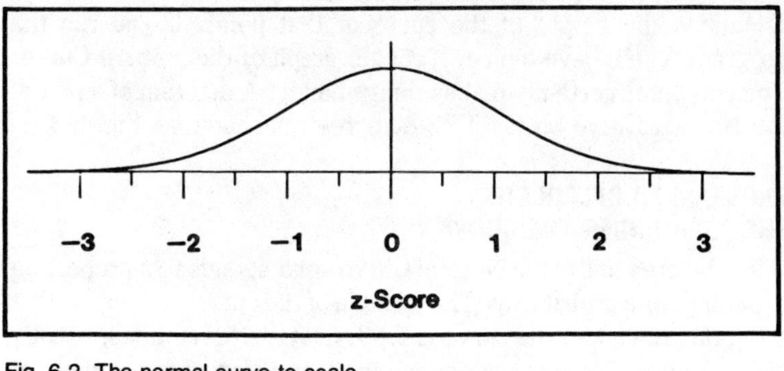

Fig. 6-2. The normal curve to scale.

to, but does not touch, the X-axis, it is said to be *asymptotic* to that axis. Is this the case with the Normal Curve? To find out, try a z-score of +5.00 (a very extreme z-score, as we'll see). The height of the Normal Curve at that point is .000001487 according to my TRS-80; that's not very high. Try a very large z-value, say 100. My TRS-80 says zero. Actually, your computer probably doesn't have sufficient accuracy to evaluate such extreme scores properly, even with double precision. The value is very slightly larger than zero. Fortunately, such scores never happen! At all events, we find that, no matter how far out we go, the y value will never be zero, and hence the curve will never touch the x-axis.

These three characteristics of the Normal Curve—that is, the fact that it has one maximum, the fact that it is symmetrical, and the fact that it never touches the horizontal axis—should be kept in mind.

You can see all of this very nicely for yourself if you use the plotting routine of the last chapter to actually plot the curve. To do this, load in or type in Program 5-2 from the last chapter, and then change or add some lines, as follows:

```
110    X(I)=I-15                     :REM SET X VALUE
120    C=1/SQR(2*3.1415927)          :REM GET CONSTANT
121    Z=X(I) :Z=Z/5                 :REM GET Z
122    Y(I)=C*EXP(-Z*Z/2)            :REM GET HEIGHT
```

These lines work as follows: line 110 adjusts the X values so that they'll start out very negative, go to zero, and increase again. Line 120 is the constant found in Program 6-1. Line 121 adjusts the value of X(I) so that it makes a pleasing graph on the screen. You may have to change the value shown in Line 121, (that is, 5) to make the graph look well, depending on the size of your screen. Line 122 calculates the height of the curve at that point. If you run the program, you'll have a nice, if rough, graph of the Normal Curve. The computer certainly makes things handy! A diagram of a plot of the Normal Curve on my TRS-80 screen is shown as Fig. 6-3.

PROGRAM TO CALCULATE THE AREA UNDER THE CURVE

The area under the Normal Curve corresponds to a proportion of people, or a probability. The formula of the curve is set up so that the total area under the curve is 1.00. In statistics courses, usually you calculate the area under the curve by reference to a table of

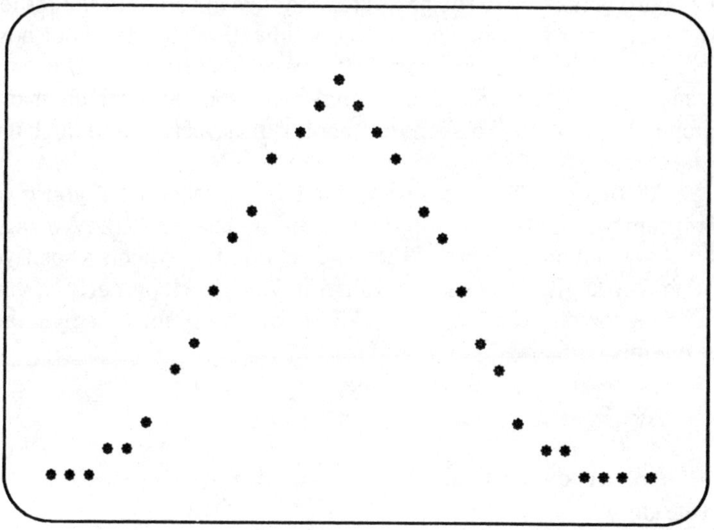

Fig. 6-3. A TRS-80 normal curve plot.

Normal Curve values, such as that provided in the table in Appendix A. However, with a microcompuer, it is perfectly possible to calculate the areas under the Normal Curve directly, without having to look in a table. The following program will do the job, and we'll use it rather often for the remainder of this chapter and the rest of the book:

```
10 REM PROGRAM 6.2. Z TO PROB NORMAL CURVE
20 INPUT Z                          :REM GET DATA
30 H=1+.0498673470*Z                :REM CALCULATE
40 H=H+.0211410061*Z*Z              :REM  APPROXIMATION
50 H=H+.0032776263*Z^3              :REM  OF FUNCTION
60 H=H+.0000380036*Z^4
70 H=H+.0000488906*Z^5
80 H=H+.0000053830*Z^6
90 H=(H^-16)/2                      :REM FINISH APPROX
100 P=0.5-H                         :REM ADJUST ANSWER
110 PRINT P                         :REM PRINT ANSWER
120 GOTO 20                         :REM CONTINUE
```

This is an approximation to the Normal Curve function, adapted from Abramowitz and Stegun (1964); but it's a rather accurate one. There are more elegant ways of writing the program, ones that will run faster, but the programming method shown above makes better sense to someone wanting to understand its operation. Essentially, a z-score is taken into the program, and then enters into the rather awesome series of calculations that follow. Don't worry about how this part of the program works! It's very complicated and uses a branch of mathematics called numerical analysis, which many people think is a good deal more complex than calculus (although it's a fascinating field).

All that you have to worry about is the fact that if you give Program 6-2 a z-value, it will tell you the area between the Average and that point on the Normal Curve. The program expects a positive value. If you give it a negative value, it won't work properly. If you can't remember that, add line 25 to the program to solve the problem.

```
25 Z=ABS(Z)
```

ABS as you recall, always makes a value positive. Let's try the program out.

SIMPLE AREA PROBLEMS

Let's try to find the area between the Normal Curve and a z-score of +1.00, as shown in Fig. 6-4. It is essential to draw a little diagram of the Normal Curve each time you work one of these problems; without the little figure, you will often become confused.

We want to find the area between the Average, and the vertical line drawn up from z=+1.00. Simply run Program 6-2, and enter +1.00. The microcomputer will say .3413. That means that 34.13

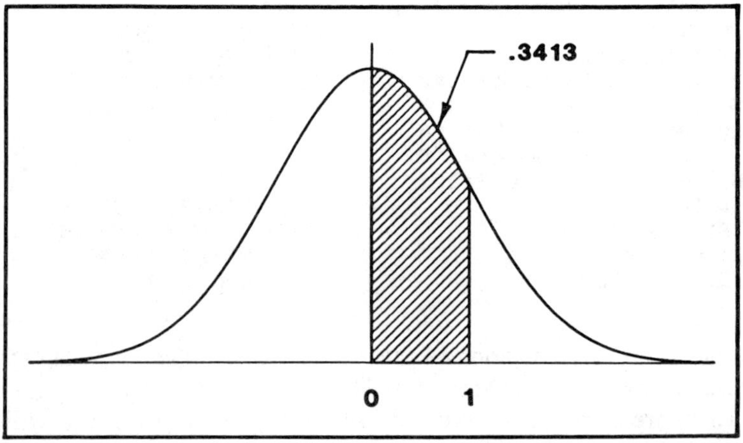

Fig. 6-4. Area to z = +1.00.

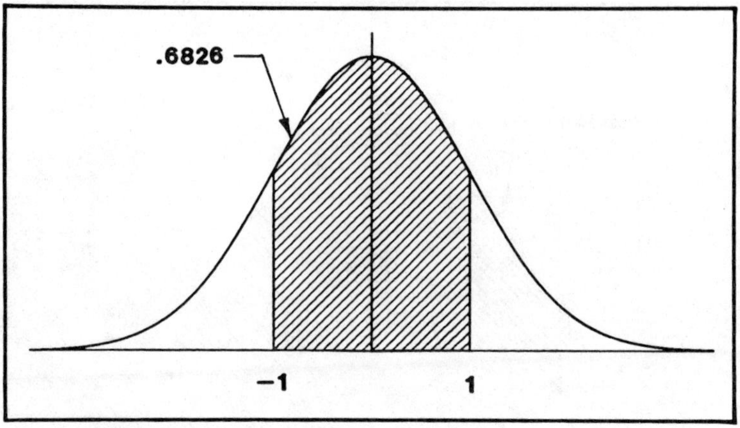

Fig. 6-5. Area between z = −1.00 and +1.00.

percent of the people fall between the Average and a z-score of 1.00. Or, since the Standard Deviation of z-scores is 1.00 (remember?!), if one goes out one Standard Deviation in one direction, one takes in 34.13 percent of the people. Now, if one goes the other direction the same distance, to z=−1.00, what will the area be between the Average and z=−1.00? Well, it's again 34.13 percent. Recall that the curve is symmetrical. In short, 68.26 percent of the people can be found between z-scores of +1.00 and −1.00, or between plus and minus one Standard Deviation. This is shown in Fig. 6-5.

Now, what percent of the people will fall between the Average and a z-score of +2.00? That's shown in Fig. 6-6. Type a z-score of

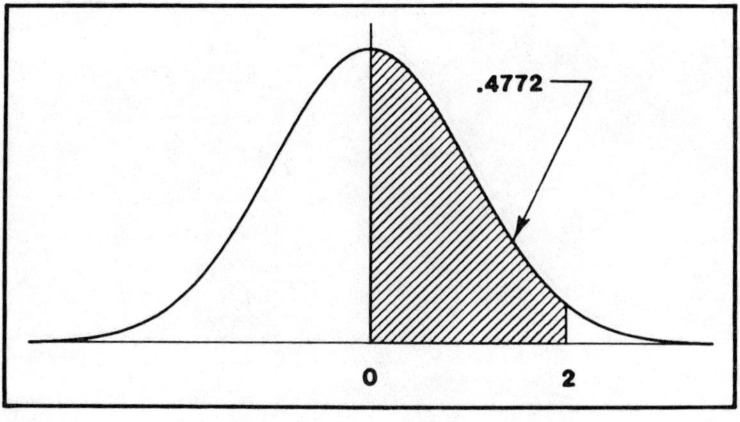

Fig. 6-6. Area to z = +2.00.

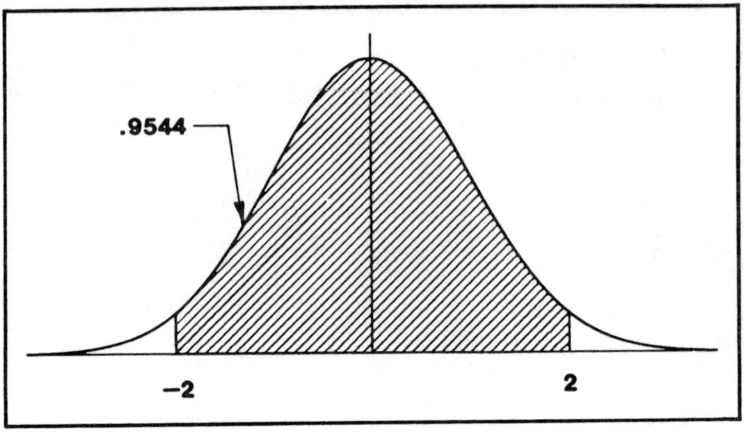

Fig. 6-7. Area between z = −2.00 and +2.00.

+2.00 into Program 6-2, and you'll find that the area out to that point is .4772. In this case, 47.72 percent of the people fall between the Average and a z-score of +2.00. The area between z-scores of plus or minus 2.00 is, then, .9544. In simple English, that means the 95.44 percent of the people will fall between plus and minus two Standard Deviations. That's important to know. This means that, with the Normal Curve, almost everyone will get scores no more deviant than two Standard Deviations (see Fig. 6-7).

A trickier problem is that of finding out what percentage of people get scores above a z-score of +1.00, as shown in Fig. 6-8. It's tricky because Program 6-2 gives the areas from the Average to

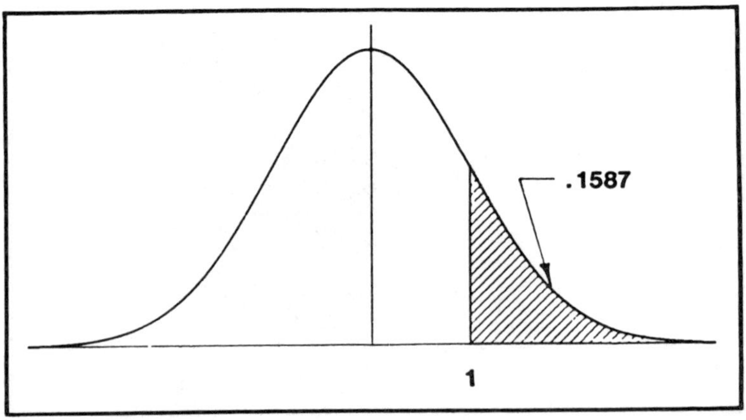

Fig. 6-8. Area above z = +1.00.

the z-score; it doesn't give the area from the z-score to the more extreme value, but the solution to the problem is not difficult. You simply type +1.00 into Program 6-2, and then subtract the answer from 0.5000. Since the total area under the Normal Curve is 1.0000, the area on the right side of the Average must be 0.5000. The area out to z=+1.00 is .3413, and the remainder, then is .5000 − .3413 = .1587. That isn't so difficult. If you like, you can change Program 6-2 so that it will give you the areas in the tails, by adding or changing the following lines:

```
105  Q=0.5-P
110  PRINT P,Q
```

And for that matter, you could make it print out the height of the Normal Curve at that point, but the height will not be used as often in this book as the other values.

GOING THE OTHER WAY

Frequently, you are given an area under the Normal Curve, and asked to find the score corresponding to the area. For example, say that you'd like to know the z-scores needed to take in exactly 95 percent of the area under the curve. In Fig. 6-5, we succeeded in getting 95.44 percent of the curve, but we'd like precisely 95 percent. How can that be done?

First of all make a drawing of what is wanted, as shown in Fig. 6-9. We'll assume that the 95 percent is spread evenly on both sides of the Average, and thus the area on each side of the curve will be

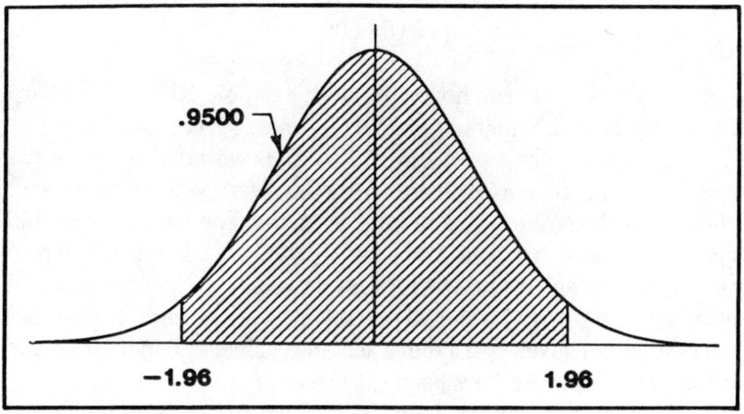

Fig. 6-9. z's for an area of 95 percent.

0.4750. What's the z-score corresponding to this value?

Program 6-2, recall, allows one to go from z-scores to areas. Now we need to go the other way, from an area to a z-score. How can this be done?

You could, of course, use Program 6-2 over and over again, trying z-score after z-score until you got as close as possible to 0.4750. But there's an easier way. Program 6-3 will yield z-score values if given an area.

```
200 REM PROGRAM 6.3. P TO Z NORMAL CURVE PGM
210 INPUT P                    :REM GET DATA
220 P=0.5-P                    :REM ADJUST IT
230 T=SQR(LOG(1/(P*P)))        :REM ADJUST AGAIN
240 U=2.515517+.802853*T       :REM TOP OF APPROX
250 U=U+.010328*T*T            :REM   CONTINUED
260 B=1+1.432788*T+.189269*T*T  :REM BOTTOM
270 B=B+.001308*T*T*T          :REM  OF APPROX
280 Z=T-U/B                    :REM GET Z-SCORE
290 PRINT Z                    :REM PRINT IT
300 GOTO 210                   :REM CONTINUE
```

This approximation again is taken from Abramowitz and Stegun (1964), and originated in work done by Cecil Hastings, in his book *Approximations for Digital Computers*, (Copyright 1955 by the Rand Corporation, published by Princeton University Press, used by permission). It's not a terribly accurate approximation, but it will do for our purposes. Type in the program, and you'll see how handy it is. Notice that the line numbers start at 200, so that both Program 6-2 and 6-3 can be in the computer at one time. Then, Program 6-3 can be run, with most machines, by typing

RUN 200

Program 6-2 is run normally, with a simple RUN. (My Sharp PC-1500 needs LN instead of LOG in line 230.)

To return to the problem of Fig. 6-9, we need to know the z-score corresponding to an area from the Average to the z-score of .4750. Run Program 6-3, and type in .4750. The computer should give you a figure of about 1.96; my TRS-80 says 1.9604 (different computers may give somewhat different values). The figure used in most statistics texts for the 95 percent area is a z-score of 1.96, but your computer gives you a more accurate value, and indeed for any value you want! See how handy it is?

Thus, to take in precisely 95 percent of the area under the Normal Curve, requires z-scores of plus and minus 1.96. This is a number that's useful to remember; it will come up again and again in this book.

A similar problem is that of finding the z-scores that correspond to an area of 99 percent under the curve. You see, statisticians like to be either 95 percent sure of things, or 99 percent sure, and so they need these values. As you can see from Fig. 6-10, the area needed on each side is 0.4950. Run Program 6-3, and type in 0.4950. You should get an answer of about 2.58 (the value usually accepted by statisticians). My TRS-80 says 2.57624. Thus, 99 percent of the people fall between z-scores of plus and minus 2.58.

We have discovered some important sets of numbers up to this point. Let's list them:

Plus or Minus z-score Limits	Total Area
1.00	68.26%
2.00	95.44%
1.96	95%
2.58	99%

All of these values are used again and again in statistics, and you should endeavour to remember them!

MORE COMPLEX PROBLEMS

With Programs 6-2 and 6-3 (or the table in the Appendix) and

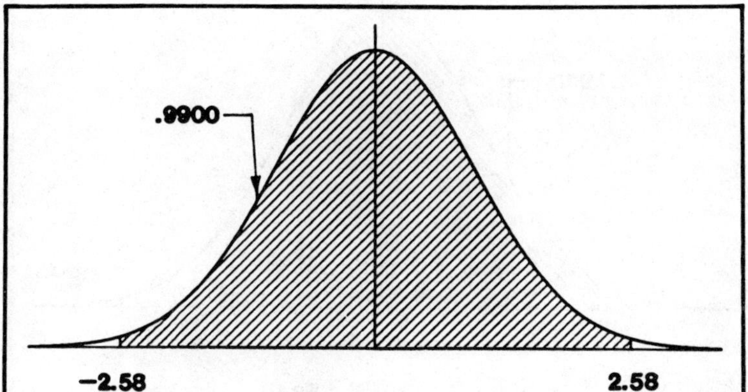

Fig. 6-10. z's for an area of 99 percent.

our knowledge of Standard Scores from the last chapter, we can work out some fairly sophisticated Normal Curve problems. Use the direct function of your computer to do the calculations, or use a small calculator. Generally, it isn't worth it to program the computer in BASIC to solve each problem, for each problem is different. Let's try one.

Problem: Find the z-score corresponging to a Percentile Rank of 79.10.

This is somewhat tricky. We haven't talked about Percentile Ranks in this book, but most people understand what they are. If someone gets a Percentile Rank of 79.10, he has done as well as or better than 79.10 percent of the people. This is shown in Fig. 6-11. From the drawing (be sure to make one whenever you do a Normal Curve problem!), we see that the area under the curve in question is .7910. The area under the curve builds up from left to right, in the usual manner of mathematical notation. Now we have to find the z-score corresponding to the top of this large area.

Recall, however, that Programs 6-2 and 6-3 deal only with areas starting from the Average. Also recall that the area for one half of the curve is 0.5000. Consequently, the area that we must supply to Program 6-3 is .7910 – .5000 = .2910. If Program 6-3 is run, and .2910 supplied to it, we find that the z-score corresponding to a Percentile Rank of 79.10 is +.81 (the sign is important here). My TRS-80 is a bit more accurate than the Normal Curve table in the Appendix, and says .809705. Both are very similar. At any rate the z-score needed is +.81.

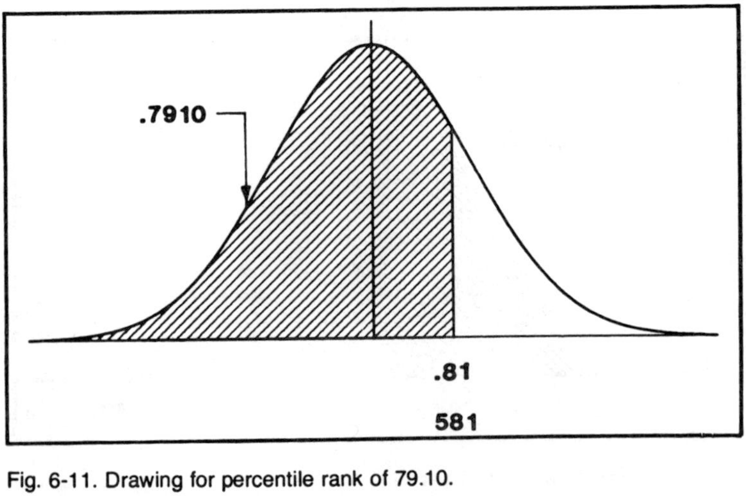

Fig. 6-11. Drawing for percentile rank of 79.10.

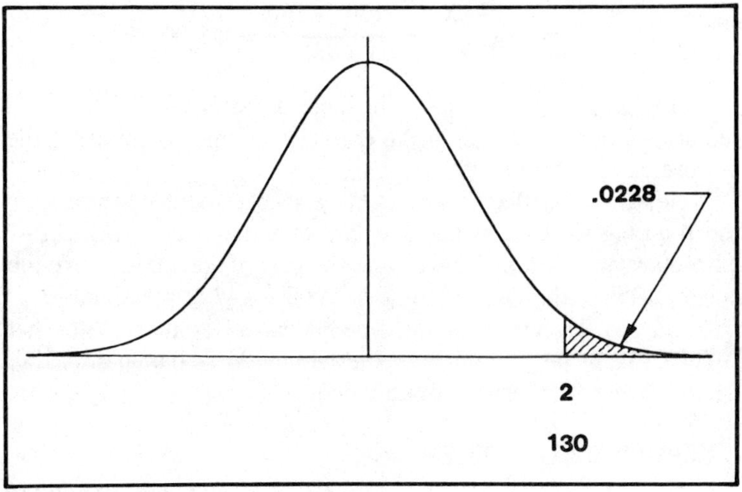

Fig. 6-12. Drawing for IQs above 130.

"But," you say, "I still don't understand z-scores. They don't make intuitive sense. Couldn't you transform them into some more understandable scores?" Nothing could be easier. You simply use Formula 4-2.

Let's assume that you want to convert the Percentile Rank of 79.01 percent to the sort of scores used for the Scholastic Aptitude Test (SAT), which have a Mean of 500 and a Standard Deviation of 100. Formula 4-2 and the proper numbers are shown below:

$$\begin{aligned}\text{Desired Score} &= (z)*(\text{desired SD})+\text{desired Average}\\ &= +.81 * 100 + 500\\ &= 581\end{aligned}$$

Consequently, the SAT-type score corresponding to a Percentile Rank of 79.10 is 581. That's not so difficult.

Problem: What percentage of the population has IQ's above 130?

Assume here that IQ scores have a Mean of 100 and a Standard Deviation of 15 (the Wechsler IQ test uses scores of this type). A score of 130 is clearly above the Average (it's the average IQ of medical doctors), and thus will look as shown in Fig. 6-12.

To solve the problem, we must first find the z-score corresponding to the score of 130. That's done by using Formula 4-1, the z-score formula, as follows:

$$z = \frac{X - \bar{X}}{SD} = \frac{130 - 100}{15} = +2.00$$

Obviously, we must know the mean and Standard Deviation of the scores before we may make the calculations. At any rate, the z-score needed is +2.00.

Next, we plug that score into Program 6-2, to find the area from the Average to the z-score. We find that this area is .4772 (my TRS-80 says .47725). But we want the part of the curve above the z-score. Thus, the desired area is .5000 − .4772 = .0228.

In short, 2.28 percent of the people have IQs above 130. That isn't very many people! If indeed the average M.D. has an IQ of 130, doctors are a very bright group indeed.

A MORE DIFFICULT PROBLEM

Normal curve problems can be made difficult, but usually the main source of confusion lies in the wording of the problem. Here's a moderately tough one.

Assume scores like those of the SAT (Scholastic Aptitude Test) type with an Average of 500 and a Standard Deviation of 100. Assume that Sam gets a score of 650 on the SAT. 86.64 percent of the people are between them. Harry does worse than Sam. What's Harry's score?

This problem is sketched in Fig. 6-13. There's nothing particularly difficult about it, if you don't let the language paralyze your

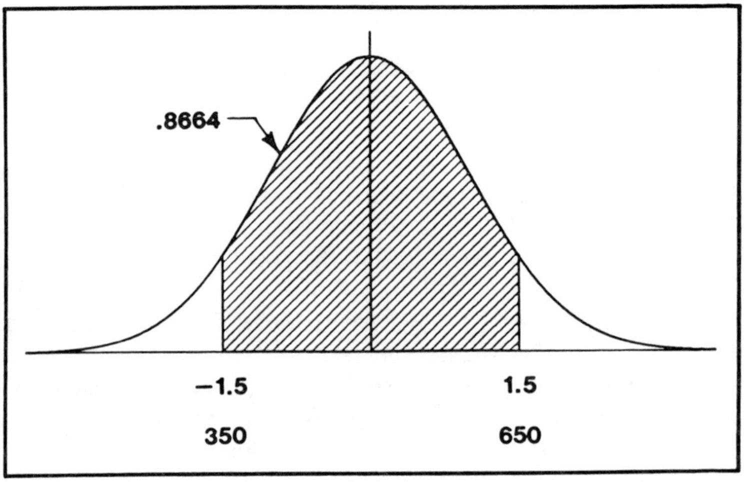

Fig. 6-13. Proportion between 350 and 650.

thought processes. Sam's score is 650. Since Harry did worse than Sam (as stated in the problem), one must first find the area between the score of 650 and the mean. To do this, calculate the z-score corresponding to the score of 650:

$$z = \frac{X - \overline{X}}{SD} = \frac{650 - 500}{100} = +1.50$$

Now, use Program 6-2, or look in the Normal Curve table in Appendix A to find the area under the curve corresponding to a z-score of +1.50. That area is .4332.

Now, the problem states that Harry did worse than Sam, and that .8664 of the people fall between them. A clever person will notice that .8664 is exactly twice .4332, the distance from the Average to the score of 650. Thus, one finds that we need an equal area (.4332) below the Average to get to Harry's score. See Fig. 6-13 if this isn't obvious. Thus, Harry's score will correspond to a z-score of −1.50; we're assuming of course that the scores are equally placed on both sides of the Average.

Thus, Harry has a z-score of −1.50. To convert this to the usual SAT score format, we simply use Formula 4-2:

Desired Score = (z)∗(desired SD)+desired Average
Desired Score = −1.50∗100 + 500

and thus Harry's score is 350. There is nothing particularly difficult about the problem except for the rather tricky wording. Drawing a diagram of the problem makes it duck soup, if you don't panic.

Problem: What percentage of the people would you expect to have IQs between 120 and 130?

Again, assume IQ test scores with a Average of 100 and a Standard Deviation of 15. The problem is sketched as Fig. 6-14. The sketch, again, makes it a fairly simple problem.

First we must find the z-scores corresponding to scores of 120 and 130. The z for 120 is (120-100)/15 or +1.33. The z for 130 is (130-100)/15 or +2.00. Now things get a bit tricky.

Look at Fig. 6-14. Program 6-2 will tell us that the area out to the z-score of 2.00 is .4772. It will also tell us that the area to a z of 1.33 is .4082. What is wanted is the shaded area. To find that, you simply subtract .4082 from .4772, yielding .0690. Thus, 6.90 percent of the people have IQs between 120 and 130. If you're a psychologist, as I am, this is a useful fact.

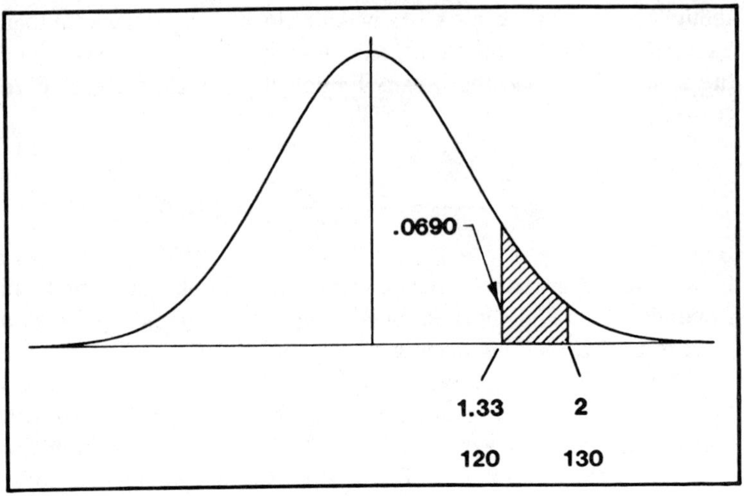

Fig. 6-14. Percentage between 120 and 130.

Problem: John has a Percentile Rank on an IQ test of 15.87. What's his score?

Our IQ test has an Average of 100 and a Standard Deviation of 15. If John has a Percentile Rank of 15.87, that means that he did as well as or better than 15.87 percent of the people who have taken the test. This situation is detailed in Fig. 6-15. Of course, we can't use the area .1587 directly, for the Normal Curve values are set up from the Average out. Consequently, since half of the people are

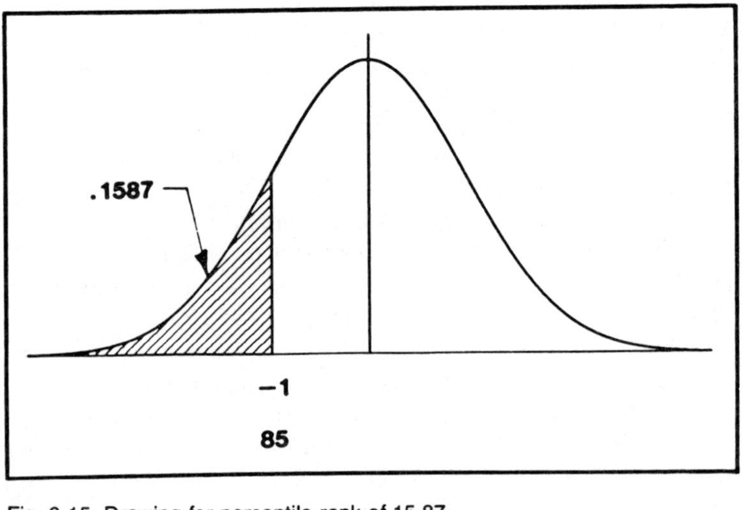

Fig. 6-15. Drawing for percentile rank of 15.87.

below the Average, we must subtract .1587 from .5000, yeilding our old friend .3413. Now, we use Program 6-3, or the Normal Curve table, to tell us the z-score corresponding to an area of .3413. This z-score is −1.00. The minus sign is very important, and must be assigned by the person working the problem. The computer, of course, always gives positive values. Consequently, whenever a z-score is to the left of the Average it must be assigned a negative value. At any rate, the z-score is −1.00.

To finish the problem, we must convert from z-score units to IQ units. This is done by the use of Formula 4-2. Let's do it:

Desired Score = (z)*(desired SD)+desired Average
Desired Score = −1.00*15+100

The answer is 85. The IQ score corresponding to a percentile rank of 15.89 is 85. That's a somewhat difficult problem because of the minus sign that must be assigned to the z-score. Many, many people forget to do that.

As you can see, Normal Curve problems are not particularly difficult, just tricky. And, except for calculation of the Normal Curve values, a computer really isn't needed. You can work out these simple problems with a calculator about as easily as you can on the computer.

Be sure that you understand the Normal Curve and ways of working problems involving it. It will come up again and again in later chapters.

Now, let's get some problems that are not psychological in nature:

Problem: Ajax Bolt Company makes bolts with an average diameter of .2586 inches, and a Standard Deviation of .0143. What percent of the bolts would have diameters between .24001 and .2443 inches, assuming that the Normal Curve is applicable?

This problem is detailed in Fig. 6-16. You should know how to solve it by now. First we must convert the scores to z-scores. For .2443, the appropriate z-score is −1.00, that is (.2443−.2586)/.0143. For .24001, the appropriate z-score is −1.30, that is (.24001−.2586)/.0143. The area out to a z-score of −1.00 is .3413, and the area out to −1.30 is .4032. The difference between them is our answer, .0619. This problem is a bit tricky simply because both z-scores are on the negative side.

Problem: Pokey Auto Company makes cars with an Average speed of 100 mph, and a Standard Deviation of 5 mph. What percent-

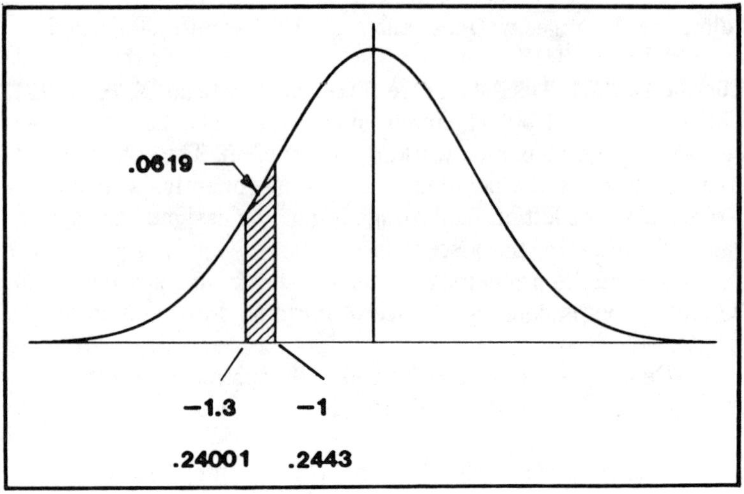

Fig. 6-16. Ajax Bolt Company drawing.

age of the cars would have speeds between 90 and 92.5 miles per hour?

This problem is precisely like the last one. A speed of 90 mph corresponds to a z-score of −2.00 (see Fig. 6-17). A speed of 92.5 corresponds to a z-score of −1.5 (figure them out, using Formula 4-1!). The areas needed are .4772 and .4332. The difference between them is .0440. However, the problem requires a percentage

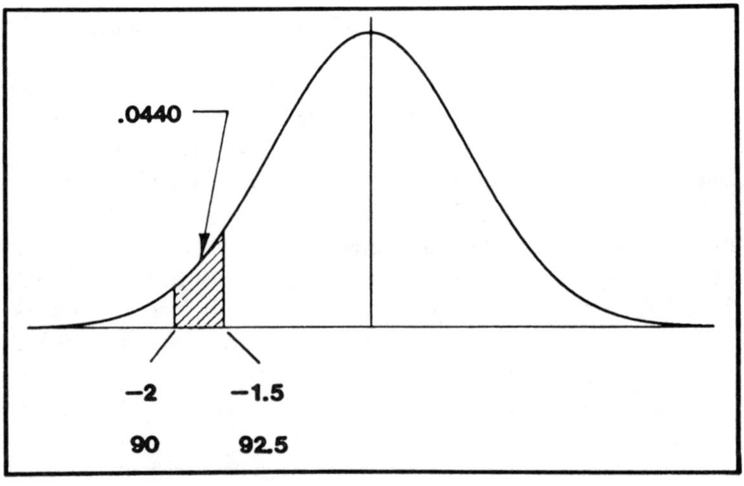

Fig. 6-17. Speeds between 90 and 92.5 mph.

value, and so the answer is 4.40 percent.

Problem: The average temperature in a room is 68 degrees, with a Standard Deviation of 10 degrees. What proportion of the time would the temperature be between 86 and 87 degrees?

This is pretty much the same sort of problem as the last one. However, the two temperatures are on the positive side of the Average, which makes things easier. For 86 degrees, the z-score is 1.8, or (86-68)/10. For 87 degrees, it's 1.9, or (87-68)/10. The two temperatures are pretty close together, and this may confuse you (see Fig. 6-18). The areas corresponding to the two z-scores are .4641 and .4713. The difference between them is .0072, the desired proportion. It's rather small, as you would expect.

Problem: Jones Tractor Company makes tractors with a Mean weight of 5000 pounds. The Standard Deviation is 1000 pounds. What percentage of the tractors will have weights between 4900 and 5500 pounds?

Figure 6-19 shows that this problem is a bit trickier than the last one. The two areas in question are on either side of the Average. Thus, the z-score for 4900 pounds is −0.1, or (4900-5000)/1000. The z-score for 5500 pounds is +0.5, or (5500-5000)/1000. Now, the area out to a z-score of −0.1 is .0390, while that to +0.5 is .1915. However, in this case, we add the areas rather than

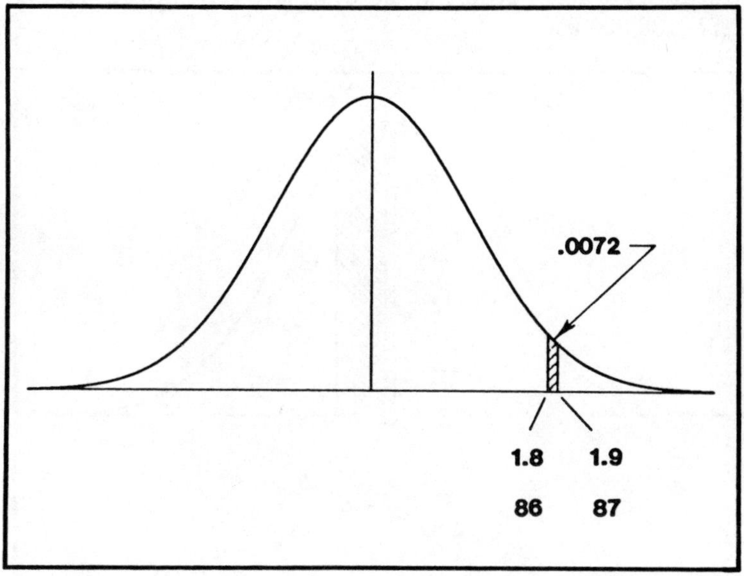

Fig. 6-18. Proportion between 86 and 87 degrees.

subtracting them (see Fig. 6-19), yielding .2313. The problem requires the percentage, so that you must multiply by 100, leading to the answer, 23.13 percent.

We have come to the end of another chapter. Notice that no new concepts have been added to our library of BASIC ideas. Indeed, our microcomputers aren't as helpful in solving these problems as they can be with certain other statistical problems. Normal Curve problems require more thought than anything else!

PROBLEMS

1. For the first set of problems, assume that you have IQ scores with an Average of 100 and a Standard Deviation of 15. Further, if needed, assume that the number of people in question is 200,000,000, roughly the population of the U.S. Answers are given in parentheses after the problems.

1a. What percentage of the people would you expect to have IQs above 145? (.13 percent) How many people? (260,000).

1b. What percentage of the people would you expect to have IQs below 70 (the mental retardation level)? (2.28 percent) How many people? (4,560,000).

1c. What percentage of the people would you expect to have IQ scores between 80 and 120? (81.64 percent)

1d. How many people in the U.S. have IQs between 85 and 115? (136,520,000)

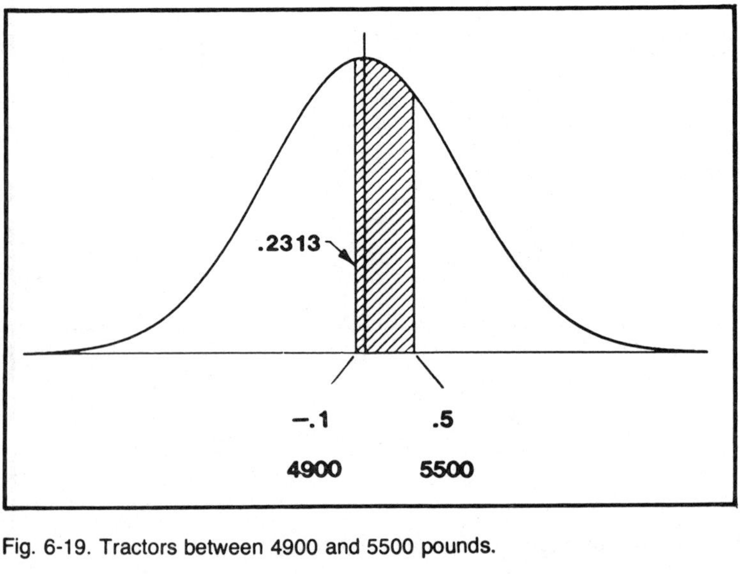

Fig. 6-19. Tractors between 4900 and 5500 pounds.

1e. Eric did better than 97.19 percent of the people. What was his IQ? (128.65)

2. For the following problems, assume MMPI Schizophrenia scale scores, with an Average of 50 and a Standard Deviation of 10. Again, there are 200,000,000 people. (The MMPI is perhaps the best known paper-and-pencil personality test used to classify mental patients.)

2a. Find the Schizophrenia score corresponding to a Percentile Rank of 70. (55.2)

2b. Find the number of people getting Schizophrenia scores between 50 and 60. (68,260,000)

2c. Schizophrenia and other MMPI scores are generally considered to be within normal limits if they fall between 30 and 70. What percentage of the people will fall into that interval? How many people? (95.44 percent) (190,880,000)

2d. Above what score would 260,000 people fall? (80)

2e. How many people would have scores between 21 and 22? (140,000)

1.00 IS A PERFECT CORRELATION!
1.00

7
Correlation

We shall now proceed to develop a way to describe the degree of similarity, or relationship, between two measures. The method used is called *Correlation.* Correlation can also be used to make predictions. *Prediction* will be covered in the next chapter, but we really can't discuss Correlation without talking about prediction from time to time.

An example here is invaluable. Assume that you want to assess the similarity between two intelligence tests—IQ Test X and IQ Test Y. Since both supposedly measure intelligence, they should be highly related to one another if the concept of IQ has any merit. Five subjects are obtained, and the two IQ tests are both given to each person (of course, a much larger sample should be used in any real situation). Suppose that the scores obtained are as follows:

Person	Test X IQ	Test Y IQ
A	120	120
B	110	110
C	100	100
D	90	90
E	80	80

Try to answer the following question: How similar are the two tests?

All the people tested earn the same score on both tests. Thus, Person A gets the highest score (120) on the Test X, and also gets the highest score (120) on the Test Y. Person E gets the lowest score on Test X (80), and also gets the lowest score on the Test Y. Thus, from the data, the tests appear to be perfectly similar in their results and it is said to be perfectly similar in their results and it is said that they are *perfectly correlated*. A plot of the points is shown in Fig. 7-1. Note that in this *scatterplot*, as it is called, all of the data points fall on a nice straight line. You can see this for yourself by using Program 5-1, and using the previous scores. Whenever there is a perfect Correlation, all of the data points fall on a straight line, as shown here. (Your microcomputer will yield a plot like that of Fig. 7-2. We shall simulate microcomputer plots for the remainder of the chapter.)

Another way of putting this is to say that one can predict perfectly from the Test X to the Test Y. If someone got the highest score on the Test X, it can be assumed that he also got the highest score on the Test Y, and so on. Of course, there are only five sets of points, so that you would probably feel uneasy about making a

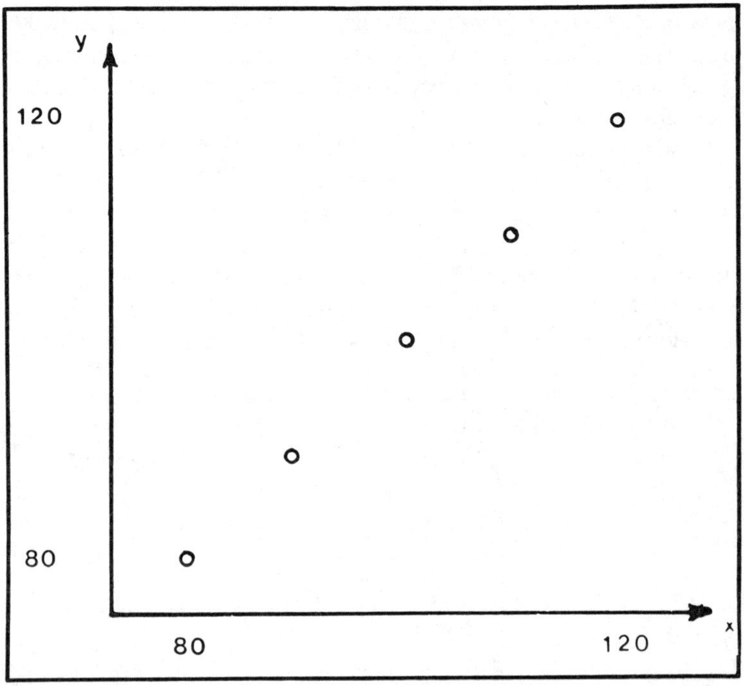

Fig. 7-1. A perfect Correlation.

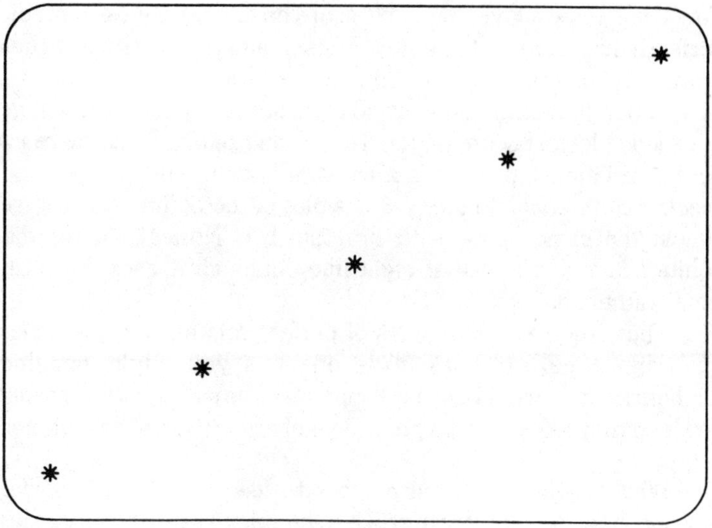

Fig. 7-2. Computer plot of perfect Correlation.

strong commitment to the idea of a perfect relationship without more data. People are better intuitive statisticians than we generally give them credit for!

Now imagine that the experiment is done again, and the results are as follows:

Person	Test X IQ	Test Y IQ
A	120	80
B	110	90
C	100	100
D	90	110
E	80	120
	---------	---------

Query: How closely related are Test X and Test Y in this set of data? You must approach this conclusion with the greatest care, for it is extremely common for people to say that, since the scores are exactly backwards, there must be no relationship whatever; some very famous scientists have made precisely this mistake. Nothing could be further from the truth. There is still a perfect relationship.

For, if we know a given person's score on one test, we can predict perfectly his score on the other. Thus, if a person gets the highest score (120) on the Test X, he will get the lowest score (80) on Test Y. The person getting the second highest score on Test X will get the second lowest score on the Test Y, and so on. Thus, there is a perfect relationship, but it's a backwards one. This is know as a *negative correlation.* Figure 7-3 is a plot of the points. You can get almost the same figure with Program 5-1. Notice that the data points again give a nice straight line, although it goes down this time, rather than up.

Thus, there are two sorts of perfect relationships, a perfect one, and a perfect backwards one. As you might imagine, mathematicians would not be happy about my awkward terminology, and they like to assign precise numbers to the various relationships.

What would be a good number to use to express a perfect relationship? Frankly, I would like to be able to use 100, since most people understand 100 percent with no trouble. However, that's not the way it's done. A perfect relationship is designated by a value of 1.00, by mathematical definition. Defining a perfect Correlation in this way helps it to agree with z-scores (Chapter 4, remember?!), which have a Standard Deviation of 1, as you recall. Consequently, a

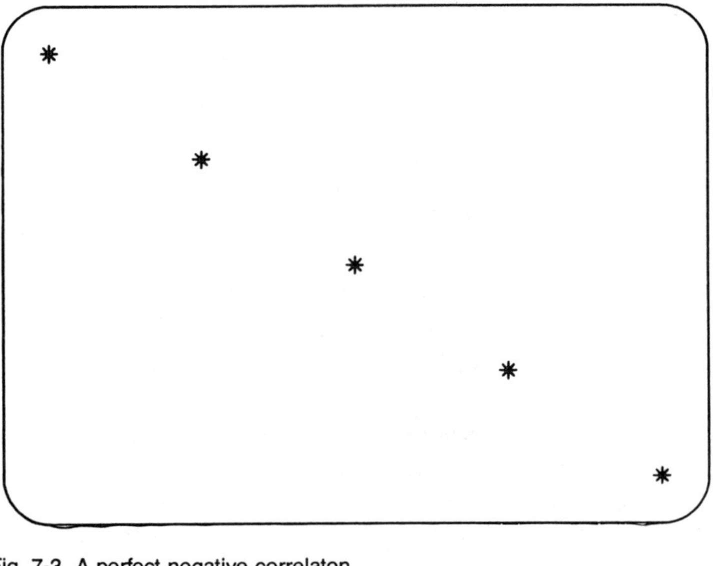

Fig. 7-3. A perfect negative correlaton.

perfect relationship is denoted as 1.00; this value applies to Fig. 7-1.

NEGATIVE RELATIONSHIPS

Now, what would be a good symbol for a perfect backwards relationship, as shown in Fig. 7-2? Be very careful, here, to avoid the temptation to guess that zero (0) would be a good choice for this number, for it is not. A perfect backwards Correlation is designed by minus 1.00 (−1.00). That makes sense, since a perfect relationship is designated by +1.00, and the perfect backwards relationship is then, by analogy, −1.00. The 1 indicates the strength of the relationship, and the sign indicates the direction. Thus, a Correlation of +1.00 indicates a perfect positive relationship, and a Correlation of −1.00 means a perfect backwards relationship, or a perfect negative Correlation, as it will be called henceforward.

Negative Correlations will probably confuse you for a while. All that is meant by them is simply that somehow or other one of the variables is scored in the opposite direction to the other. There are many examples of negative Correlations in real life. For example, a good golf player hits the ball a long distance. But his overall score is low. Thus, if one computes the Correlation between length of drive, and golf scores, a negative Correlation results. There is also a negative Correlation between the number of cigarettes smoked and the number of years that one has left to live. There is a perfect negative Correlation between the number of days gone in a year with the number of days left. There is a strong negative Correlation between altitude and air pressure. All that any of these relationships means is simply that somehow one scale of measurement is arranged in a direction opposite to the other.

Thus, a Correlation of +1.00 indicates that two variables are perfectly related, and vary in the same direction. A Correlation of −1.00 means that there are two variables, perfectly related, but varying in the opposite direction. However, one definition remains. What shall we call a total lack of relationship? It is not easy to develop an intuitively obvious, simple example of this situation, and we shall wait to give an example until later. However, the number used to denote no relationship at all, is, appropriately, 0.00 (zero). Thus:

A perfect positive relationship is	+1.00
A lack of relationship is	0.00
A perfect negative relationship is	−1.00

A BETTER INDEX: THE z-SCORE FORMULA

Thus far, our examples have been so lovely that we have established the basic concepts without any need for calculations. However, a slight change in the data we've been using will ruin things completely. Let's take the first example, and interchange a couple of numbers.

Person	Test X IQ	Test Y IQ
A	120	120
B	110	90
C	100	100
D	90	110
E	80	80
	------	------

The lovely, simple inferences here must come to an end. There is no longer a perfect relationship, although three out of the five pairs retain the possibility of perfect prediction. But Persons B and D ruin things. It can no longer be said that the Correlation is +1.00 or −1.00 simply by looking at the data. A mathematically elegant method of solving the problem is needed.

However, let's try to get some insight into the likely answer. Since three out of five pairs still agree perfectly, it would seem that there is still a strong positive relationship, but it is no longer perfect. Thus, it must be less than +1.00, but not so low as 0.00. Perhaps the relationship is on the order of +.7 or so; there is no way of knowing at this point. Note also that the little scatterplot in Fig. 7-4 no longer shows a straight line. Two of the five points have moved off the line.

Developing the proper formula is not as difficult as you might think. Indeed, if it is done properly, the development is almost intuitively obvious.

To assess the similarity between two sets of scores, we must compare the sets of scores. What must we do before two sets of scores are compared? In Chapter 4, you learned that before two sets of scores are compared, they must first be reduced to Standard Scores. That is, the sets of scores must have the same Average and Standard Deviation, at minimum. We cannot be certain that Test X and Test Y have the same Average and Standard Deviation, and so we must convert them to Standard Scores. The simplest sort of

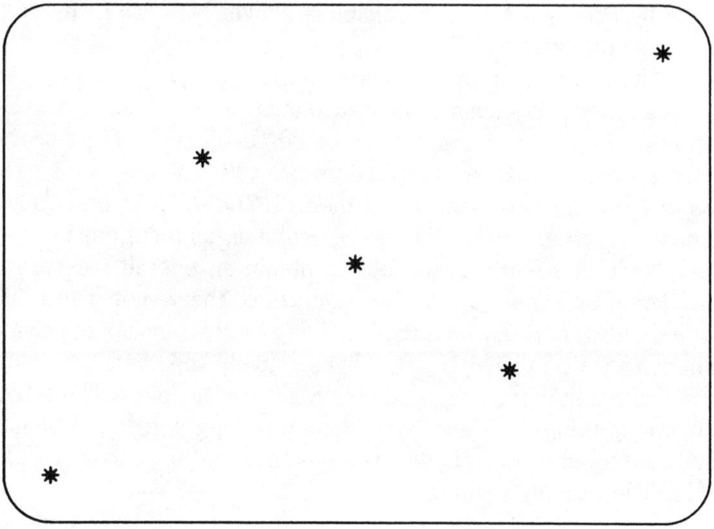

Fig. 7-4. An imperfect Correlation.

Standard Scores are z-scores, and we have our little microcomputer to do all the work for us. Consequently, run Program 4-1, the z-score program, and find the z-scores for this example. The z-scores will be found to be as follows:

Person	Test X IQ z_x	Test Y IQ x_y	$z_x\ z_y$
A	1.4142	1.4142	2.0000
B	.7071	−.7071	−.5000
C	.0000	.0000	.0000
D	−.7071	.7071	−.5000
E	1.4142	1.4142	2.0000
		Sum =	+3.0000

It would be good practice for you to work out a few of these numbers by hand.

To repeat, before two sets of scores can be compared, both sets must be reduced to comparable Standard Scores, that is, scores with identical Averages and Standard Deviations, and that has been done by Program 4-1 in the example above. Now we're ready to compare the scores.

The comparison of the scores is made simply by multiplying them together, as shown in the last column above. Thus, Person B's score on Text X is 110, and his score on Test Y is 90. The z-scores corresponding to these two scores are .7071 and −.7071. The product of the two z-scores is then .7071*(−.7071) or −.5000. These scores are shown in the last column, as mentioned.

Next, it seems reasonable to obtain an overall measure of similarity by adding up all the products of the z-scores (actually called *cross-products*), and then dividing by the number of pairs of numbers (to be called N_{pairs}). That's roughly the same procedure used with the Average, and also the Standard Deviation. If you keep adding up things, the sum should keep getting bigger and bigger, and consequently one divides by the number of things to correct for this. The formula is then:

$$\text{Correlation} = r = \frac{\Sigma z_x z_y}{N_{pairs}} \qquad \textbf{7-1}$$

In the present case, the sum of the cross-products is +3.0000 (see little table above). Therefore the Correlation is +3/5 or +.60; the degree of relationship has been pinned down much more accurately than it was before. All that was known originally was that it had to be somewhere between 0.00 and +1.00, and now it is put at precisely +.60.

Thus, we have the formula for the Correlation coefficient. The symbol for Correlation is "r," which stands for "regression;" we'll talk about regression in the next chapter.

Notice how simple Formula 7-1 is. It involves only an averaging of products. The z-score is obtained for each score, multiplied together, and then the products are averaged. The process could not be simpler. Putting correlational problems in z-score form makes them duck soup.

A PROGRAM TO CALCULATE r's

Our microcomputers make the computation of Correlation very easy indeed, even though a substantial amount of work is needed. You see, one needs to get the Average and Standard Deviation for each variable, so that the z-scores can be calculated. Then, the z-scores must be obtained, multiplied together, the products added up, and the average taken. It's a lot of work. If you don't believe me, try it by hand. Program 7-1 will make things a good deal simpler for you.

```
10 REM PROGRAM 7.1. Z-SCORE CORRELATION
20 READ X,Y                         :REM GET DATA
30 IF X=-1 THEN 100                 :REM TEST FOR END
40 SX=SX+X  :SY=SY+Y                :REM GET SUMS
50 QX=QX+X*X:QY=QY+Y*Y              :REM GET SQR SUMS
70 N=N+1                            :REM COUNT ENTRIES
80 GOTO 20                          :REM CONTINUE
100 AX=SX/N:AY=SY/N                 :REM GET AVERAGES
110 VX=QX-SX*SX/N                   :REM GET TOP OF SD
120 VY=QY-SY*SY/N                   :REM FORMULA FOR X,Y
130 DX=SQR(VX/N)                    :REM GET SD FOR X
140 DY=SQR(VY/N)                    :REM GET SD FOR Y
200 RESTORE                         :REM RESET DATA
210 FOR I=1 TO N                    :REM R CALC LOOP
220 READ X,Y                        :REM GET DATA AGAIN
230 ZX=(X-AX)/DX                    :REM GET Z FOR X
240 ZY=(Y-AY)/DY                    :REM GET Z FOR Y
250 S=S+ZX*ZY                       :REM GET SUM ZX*ZY
260 NEXT I                          :REM END LOOP
270 R=S/N                           :REM CALCULATE R
280 PRINT "CORREL IS";R             :REM PRINT RESULT
300 DATA 120,120                    :REM DATA TO USE
310 DATA 110,90
320 DATA 100,100
330 DATA  90,110
340 DATA  80,80
350 DATA  -1,0
```

Notice that in this program, variable names of two letters are used for the first time in this book. Almost all microcomputers accept such variable names, and they make naming variables much easier for us. If only one letter is used, as in the past, it is very difficult to refer to variables in any sensible way, but the use of two letters makes more sense. If you have a tiny computer that allows only single-lettered variables, you can easily convert all the variables to single letters. But we shall use two letters often, just because they make better sense to most people.

Program 7-1 functions just as you would expect from an inspection of Formula 7-1. You go through the data once to get the Averages and Standard Deviations, and then through the data a second time to calculate the z-scores and their products. Finally, you calculate the r. The procedure is so complex that the flowchart in Fig. 7-5 can show only a part of it. Let's go through the program step-by-step.

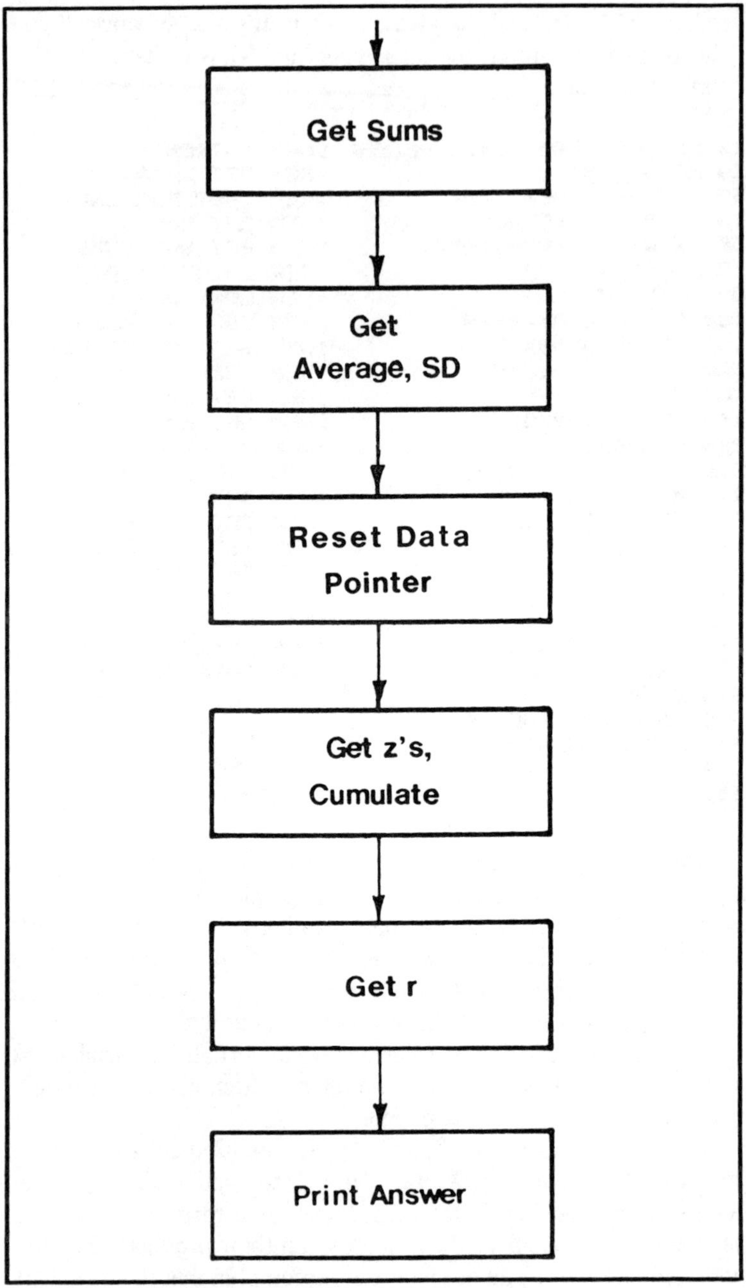

Fig. 7-5. Flowchart for the z-score Correlation program.

Lines 10-80 are almost an exact duplicate of Lines 10-70 of Program 4-1 (the z-score program). Indeed, the whole program mirrors Program 4-1. There is a difference, however, for we must calculate the z-scores on two variables now, X and Y. Consequently, each variable is handled separately. Line 20 reads the data, and, as always, Line 30 checks for the end of the data. Line 40 gets the sums for X and Y, called "S" for sum. Line 50 gets the sums of the squared numbers for X and Y, called "Q" for square. Line 70 counts the number of pairs entered. Lines 100 to 140 calculate the Averages and Standard Deviations for the two variables, just as was done in Lines 80-100 of Program 4-1.

Line 200 resets the data pointer, so that the computer can reread the data. Then the loop in lines 210 through 260 calculates the z-scores (lines 230-240), multiplies them together, and sums them in variable S (Line 250). The Correlation is calculated, as shown in Formula 4-1, in line 270, and the result printed in line 280.

The program isn't all that complex, but notice that it's a rather long one. As mentioned above, there's a lot of calculation going on: Averages and Standard Deviations must be calculated, then z-scores, then their products, these must be summed, and finally the Correlation itself must be computed. But the operation of the program itself is no more complicated, at bottom, than the z-score program, Program 4-1.

Run the program. It will tell you that the Correlation equals +.60. If you have a lot of stamina, work the problem out with only a hand calculator. Then you'll really appreciate your microcomputer.

EXPLORATIONS WITH THE CORRELATION PROGRAM

With Program 7-1, and the plot program, 5-1, you can learn a lot about how the Correlation Coefficient works. Simply by changing the DATA statements you can work out a whole new problem, and plot the results to give you a feel for the data. When I was a student, it took hours, literally, to work a single problem, and therefore, no one could work enough problems to get a feel for the statistic. Microcomputers have changed all that.

First, try the data for the perfect positive and negative Correlations given above (Figs. 7-2 and 7-3). As you can see, such Correlations give completely straight lines. The Correlation is always 1.00 in such circumstances: a perfect Correlation.

The following data will yield a far different result:

Variable X	Variable Y
−3	9
−2	4
−1	1
0	0
1	1
2	4
3	9

Put the data above into DATA statements, and run both the correlation program and the plot program. You can put both of them together in the computer at one time, but you'll have to change the numbers on one (starting, say, at 5000), and put an END statement at the end of the first one. Also, notice that since one of the data points is minus one (−1), and since the computer expects the data to end when it encounters a minus one, you'll have to change Program 7-1 slightly to make things work. You could use, say, −99 to indicate the end of the data. Then you'd have to change the −1 in line 30 of the program to −99.

Running the correlation program will show you that the Correlation between the two variables is 0.00. Thus, there would seem

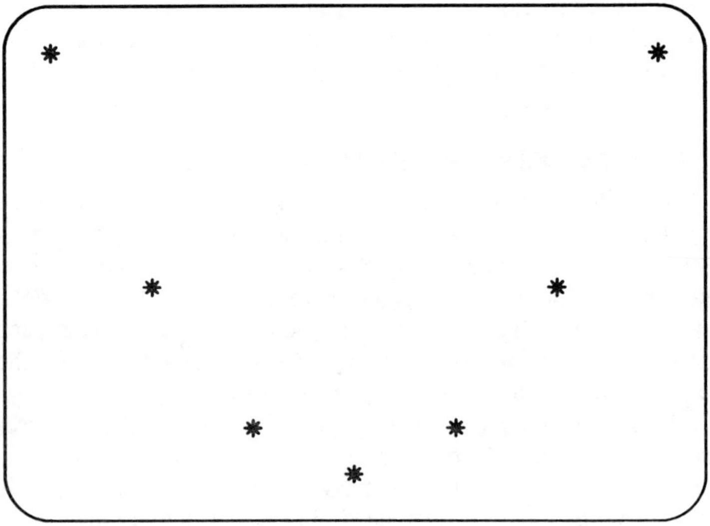

Fig. 7-6. A parabola.

to be no relationship between variable X and variable Y, but the plot program produces a plot similar to that of Fig. 7-6. Notice that there is, in fact, a strong relationship between X and Y, but, as you can see from the plot, the relationship is a parabola. The equation for this curve is as follows:

$$y = x^2 \qquad \textbf{7-2}$$

Since the two variables are so highly related in a parabola, why is the Correlation zero? That doesn't make sense.

The Correlation is zero because the Correlation formula tries to put a straight line through the points. Of course, you can't put a straight line through the points of a parabola. You can only put a parabola through those! Thus, the computer says that the relationship is zero. Another way to look at it is to note that while you can predict the Y values from X, you can't predict the X values from Y. Thus, if you know that the value of Y is 9, the value of X may be either plus 3 or minus 3. The correlation program simply reflects the ambiguity of the situation.

Now, we'll present a series of plots, just like the ones that you'd get on your microcomputer with Program 5-1, and your job will be to guess the Correlation in each case. In this way, you may get a feel for the Correlation Coefficient.

In Fig. 7-7, readings were taken on varying electrical voltages by two inexpensive voltmeters. What's the Correlation?

This is a fairly easy one. Figure 7-7 looks a lot like Fig. 7-2, the perfect Correlation at the beginning of the chapter. However, the relationship is by no means perfect, as is shown by fairly marked variations in the line. Surprisingly, the Correlation is .9938! That's very nearly perfect. You'll notice a tendency of the Correlation Coefficient to overestimate the relationship in future examples. Obviously our meters are far from perfect!

Figure 7-8 is plotted from purely fictitious data. The plot looks a lot like Fig. 7-3, the perfect negative relationship. Remember that the Correlation Coefficient tries to put a straight line through the data, and it is fairly easy to see where the line should go. It will slope down from the upper left hand corner. Thus, we'd expect a negative Correlation. However, the dots are scattered a good deal, and to my eye, one wouldn't expect a very high r. Surprise! The Correlation is −0.9773. Again, that seems much higher than Fig. 7-8 would indicate.

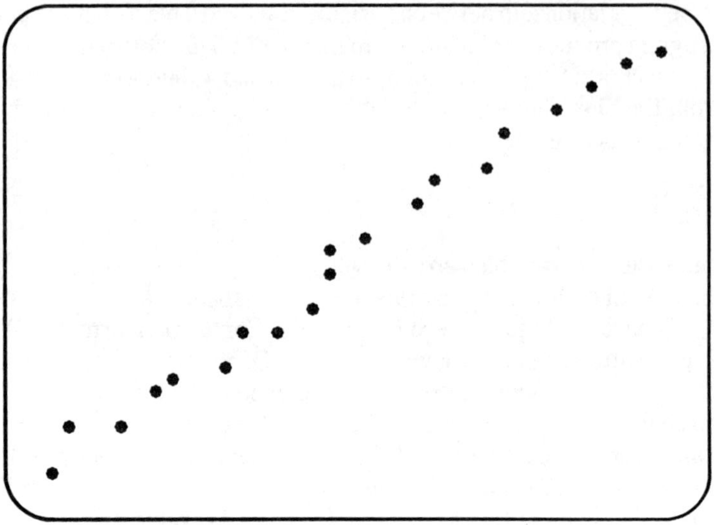

Fig. 7-7. Scatterplot of two inexpensive voltmeters.

Look at Fig. 7-9. This plot is perfectly even in both directions across the screen. There seems to be no relationship whatever between the variables—and such is the case! The r is zero with these variables. Frankly, since the plot is a bit longer than it is wide,

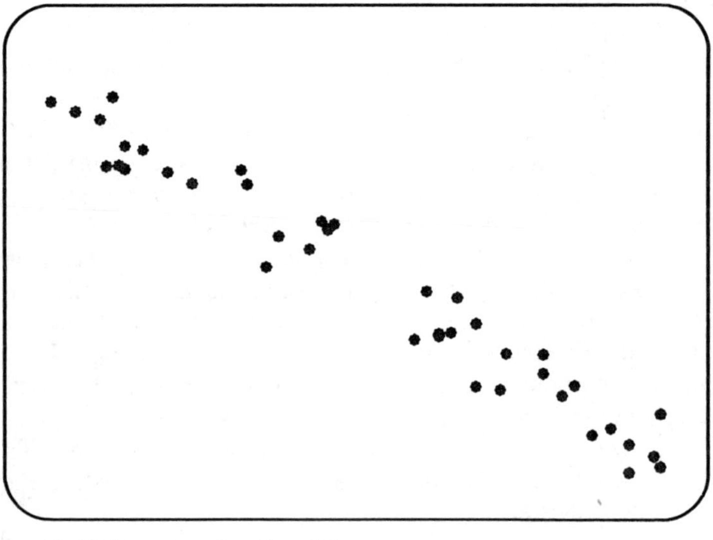

Fig. 7-8. Fictitious negative Correlation.

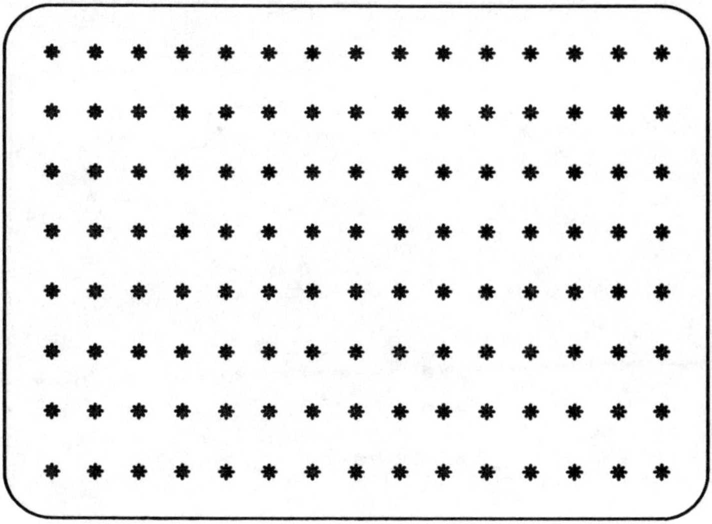

Fig. 7-9. Uniform scatterplot.

I had a hunch that the Correlation might be a bit different from zero, but the formula proves me wrong! Incidentally, it's very, very difficult to find really random data that will give an r of zero. If you don't believe me, try to find some!

Figure 7-10 is a good example of the sort of rs that we typically get in my field of Psychology. It's based on real data, and is a plot of two Wechsler IQ test scores: the Information Test score and the Full Scale IQ. Notice that the points vary all over the place. You would not expect much of a relationship at all, and yet the r is +.4942! A good rule of thumb is that if you can see any way of putting a straight line through the data, the Correlation is fairly high. With much imagination, you could put a straight line from the bottom left to the top right. Certainly, predictions made in such a situation would not allow very accurate predictions. If only we could have tests in my field that were as accurate as the electrical meters in Fig. 7-7!

Figure 7-11 again shows a plot of two Wechsler IQ test scores, the Verbal IQ and the Full Scale IQ. What's the Correlation in this case?

Clearly, Fig. 7-11 shows a stronger relationship than Fig. 7-10. It's easy to see the direction of the straight line—up to the right, but there is still a very large amount of scatter. However, the r is +.8587. That's a substantial value in my field, one which, quite

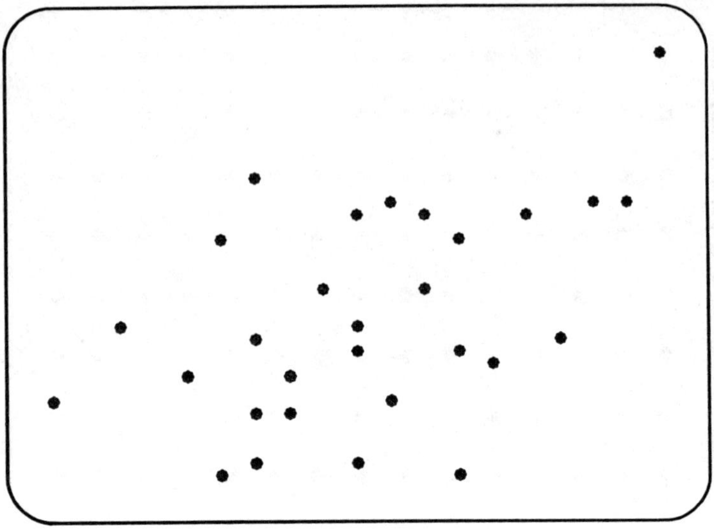

Fig. 7-10. Scatterplot of two intelligence tests.

frankly, is almost never attained. The r is this high in this case simply because the Full Scale score contains the Performance Score in its computation.

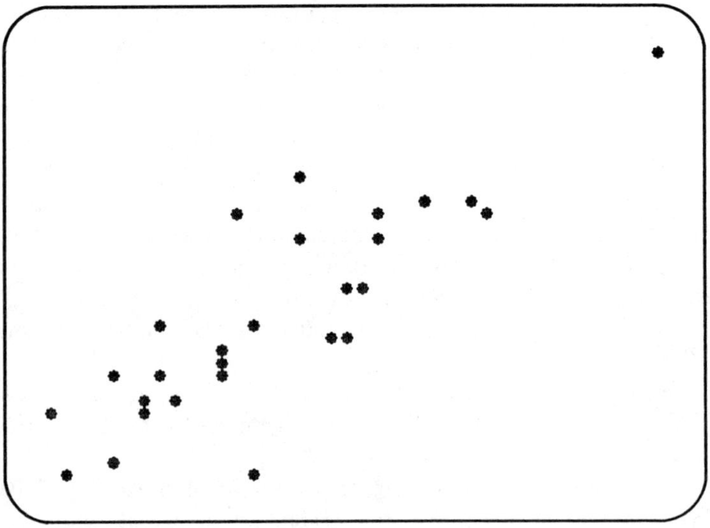

Fig. 7-11. Verbal vs. full-scale IQ.

Figure 7-12 is purely fictitious data, actually composed on the screen of my TRS-80, and then transferred to the plotting program. Clearly, a line put through the data would slope down from the upper left. It's not difficult to see where the line would go. What's the Correlation?

It's −0.8916. Again, the scores seem to spread out more than you would expect from that value. −0.8916 seems a bit too high. Once more, the r is deceptive.

Figure 7-13 again plots two Wechsler IQ test scores—the Information subtest and the Comprehension subtest. The diagram seems to slope up to the right. What's the r?

It's +0.6769. That seems reasonable. It's not difficult to see where the line should go, and if it weren't for the point at the center of the diagram, in the middle, it would seem likely that the Correlation would be higher.

Figure 7-14 should be easy for someone who's already seen Fig. 7-6. It was made up by taking some data, transforming it, and putting it into the equation for the parabola given earlier in the chapter. However, I didn't quite succeed in making a symmetrical parabola! Consequently, the Correlation is not zero, as you'd ex-

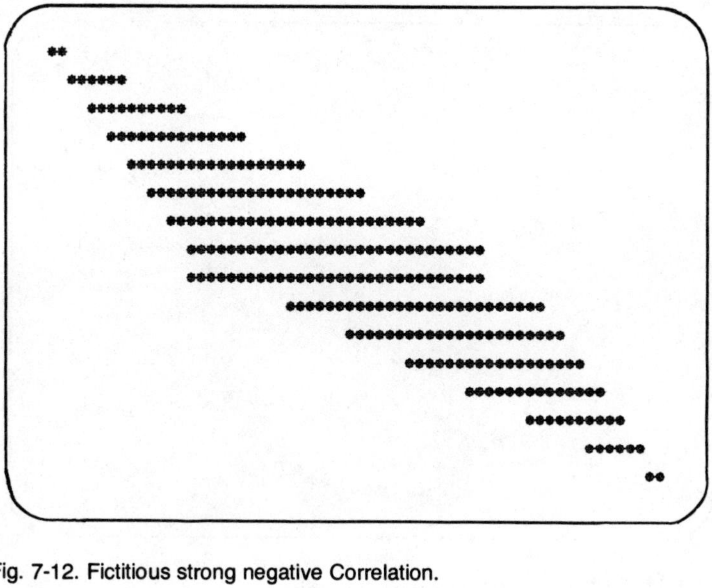

Fig. 7-12. Fictitious strong negative Correlation.

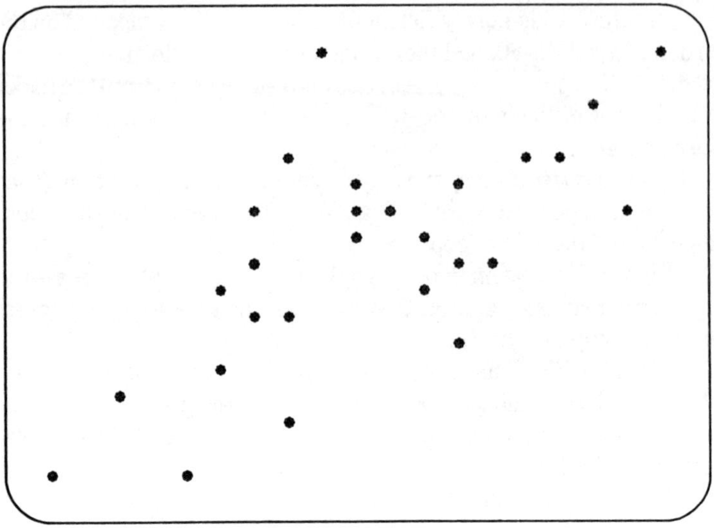

Fig. 7-13. Information and comprehension IQ tests.

pect, but, rather, +0.0833. However, the effect of the parabola is still there. There's a nearly perfect relationship, but it doesn't fit a straight line at all.

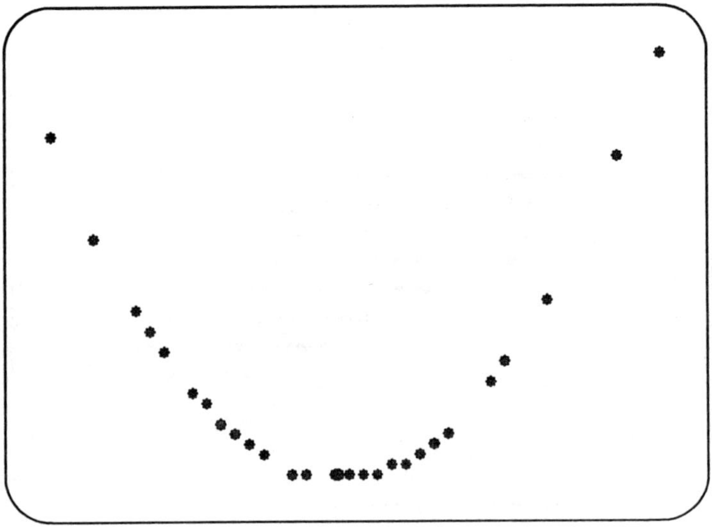

Fig. 7-14. A parabola.

Figure 7-15 is of great historical interest. It's a plot of the data collected by Sir Francis Galton over a hundred years ago, on the heights of parents versus the heights of their children. Galton is the man who did the original work on Correlation. The data should look peculiar simply because it's grouped into rough categories. But, to make it look better, I used a random number generator to spread the points out a bit. The Correlation for Fig. 7-15 is +0.4635. We'll talk about Galton more in the next chapter.

Figure 7-16 is a plot of two scales of the Sixteen Personality Factor Questionnaire, a personality test. The scales plotted are Boldness and Guilt. What's the Correlation?

You should be getting the hang of things by now. The diagram seems to slope down from the upper left hand corner, implying a negative Correlation. The points are widely scattered, but you can indeed get a slight idea of the location of the straight line that the correlation formula will put through the points. Thus, you'd expect a moderate Correlation, and it is, with a value of −0.4115. Negative Correlations in psychology occur mainly in the field of personality testing, such as this one.

Figure 7-17 is a plot of the points listed on the following page.

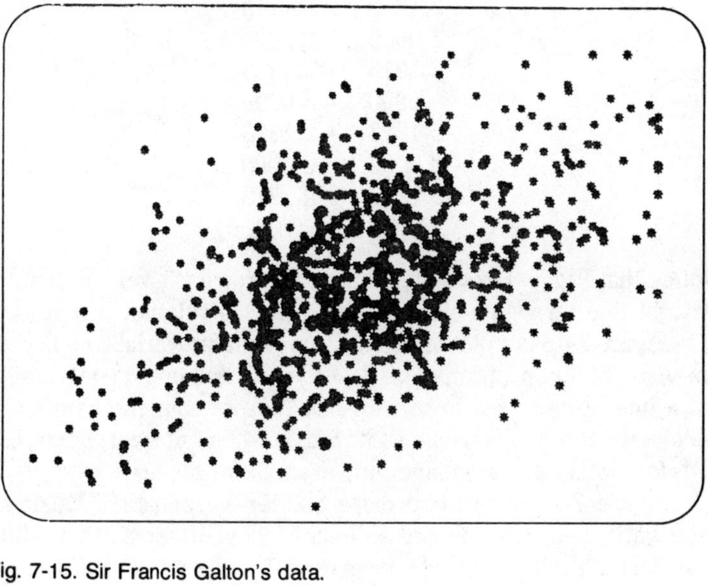

Fig. 7-15. Sir Francis Galton's data.

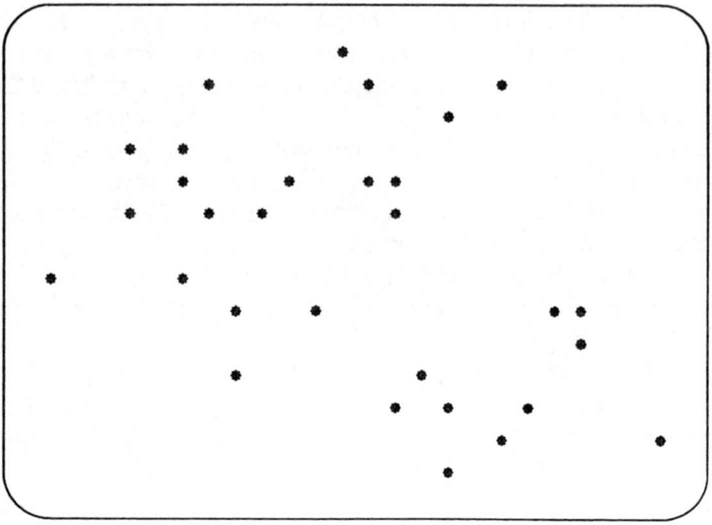

Fig. 7-16. Boldness vs guilt scores.

X	Y
0.0000	3.0000
2.1213	2.1213
3.0000	0.0000
2.1213	−2.1213
0.0000	−3.0000
−2.1213	−2.1213
−3.0000	0.0000
−2.1213	2.1213

Notice that Fig. 7-16 is a plot of a circle. You can't very well put a straight line through a circle! And so the Correlation is zero!

Figure 7-18 is a plot of two more personality variables: masculinity and family position. The data is real. You could conceivably put a line up from the lower left hand corner, but the points are widely scattered. The r shows it: it's +0.4736, not very high, but high for Correlations obtained in research in this area.

Figure 7-18 shows two more Sixteen Personality Questionnaire variables—warmth and self-sufficiency. It seems that a line could be put through the data down from the upper-left hand corner.

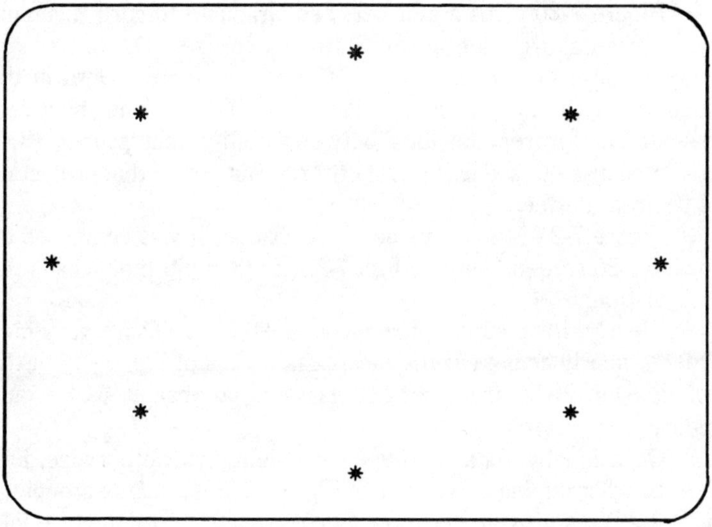

Fig. 7-17. Scatterplot of a circle.

The Correlation of -0.5704 shows this to be the case. This is a fairly high Correlation for psychological research, and notice that it's fairly easy to visualize the line.

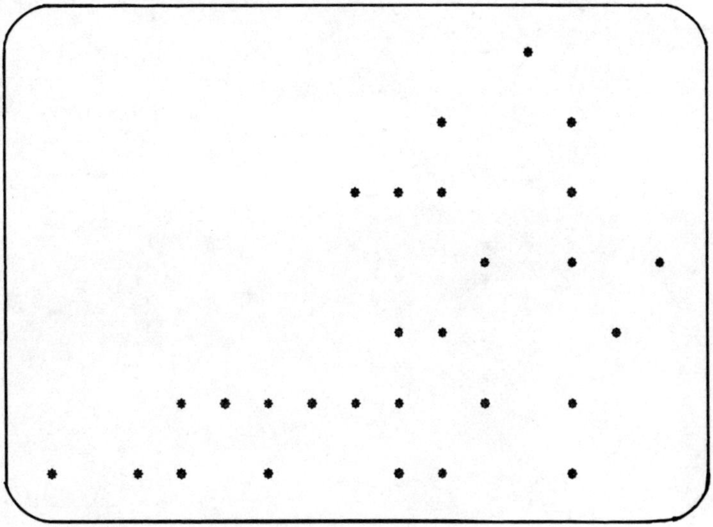

Fig. 7-18. Masculinity vs family position.

Figure 7-20 gives a plot between Graduate Record Examination scores and grades in my Statistics course; The scores are widely scattered, but a line would seem to slope upward in the positive direction. The Correlation is +0.5149. This is about representative for relationships between ability test scores (IQs, GREs, and so on) and real world criteria. But notice that prediction is far from perfect.

Figure 7-21 is an easy one, for a change! It was composed on my TRS-80 screen, just like Fig. 7-8, and then plotted. What's the Correlation?

Clearly, it's positive. The value is +0.9701. That's very high indeed, nearly as high as the electrical meters of Fig. 7-7, but the relationship is far from perfect, far less so than 0.9701 would indicate.

Occasionally, data is grouped, accidentally or otherwise, into very rough groupings, as shown in Fig. 7-22. Such coarse groupings do not always provide very low Correlations, as you might think. The r in Fig. 7-22 is −0.4726, which is surprisingly high, under the circumstances.

Finally, Fig. 7-23 is a very difficult one to estimate by eye. The variables are real ones, derived from Sigmund Freud's theories. It's almost impossible to judge where the line will go, and the points are

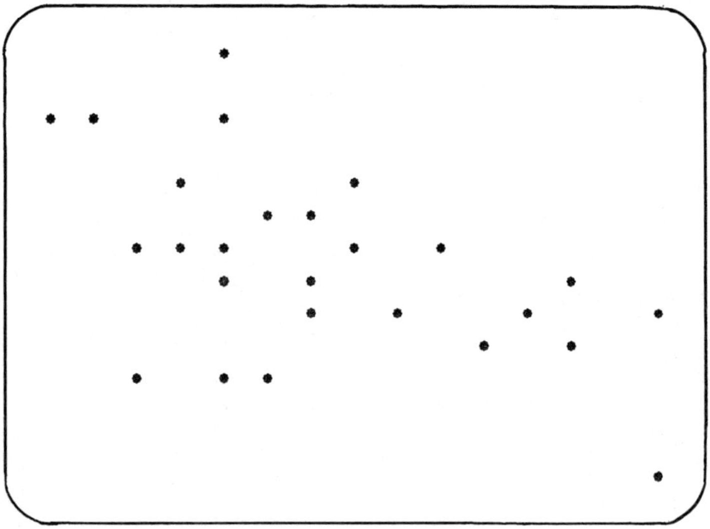

Fig. 7-19. Scatterplot of warmth vs self-sufficiency.

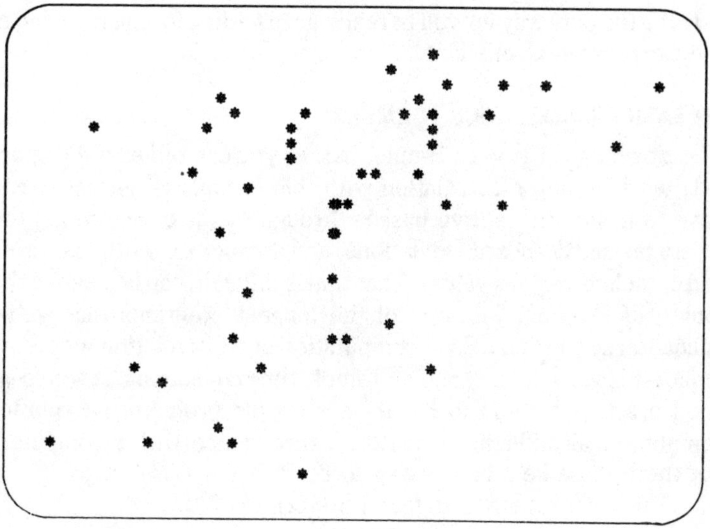

Fig. 7-20. GRE Scores and statistics grades.

scattered all over the place. But the r is +0.3662, which is a great deal higher than I would have guessed.

Try some more data on your computer with Programs 5-1 and

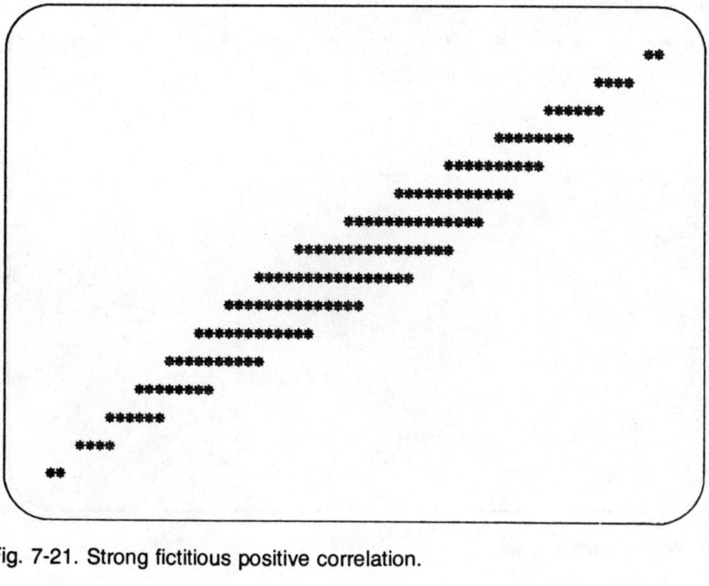

Fig. 7-21. Strong fictitious positive correlation.

7-1. It's the only way for you to really get a feeling for the meaning of the Correlation Coefficient.

AN EASIER CORRELATION FORMULA

Formula 7-1 is very simple, but, as you can tell from Program 7-1, working out a Correlation with this formula is not easy. You have to make at least two passes through the data, one to get the Averages and Standard Deviations, and another to get the z-scores and cumulate various values. That this is difficult, can be seen in the length of Program 7-1, one of the longest programs that we've encountered thus far. Hand computation of a Correlation with Formula 7-1 is very difficult and extremely time consuming. It would be good practice for you to try it on a simple problem. Of course, computational difficulty is of little consequence with a computer, but there must be a better way to do it.

The better method is found in Formula 7-2:

$$r = \frac{\Sigma XY - \dfrac{\Sigma X \, \Sigma Y}{N}}{\sqrt{\left(\Sigma X^2 - \left(\dfrac{(\Sigma X^2)}{N}\right)\right)\left(\Sigma Y^2 - \left(\dfrac{(\Sigma Y^2)}{N}\right)\right)}} \qquad \textbf{7-2}$$

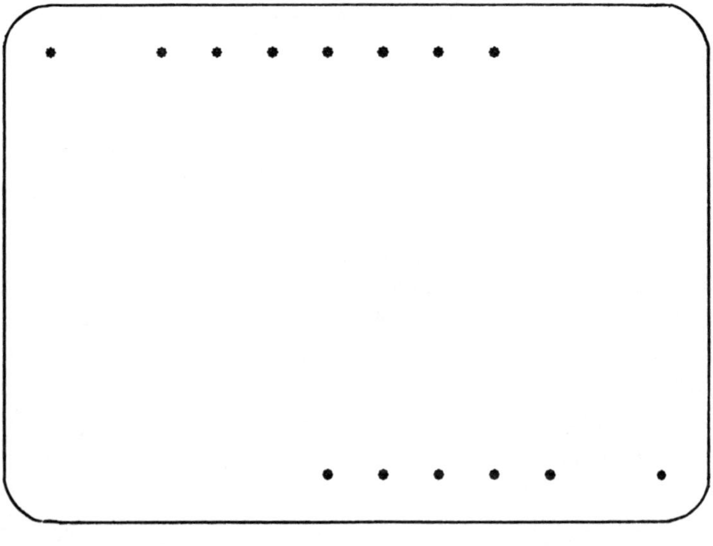

Fig. 7-22. Grouped data.

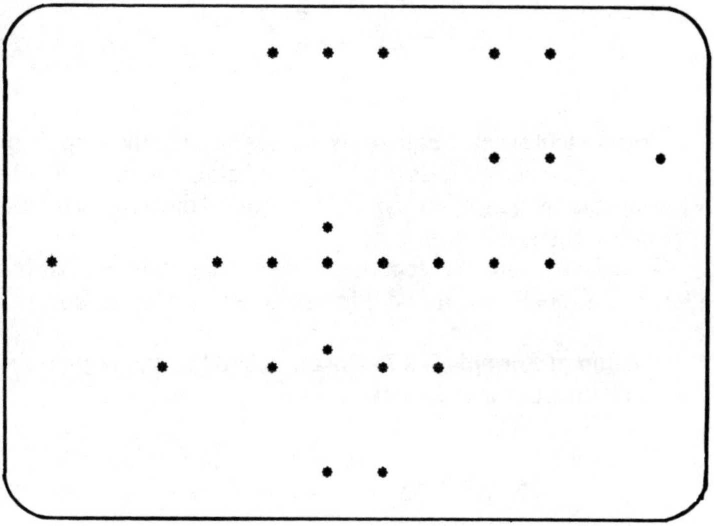

Fig. 7-23. Scatterplot of Freudian variables.

"Good Grief!" you say. "How can such an awesome-looking formula be simpler?! It looks three times as hard!"

It is, however, a much simpler formula. In fact, the formula breaks down into the three sections, which are largely identical. Thus, the section on the bottom left of the formula is:

$$\Sigma x^2 = \Sigma X^2 - \frac{(\Sigma X)^2}{N} \qquad \textbf{7-3}$$

You've seen this formula before. It's the top of the Standard Deviation formula, Formula 3-5. This expression comes up again and again in statistics, and it has a confusing name. It's called the *sum of squares.* That term is confusing because of course ΣX^2 and $(\Sigma X)^2$ are also sorts of sums of squares. But the term has been used for decades. Worse, it's denoted by Σx^2, that is, the sum of little **x** squared! Consequently you have to keep track of little xs and big Xs and it's all very confusing. But in practice one seems to get used to it.

The right-hand section of the bottom of Formula 7-2 is almost identical. However, the letter Y is used whenever X was used previously.

$$\Sigma y^2 = \Sigma Y^2 - \frac{(\Sigma Y)^2}{N} \qquad \textbf{7-4}$$

Both formulas say, "Square the values and add them up. From this subtract the sum of the values, squared after summing, divided by the number of things." In short, the bottom of the new correlation formula has the same term twice.

This sum of squares term is extremely important in what is to follow in this book. You'll find it in almost every chapter from now on.

The top of Formula 7-2 is closely related to the bottom sections. Let's look at it in isolation:

$$\Sigma xy = \Sigma XY - \frac{(\Sigma X)\,(\Sigma Y)}{N} \qquad \textbf{7-5}$$

If X is the same variable as Y, then Formula 7-5 reduces to either Formula 7-3 or 7-4, depending upon which way one alters things. Formula 7-5 says, "Multiply the original scores together, then add them up. From that value, subtract the sum of the X values times the sum of the Y values, divided by N. At any rate, it's not so difficult.

In sum-of-squares form, Formula 7-2 reduces to the following expression:

$$r = \frac{\Sigma xy}{\Sigma x^2\ \Sigma y^2} \qquad \textbf{7-6}$$

If you're the sort of person who doesn't like mathematics a lot, you probably don't think that the recent manipulations have improved things a great deal. But look again at Formula 7-2. All the values there can be obtained with one pass through the data. ΣX^2 and ΣY^2 say to square each value, and then sum them. (ΣX) and (ΣY) simply mean to sum the values, and ΣXY says to multiply the values together and then sum them. All of these things can be obtained the first time through the data. Program 7-2, which translates Formula 7-2 for the computer, shows this very clearly:

```
10 REM PROGRAM 7.2. RAW SCORE CORRELATION
20 READ X,Y                          :REM GET DATA
30 IF X=-1 THEN 110                  :REM TEST FOR END
40 SX=SX+X   :SY=SY+Y                :REM GET SUMS
50 QX=QX+X*X:QY=QY+Y*Y               :REM GET SQR SUMS
60 XY=XY+X*Y                         :REM GET CROSS PROD
70 N=N+1                             :REM COUNT ENTRIES
80 GOTO 20                           :REM CONTINUE
110 VX=QX-SX*SX/N                    :REM VALUES FOR
120 VY=QY-SY*SY/N                    :REM  BOTTOM OF 7.2
130 CP=XY-SX*SY/N                    :REM TOP OF FORM 7.2
140 R=CP/SQR(VX*VY)                  :REM CALCULATE R
150 PRINT R                          :REM PRINT ANSWER
160 DATA 120,120                     :REM DATA TO USE
170 DATA 110,90
180 DATA 100,100
190 DATA  90,110
200 DATA  80,80
210 DATA  -1,0
```

This program calculates the Correlation with one pass through the data, and in fewer lines than did Program 7-1. Thus, it must be a great deal more efficient (see Fig. 7-24). Lines 10 through 80 are virtually identical to the same lines in Program 7-1. However, in Line 60, the sum of XY is obtained; that's not in the previous program. Lines 110 through 130 calculate the three elements of Formula 7-2 (CP stands for cross-product), and line 140 actually calculates the Correlation.

If you don't understand Formula 7-2, don't worry about it. Program 7-2 will do the job for you, and is very fast and efficient. Program 7-2, or some variation of it, is the way in which the Correlation is almost always calculated in a computer. Try various values both with this Program, and Program 7-1. You'll find that you get the same answer with both, but Program 7-2 is faster, and shorter.

INTERPRETING CORRELATIONS

It's an unfortunate fact that the Correlation Coefficient, r, can be rather misleading. We saw that while examining scatterplots earlier in the chapter. You would think that if you have a Correlation of +.90, say, that the two variables involved are about 90% similar to one another. Or, putting it another way, you might think that you could predict accurately about 90 percent of the time from X to Y. This is not at all the case. There is no exact agreement about the

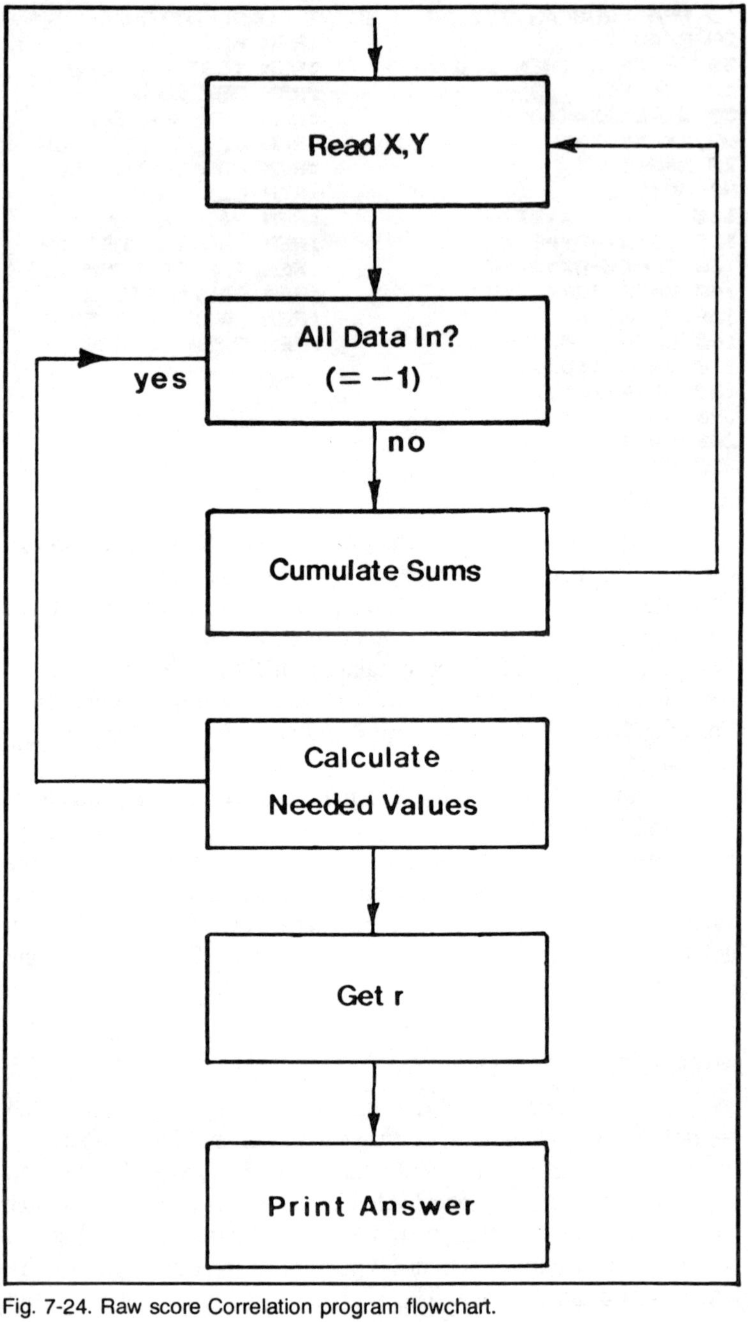

Fig. 7-24. Raw score Correlation program flowchart.

proper method for interpreting Correlations, and several systems are used, most more stringent than the one shown here, but the one to be presented is the simplest.

To interpret a Correlation Coefficient, you simply square it. Since Correlations are always decimal numbers, this means that the value will always become smaller. Take, for example, a Correlation of +.70. This looks fairly substantial, and you would probably be rather happy with such a value. However, 0.70 squared is 0.49. This is not nearly as impressive as 0.70, but 0.49 is a much better indication of the meaning of the Correlation. Loosely, you can say that if the Correlation is 0.70, then the two variables are 49 percent similar, or that you can predict from X to Y correctly about 49 percent of the time. The point is that Correlations are never as impressive as they look, unless they happen to be 1.00. It's possible to make up a little table of various r values and their squares:

r	**$r^2 \times 100$**
1.00	100%
.90	81%
.80	64%
.70	49%
.60	36%
.50	25%
.40	16%
.30	9%
.20	4%
.10	1%

This table is terribly depressing, especially when you have become accustomed to Correlations of the size generally found in my field, Psychology, where Correlations almost never exceed .70. A Correlation of .70, from the table above, does not imply 70 percent accuracy, but, rather, only 49 percent accuracy. A Correlation of .60 looks very respectable at first, and is about the value that one gets when relating IQ test scores to school grades, but it implies only 36 percent accuracy. This method of interpretation of Correlations can be called the "proportion of variance accounted for" method, and perhaps is the most liberal of several systems used. In any event, the important thing to remember is that Correlations are rarely as impressive as they look at first glance.

CORRELATION AND CAUSATION

It is known that there is a substantial negative Correlation between the amount of illumination at a busy intersection and the number of automobile accidents that occur there. From this fact can we conclude that lower illumination causes more accidents? If you say that it does, you may be making one of the most common mistakes that people make about interpreting Correlation Coefficients. For Correlations only indicate that two things vary together. They cannot tell you anything about causation. Indeed, you can make the example just given rather ridiculous by making it go in the other direction. Since Correlations may go either way, can we say that if we were to cause many more accidents at that intersection, it would be certain that the illumination would grow dimmer? Certainly not. All that the Correlation can tell us is that the two things vary together, not that they cause one another.

Another example: It is well known, and true, that the length of one's thumb correlates very highly, say +.90, with mental age. Does that mean that stretching your thumb would make you brighter? Or, on the other hand, if you had an accident and received brain damage that impaired your intelligence, would you expect your thumb to shrink? That's ridiculous! Causation cannot be inferred in either direction. The Correlation in this example is being induced, as it usually is, by a third factor—simple age. As children grow older, their thumbs grow longer, and their mental ages also increase. Hence the high Correlation. The danger of something else causing a Correlation is always present.

It is also no doubt a fact that there is a substantial relationship between the amount of pornography being read in any given year and the number of people who go to church. The relationship would certainly be at least +.90 or so. Does this mean that an increase in pornography leads to an increase in churchgoing? That somehow seems reasonable. However, because of the bidirectional nature of Correlations, it might also mean that going to church more often has led to greater consumption of pornography. That doesn't make sense. What's going on?

Before answering, let us give another ridiculous, but no doubt true example. There is undoubtedly a strong Correlation between the price of brass doorknobs and the salary of the Governor of Michigan. Does that mean, somehow, that increasing the price of brass doorknobs would increase the Governor's salary? Certainly there is something wrong here.

The problem in this example, and the one before it, is that almost everything increases over a long period of time. If you just wait long enough, prices on everything rise. Thus, it is the passage of time, and inflation, that causes the relationship between doorknob prices and the Governor's salary. Similarly, there are more people now then there ever were, and the world population keeps growing. That means that more people are reading pornography, and also more people are going to church. Of course, there will be small variations in things over a short period of years, but if you take a long period of time, especially when measured in centuries, things go up drastically. This is a little like the argument that Columbus is responsible for the destruction of Hiroshima by the atomic bomb, because, if Columbus hadn't discovered America, there would have been no World War II, etc., etc. Thus, you can blame everything on Columbus, or, to be more general, on Adam and Eve.

Another example will be of use. There is a definite relationship between the price of the car that a family drives and the children's chances of going to a good college. The Correlation is substantial. Does that mean that if your father goes out and purchases a Cadillac, that you'll have a better chance of going to a fine school? Or, to take the sillier side, does it mean that if you get into college, your father will suddenly be given a fancier car? Of course, in this example the third factor influencing all variables is overall family income and social status. People from rich families have a much greater change of getting into college; they make donations to endowment funds, and they know how to play the game of getting into good schools better than people without these advantages. Money breeds money, as they say.

Finally, notice that all of these things tend to increase with the passage of time:

- Lower incidence of disease.
- Greater use of contact lenses.
- Salaries of State Highway Patrol officers.
- Profits of Nevada gambling halls.
- Use of x-ray machines.
- Greater use of pocket calculators.

From this list, all of which are dependent on the passage of time as the real causative factor, you could conclude, erroneously, that greater use of contact lenses leads to higher profits in Nevada gambling halls. Or the greater use of pocket calculators will lead to higher salaries for Highway Patrolmen.

In short, Correlation cannot establish causation. Causation may be established by carefully controlled experiments, in which there is control over an independent variable. If all other factors are kept constant, one can then observe the effect of the experiment on a dependent variable. Even here, a third, uncontrolled, outside variable, may lead to erroneous conclusions, even in the purely experimental situation.

PROBLEMS

1. Two different voltmeters, both inexpensive, are connected to a variable voltage supply. The voltage is varied to see how well the two meters agree. Calculate the Correlation between the readings of the two voltmeters: (+.9992)

X	5.0	8.5	9.6	14.5	11.0
Y	5.4	8.9	10.1	14.6	11.5

2. Scores on a personality test scale for hypochondriasis (X) are to be compared with scores on the schizophrenia scale (Y). What's the Correlation? (+.7350)

X	22	28	45	33	21	33
Y	25	31	78	26	44	49

3. This example isn't a real one, but notice that the Ys are equal to the X's squared. Whats the Correlation? (+.9776)

X	1	4	3	2	7	8	4
Y	1	16	9	4	49	64	16

4. Here assume that we're comparing golf scores (X) with driving distance scores (Y) in yards. The data are obviously fictitious! What's the Correlation? (−.9984)

X	81	61	41	71	51
Y	390	610	790	490	690

5. Here's an interesting one. Calculate the Correlation:

X	44	55
Y	33	79

Notice that the Correlation is perfect. That's because two points, no matter which ones, always define a straight line. And the Correlation Coefficient always tries to put a straight line through the data. With two points it can always do so perfectly!

THE CORRELATION IS THE TANGENT OF THE ANGLE !

8 Prediction

We're now ready to face the problem of making accurate predictions. Prediction is the very heart of science; one does not usually consider a branch of science to have made progress until its predictions are quite accurate. Thus, Einstein predicted that a certain mass of uranium would explode. But he could not be taken completely seriously until the prediction was borne out when the first atomic bomb worked. It this chapter, the method most generally used to make simple predictions will be developed. Its technical name is *Regression.*

Let's take an example similar to the one from the last chapter, but with somewhat different numbers. We'll assume that we're going to predict grade point average (GPA) from scores much like the scores on the widely-used Scholastic Aptitude Test (SAT). The values look as follows (of course the data are completely fictitious):

Person	**Aptitude X**	**GPA Y**
A	700	4
B	600	1
C	500	2
D	400	3
E	300	0
	------	------
Average	500	2
SD	141.42	1.4142

The Correlation between the Aptitude Test and GPA is +0.60, just as in the example of the last chapter. However, the numbers have been changed to make a better example for this chapter.

Assume that you want to predict GPA from Aptitude Test scores. In z-score form, the method could not be simpler. To predict values of Y from values of X, simply use this formula:

$$z_y' = rz_x \qquad \textbf{8-1}$$

The left term, pronounced z-y-prime, refers to the predicted values of y, that have been predicted from the z-score values of X. With an Aptitude Test score of 700, what would you predict the GPA to be?

First, you must convert the Aptitude Test score to a z-score, using Formula 4-1, as follows:

$$z_x = \frac{X - \overline{X}}{SD_x} = \frac{700 - 500}{141.42} = +1.4142$$

Then, to find the z-value of Y to be predicted from the z-value of X, use Formula 8-1:

$$z_y' = rz_x = +0.60 * +1.4142 = 0.8485$$

Thus, the predicted value of Y (in z-score form) to be expected from the X value of 700 is +0.8485.

If all possible predictions are made from all of the z-score X values to all the z-score Y values, we obtain the following table:

X	z_x	z_y'
700	1.4142	.8485
600	.7071	.4243
500	.0000	.0000
400	−.7071	−.4243
300	−1.4142	−.8485

To get the values in the last column, we simply multiply the z-score in the second column by +0.6. The result is predicted value for Y, in z-score form. The original data points, and the line showing the predicted values are plotted in Fig. 8-1. To plot the line, first find the z_x value of 1.4142. Then go up to the z_y value of .8485, and plot a point. Next, plot any of the other points in the little table above, say

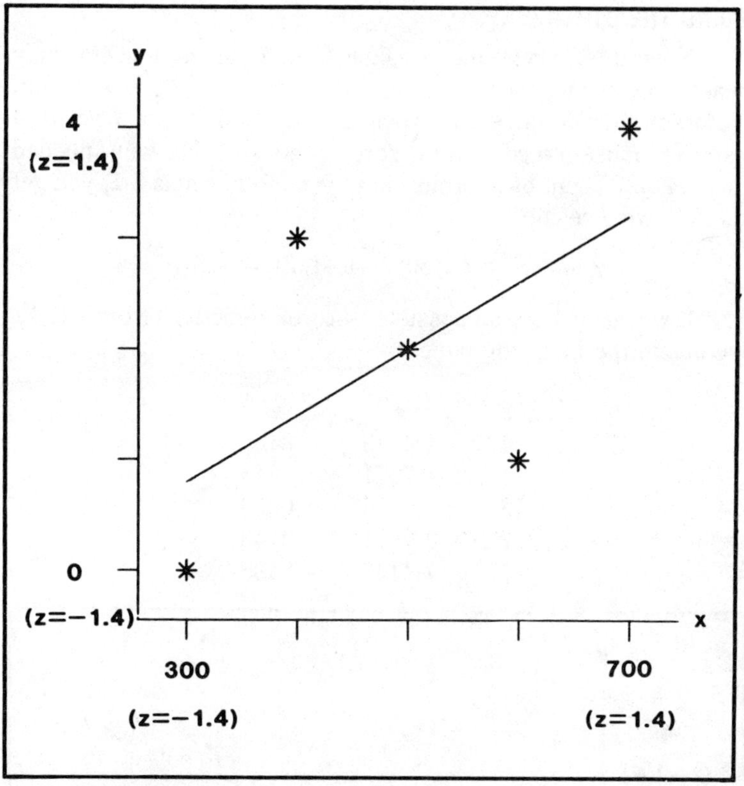

Fig. 8-1. A Regression line.

$z_x = -1.4142$ and $z_y = -.8485$. Finally, connect the two points with a line. The line in Fig. 8-1 is known as the line for the Regression of Y on X; this peculiar term will be explained shortly. In Fig. 8-1, since the z-scores and the original scores are linearly, or directly related, the same plot will serve for both sets of values.

"Whatever do the predictions mean?" you say. Alas, the usual problem with z-scores occurs, the one already dealt with in Chapter 4, namely that z-scores are very difficult to understand. Therefore the z-score must be converted back to GPA units. The method for doing this will be shown in the next section.

Nothing could be simpler than this method. The relationship used is $z_y' = rz_x$. To make predictions from one z-score to another, you simply multiply the predictor by the Correlation Coefficient, and You can prove that this method will make the best prediction possible in the situation. It seems too good to be true that such a simple method could do so much.

GOING THE OTHER WAY

Now, if we can predict Y values from X values, it seems only reasonable that we can just as well predict in the opposite direction. To predict the X value that would be expected from a Y value of zero, you must first convert the zero to a z-score. The value needed is −1.4142. Then, by a formula analogous to Formula 8-1, you get the following result:

$$z_x' = rz_y = (+.60)\ (-1.4142) = -.8485$$

If we again make all possible z-score predictions from Y to X, we obtain the following table:

Y	z	z′
4	1.4142	.8485
1	−.7071	−.4243
3	.0000	.0000
2	.7071	.4243
0	−1.4142	−.8485

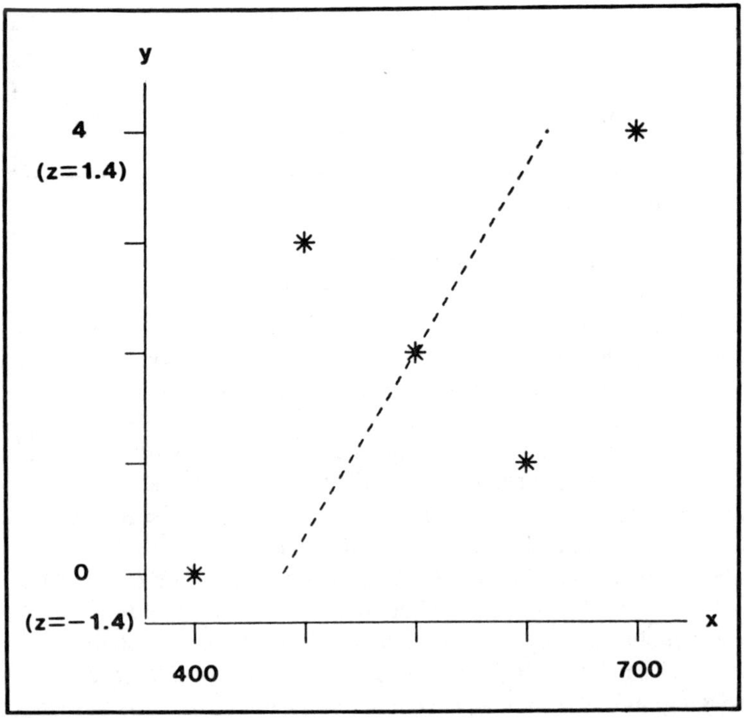

Fig. 8-2. The Regression of X on Y.

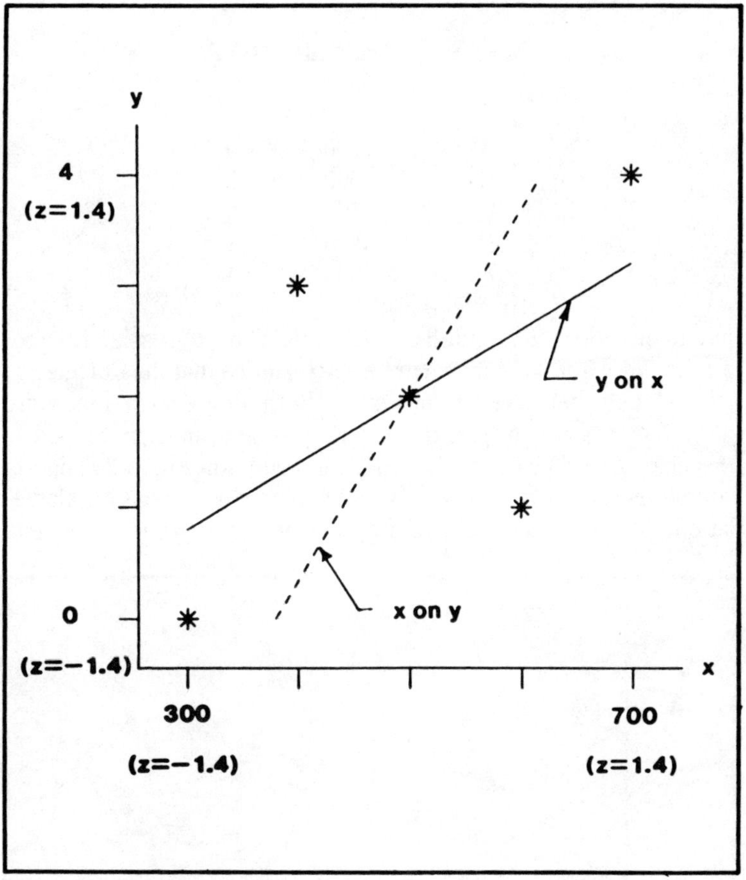

Fig. 8-3. Both Regression lines.

These points are plotted in Fig. 8-2, as the dotted line, in precisely the same way as the line in Fig. 8-1, with X and Y reversed, of course. This is the line for the Regression of X on Y, or the line for predicting X from Y. For the remainder of this chapter, solid lines will be used for line showing the Regression of Y on X, and dotted lines for the Regression of X on Y. Check the numbers against the plots to be certain that you see how they were arrived at.

Both of these Regression Lines are plotted in Fig. 8-3. In every problem of this sort, there will be two Regression Lines—one for Y on X, and one for X on Y.

It is of course possible to make predictions when the Correlation between two variables is negative. Take the following fictitious data:

Person	X Aptitude Test	Y GPA
A	700	0
B	600	3
C	500	2
D	400	1
E	300	4

In this situation, the Correlation is −0.6. The Regression Line for predicting Y from X is shown in Fig. 8-4. Notice that the Correlation is obviously negative in this case. Both Regression Lines are plotted in Fig. 8-5. Notice that the Regression Lines are flipped so that they are precisely backward from the plot in Fig. 8-3. That, of course, is precisely what you would expect with a negative Correlation. There would be two Regression Lines—one to go one

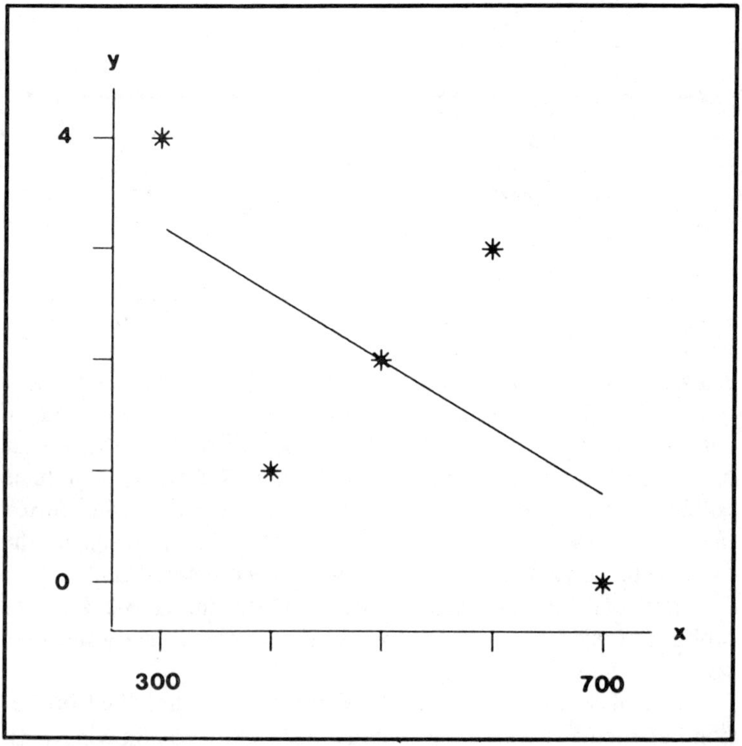

Fig. 8-4. Prediction of GPA from test scores.

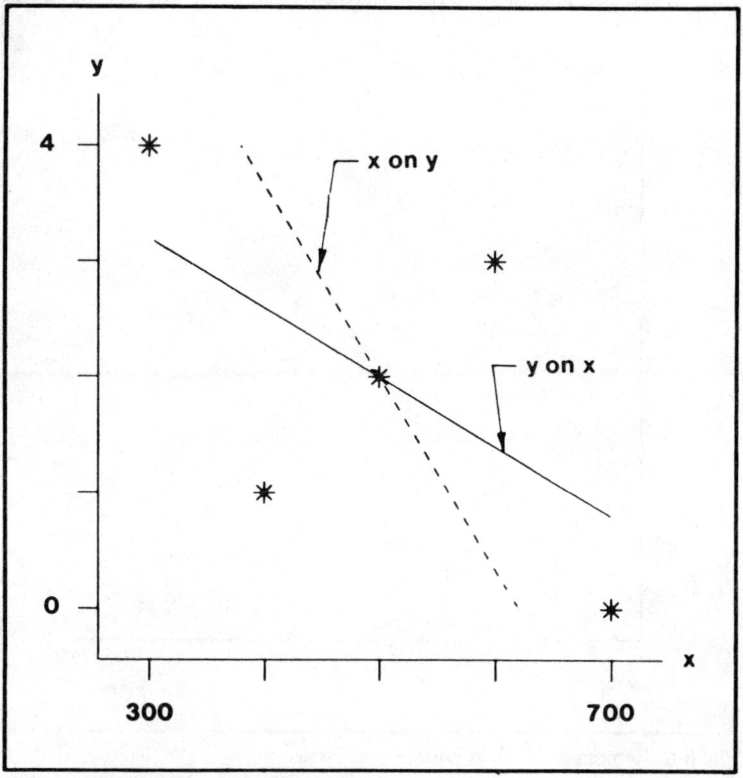

Fig. 8-5. Both GPA-test prediction lines.

direction, and another to go the other direction. This fact has puzzled generations of students of statistics!

Figure 8-6 shows the situation in which the Correlation is perfect—+1.00. Notice here that there is really only one Regression Line. In Fig. 8-6 the lines have been shifted just a bit so that you can see both, but with a perfect relationship, the two lines are really on top of each other.

To summarize, then, to predict Y from X, or X from Y, in z-score terms, simply multiply the z-score of the predictor by the Correlation. This leads to two formulas:

$$z_y' = rz_x \qquad 8\text{-}1$$
$$z_x' = rz_y \qquad 8\text{-}2$$

This is an amazingly easy prediction strategy. It just couldn't be simpler. In this case, just as with the calculation of the Correlation

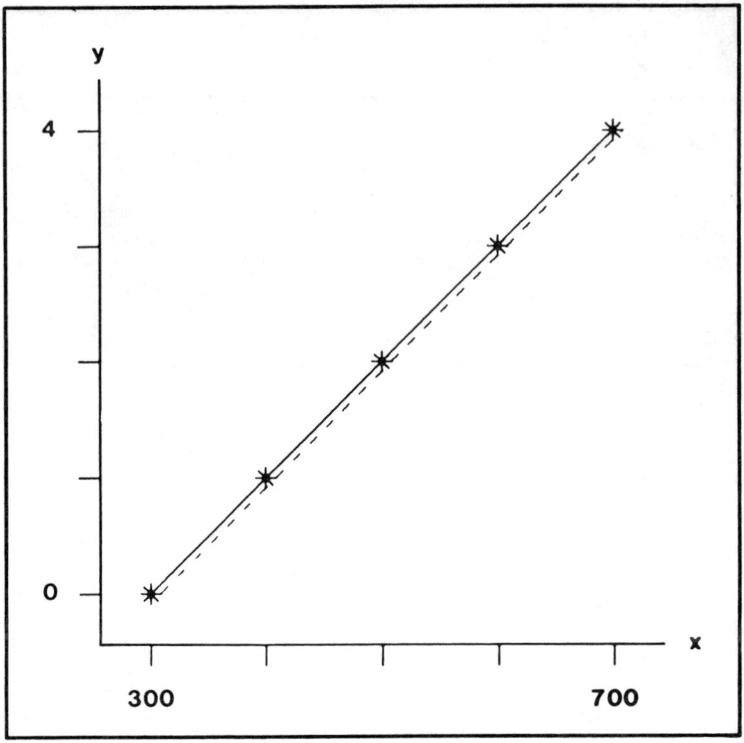

Fig. 8-6. Regression lines in perfect prediction.

Coefficient, z-scores make things very simple indeed. Unfortunately, z-scores present problems of their own, for most people have great difficulty in understanding them. Therefore, just as in Chapter 4, we must develop methods for going directly from one set of real scores to another, without having to go through the intermediate step of z-scores.

PREDICTION FORMULAS IN THE PROPER UNITS

If you take Formula 8-1 and, with a little algebra, replace the z-scores with the Averages and Standard Deviations you'll get the following formula in real score units. I'd go through the derivation, but it might scare you to death!

$$Y' = r \frac{SD_y}{SD_x} (X - \overline{X}) + \overline{Y} \qquad \textbf{8-3}$$

I told you it would be a bit frightening! But it isn't difficult. It's merely the z-score formula rearranged, with the z-scores converted back to the proper units. Notice, for example, that the z-score formula for X is still in there. See the $(X - \overline{X})/SDx$ in the middle of the formula? All that the equation does is to put back in the units that the z-score equation purposely leaves out. Let's try Formula 8-3, predicting from an Aptitude Test score of 700 to a GPA:

$$Y' = (+0.6)\ \frac{(1.4142)}{(141.41)}\ (700 - 500) + 2$$

$$= (+.006) * (700 - 500) + 2$$

$$= 3.2$$

In other words, if someone gets a score of 700 on the Aptitude Test, we would predict his GPA to be 3.2, and that can be proven mathematically to be the best possible prediction that one can make under the circumstances.

If you're going to do a great many predictions, it's convenient to work out the equation without putting in the value for X.

$$Y' = (+0.6)\ \frac{1.4142}{141.42}\ (X - 500) + 2$$

$$= +0.006 * X - 1$$

An equation in this form is often handy to have. It's called the Regression Equation for predicting Y from X. Say that X is 300. Putting that value in the formula,

$$Y' = +0.006 * 300 - 1 = 0.8$$

Thus, you'd predict a GPA of 0.8 from the Aptitude Test SAT score of 300. Notice again that this score is more in the direction of the Average than the real GPA score in the pair of 300 and 0.

All the possible predictions in the X to Y direction look as follows:

X Aptitude	Y′ GPA′
700	3.2
600	2.6
500	2.0
400	1.4
300	0.8

As mentioned, the plot of these predicted points will fall precisely in the same position on the graph as the z-score plot in Fig. 8-1.

These predicted values are interesting. Notice that, in general, the predicted values tend to be closer to the Average than the values that you'd expect if the relationship were perfect. This is the phenomenon of Regression, to be covered later in this chapter. The only prediction that remains the same as the original data is the prediction made from the value of 500. That still predicts the GPA of 2.

A formula for predicting X from Y in real score units can also be developed, and is shown as Formula 8-4:

$$X' = r \frac{SDx}{SDy} (Y - \overline{Y}) + \overline{X} \qquad \textbf{8-4}$$

To make a prediction of X from Y, the following formula would result:

$$X' = (+0.6) \frac{141.42}{1.4141} (Y - 2) + 500$$

$$X' = 60Y + 380$$

This is the Regression Equation for predicting X from Y, in real score units. Thus, a GPA of 0 predicts an Aptitude Test score of 380 (60 * 0 + 380 = 380). A GPA of 4 predicts an Aptitude Test score of 620.

All the possible predictions from Y to X, then, may be shown as follows:

Y GPA	X′ Aptitude′
4	620
1	440
2	500
3	560
0	380

Notice that the predicted values tend to be less extreme than the original Aptitude Test values, again because of the phenomenon of Regression.

To summarize up to this point, there are two Regression Equations that may be obtained from our set of data.

$$Y' = +0.006X - 1$$
$$X' = 60Y + 380$$

These look rather frightening, but they're not. They're really Formulas 8-1 and 8-2, adjusted for the different Averages and Standard Deviations of the Aptitude Test and GPA scores. To predict GPA scores (Y), which are small numbers, we multiply the large Aptitude Test scores by a small number (.006). To predict the large Aptitude Test scores (X) from the tiny GPA scores, we multiply by a large number (60). Finally, the value added in at the end corrects for the difference in Averages.

Still, Formulas 8-3 and 8-4 are complex, and need Averages, Standard Deviations, and Correlations which are not easy to calculate by hand. As you would expect in this book, your microcomputer will be called on to do all the dirty work!

A PREDICTION (REGRESSION) PROGRAM

The following program will calculate the prediction equation for you, and let you make new predictions to boot.

```
10 REM PROGRAM 8.1. PREDICTION PROGRAM
20 READ X,Y                        :REM GET DATA
30 IF X=-1 THEN 100                :REM TEST FOR END
40 SX=SX+X  :SY=SY+Y               :REM GET SUMS
50 QX=QX+X*X:QY=QY+Y*Y             :REM GET SQR SUMS
60 XY=XY+X*Y                       :REM GET CROSS PROD
70 N=N+1                           :REM COUNT ENTRIES
```

```
80 GOTO 20                          :REM CONTINUE
100 AX=SX/N:AY=SY/N                 :REM GET AVERAGES
110 VX=QX-SX*SX/N                   :REM GET SSQ FOR X
120 VY=QY-SY*SY/N                   :REM GET SSQ FOR Y
130 DX=SQR(VX/N)                    :REM X STD DEV
140 DY=SQR(VY/N)                    :REM Y STD DEV
150 CP=XY-SX*SY/N                   :REM GET CRS PRD
160 R=CP/SQR(VX*VY)                 :REM CALCULATE R
170 M=R*DY/DX                       :REM GET SLOPE
180 B=AY-M*AX                       :REM GET INTERCEPT
200 PRINT "SLOPE IS",M              :REM PRINT RESULTS
210 PRINT "INTERCEPT IS",B
300 INPUT "NEW X=";X                :REM GET NEW VALUE
310 Y=M*X+B                         :REM NEW PREDICTION
320 PRINT Y                         :REM PRINT NEW VALUE
330 GOTO 300                        :REM CONTINUE
400 DATA 700,4                      :REM DATA TO USE
410 DATA 600,1
420 DATA 500,2
430 DATA 400,3
440 DATA 300,0
450 DATA -1,-1
```

If you type in the program, and run it, you'll find that it gives you the prediction equation. Then, at the end, it lets you enter a new X value, and it will make a new prediction from it again and again. Let's look at it (see Fig. 8-7).

Lines 10 through 80 are identical with those of Program 7-2, the last Correlation program in Chapter 7. They cumulate the sums, sums of squared numbers, and the cross products. Line 100 gets the Averages, and lines 110 and 120 the sums of squares, the top of the Standard Deviation formula for both X and Y. Lines 130 and 140 finish the calculations for the Standard Deviations. Line 150 gets the value needed for the top of the Correlation formula, which is finished in line 160. Thus there is nothing new through Line 160.

Line 170 is new. This calculates the slope of the Regression Line. The slope is the number you multiply X by to make a prediction. Line 180 calculates the *intercept,* as it is always called in regression analysis. The intercept is the value added in at the end of the equation. An example will help here.

$$Y' = +0.006X - 1$$

The slope is +0.006, and the intercept is −1. The terms *slope* and *intercept* come from algebra, and refer to the tangent of the angle that a straight line makes with the X-axis, and to the place where the line cuts the Y-axis. These terms are so universally used in talking

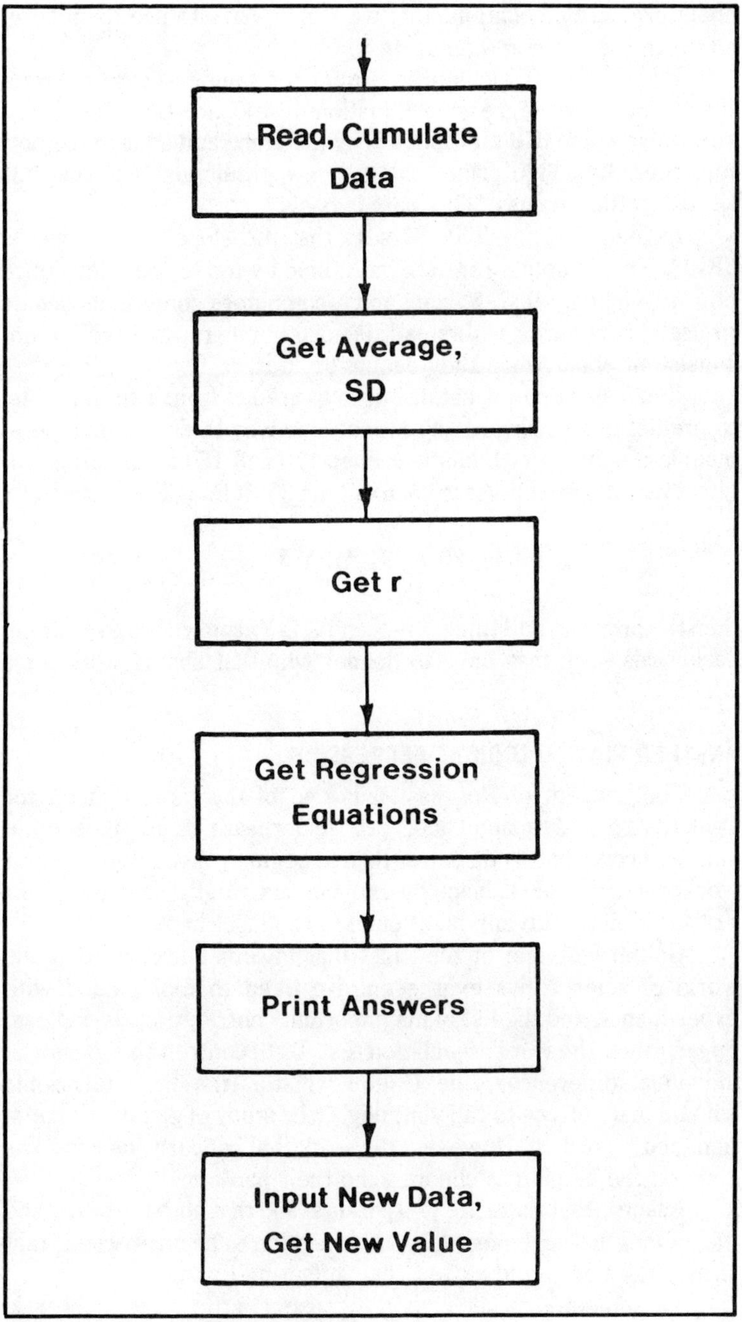

Fig. 8-7. Prediction program flowchart.

about prediction equations that we simply have to use them here, although they're very confusing.

Lines 300-320 calculate new predicted values, in case you want them, using the Regression Equation developed. Line 300 gets a new value, Line 310 mulitplies it by the slope and adds in the new intercept. Line 320 prints out the new prediction, and line 330 keeps up the process. This part is easy.

Incidentally, my TRS-80 says that the slope in this case is 6E-03. This simply means six multiplied by ten to the minus third power, which is .006. Various microcomputers sometimes decide to use this notation with small decimal numbers, and they're not consistent about when they decide to do it!

You may wonder what one does to predict from Y to X, that is, to predict in a backwards direction. One way to do it would be to include a new set of Lines like lines 170 and 180 in the program. However, it's easier just to change Line 20. If line 20 is changed to

```
20   READ Y,X
```

then the program will think Y is X and X is Y, and will do everything backwards. You may have to ponder why that should work for a while.

ANOTHER WAY TO LOOK AT REGRESSION

The concept of Regression is one of the most difficult for students to understand, and, yet, underneath it all, it is quite simple. I could hardly do better than to begin by discussing the first work done on the subject by the famous English scientist, Sir Francis Galton, already mentioned in the last chapter.

Galton was one of the most magnificent putterers that the world of science has ever seen. He liked to fool around with experiments, and the list of his important contributions is endless: fingerprints, the word association test, birth control, the science of individual differences, the Galton whistle (the little ultrasonic whistle that you use to call your dog!), the study of genius, Correlation, and on and on. However, the work that concerns us here was done on the heights of children and their parents.

Assume that there are two parents, and that both the father and the mother are seven feet tall. The question to be answered is that of how tall you would expect their offspring to be.

To someone as naive as the present writer, the answer is obvious: Since both parents are far above average height, it makes sense that the two extreme tendencies would be accentuated, and that the children of this marriage should be even taller still, perhaps 7½ feet tall, on the average. This solution is shown in Fig. 8-8. This conclusion makes such excellent sense that few people questioned it until Galton's time.

Or, take the opposite example. If one had two parents, both 3 feet tall, how tall would you expect their offspring to be, on the average? Again, I'd expect that their children would be even shorter than the parents were (see Fig. 8-9). You'd think that the extreme tendency would be magnified, and that their children might be 2½ feet tall.

Galton set out to investigate this situation experimentally. He did so by offering a large prize to the person who sent him the best family history, giving especially the heights of all the parents in the

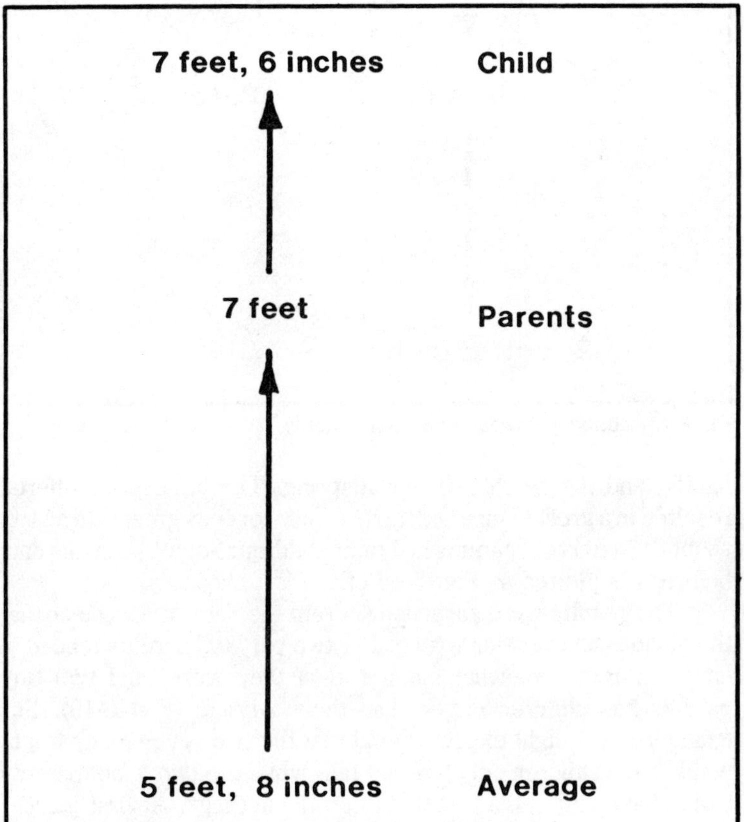

Fig. 8-8. Incorrect prediction from tall parents.

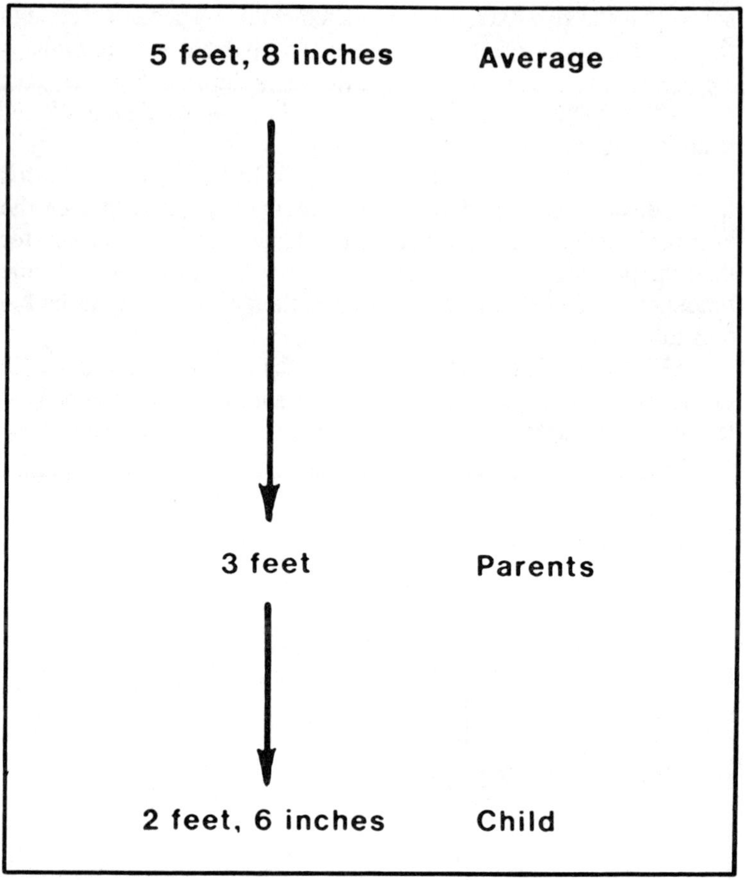

Fig. 8-9. Incorrect prediction from short parents.

family, and the heights of the offspring. The large prize offered resulted in a great number of entries, and, hence, a great deal of data about the heights of parents and their children. You've seen this data before. It's plotted as Fig. 7-15 of the last chapter.

The results were surprising. From the data, it was clear that the obvious answer was wrong! For two very tall parents tended to have children somewhat smaller than they were, and two tiny parents had children larger than their average (Fig. 8-10). For example, you might expect to find that the two seven foot parents would have children only 6½ feet tall, while the two 3 foot parents would have children 3½ feet tall, on the average. At first glance, this finding is rather amazing. Yet it shouldn't be.

For what would happen if tall parents produced even taller

children, and small parents produced smaller children? Well, tall people tend to marry tall people, and short people tend to marry short people. It is not very likely that an 8-foot man would marry a 3-foot female! Now, since talls marry talls, and shorts marry shorts, if the tendency toward extremes were operating, the 7-foot parents would produce 7½-foot children. These 7½-foot children would have to marry other people of similar height, and their offspring would be 8 feet, and it would go on and on like that. Similarly, the 3-foot people would produce successive generations of 2½, 2, 1½, ½ and perhaps 0 feet! Of course, things would not work out that rapidly, but it should be obvious that over a period of thousands of years there would be a race of enormous giants, and another race of very, very tiny people, and the giants would have to be very careful about where they walked, to avoid stepping on their little friends!

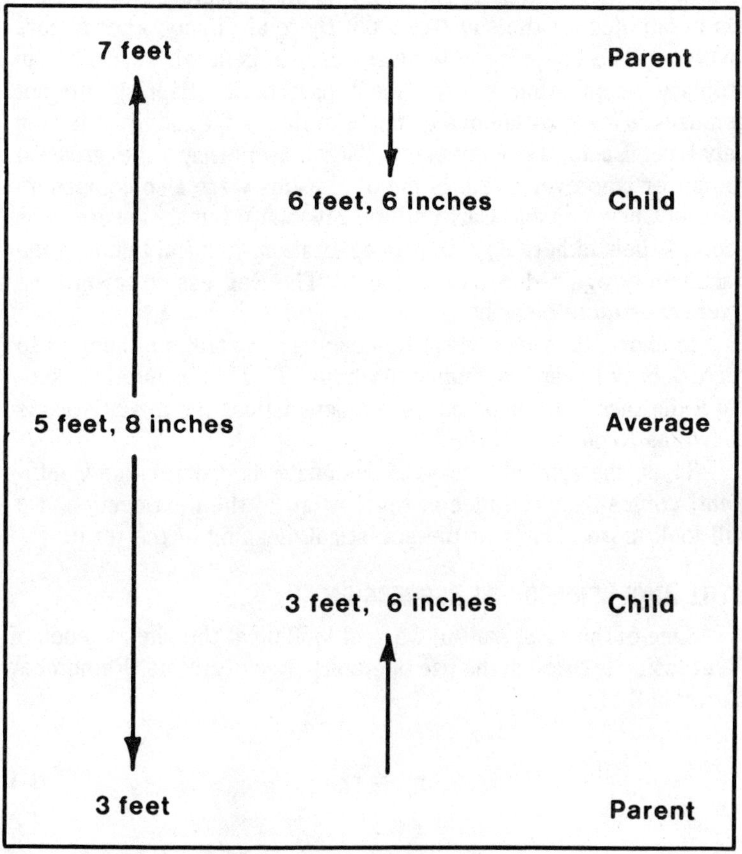

Fig. 8-10. Galton's explanation of Regression.

Thus, it can be seen that it is an absolute necessity for characteristics in human beings to avoid the tendency toward extremes.

Put another way, extreme characteristics in parents tend to be modified, to go back toward the Average of the race. A common term in science denoting going back is regression. Consequently, Galton used the term Regression to describe the tendency of human characteristics to level out in heredity, the tendency to go back to the Average in hereditary transmission of the human race.

This trend may be seen in many fields. Very intelligent men tend, on the whole, to marry very intelligent females. And duller people also tend to mate. Consequently, without the general concept of Regression, the human race would arrive at a race of immensely bright people, and another race of people with profound mental retardation. Yet this does not happen. Bright people do pop up now and then, and there can be no question that intelligence does run in families for the short run, but there are many other factors involved. The offspring of geniuses are, in general, smarter than ordinary people, but for the most part their offspring are not geniuses. The most famous example of this is the family of Johann Sebastian Bach, the composer. Bach was perhaps the greatest composer who ever lived. Some of his sons were also composers (Johann Christoph and Carl Phillipp Emanual), but they weren't as good as their father. By the next generation, musical talent in the Bach family was reduced enormously. The Regression toward the Average is quite clear here.

In short, there is a great tendency among traits in humans to regress back toward the human Average. This is absolutely essential if the race is to maintain its characteristics, and in statistics is known as *Regression to the Mean.*

Thus, the symbol r, used to designate the Correlation Coefficient, comes from the idea of regression. In the next section, we will look more closely at the statistical meaning of the term.

STILL ANOTHER LOOK AT REGRESSION

One of the most fruitful ways of looking at the phenomenon of Regression is through the use of graphs or scatterplots. Remember Formula 8-1.

$$z_y' = rz_x \qquad \textbf{8-1}$$

Let's assume that the Correlation between X and Y is 1.00, and make prediction from X to Y. If r=+1.00, Formula 8-1 becomes $z_y' = z_x$, and selected values would be as follows:

z_x	z_y'
1.4142	1.4142
.7071	.7071
.0000	.0000
−.7071	−.7071
−1.4142	−1.4142

If we plot these points, we find that they make a nice straight line, at 45 degrees to the x-axis (refer back to Fig. 8-6). It's interesting to note that, when z-scores are used, the Correlation is the tangent of the angle the line makes with the x-axis. With a perfect Correlation, the angle is 45 degrees, and the tangent of 45 degrees is +1.00!

Now, assume that the Correlation is +0.6, as in our earlier examples. Formula 8-1 becomes $z_y' = 0.6z_x$. Selected predictions would be as follows:

z_y'	z_x
1.4142	.8485
.7071	.4243
.0000	.0000
−.7071	−.4243
−1.4142	−.8485

These values were plotted as the line in Fig. 8-1, earlier in the chapter. In going from Fig. 8-6 to Fig. 8-1, notice that what happened is that the Regression Line has moved down. In the Fig. 8-6 it made an angle of 45 degrees with the horizontal, and now it's only 31 degrees (the angle whose tangent is 0.60). The effect of the imperfect relationship between X and Y, then, is to make the Regression Line move down more toward the horizontal axis.

What is the horizontal axis? Well, it corresponds to a z-score of zero. You've probably forgotten, but a z-score of 0 is the Average of the z-scores.

What all of this means is the following: If there is any doubt or error in a prediction, as shown by a Correlation that is less than 1.00, the prediction strategy we're using here says to make the line more horizontal. The more doubt about the prediction, the more horizontal the line becomes. If the Correlation is zero, that is, if there is total doubt about the prediction, Formula 8-1 becomes $z' = 0.00z$, and the following table results:

z_x	z_y'
1.4142	0
.7071	0
.0000	0
−.7071	0
−1.4142	0

The Regression Line will be horizontal. No matter what the X value, the value of Y that is predicted is zero. Notice carefully what this means. If there is complete doubt about a prediction, as shown by a Correlation of zero, you simply predict the Average every time for Y. In short, you can't do worse than predicting the Average.

Further, with increasing doubt about predictions, the line becomes more and more horizontal, and one predicts values that are closer and closer to the Average. In short, the strategy shown here says that the more doubt you have about a prediction, the more you should predict the Average. With total doubt, just predict the Average.

The limiting situation would be that in which r really was zero. Such a situation can be found when the points form a circle, as shown in Fig. 8-11. Notice that with the circle, in which the variables are completely unrelated ($r=0$), the line for the Regression of Y on X is horizontal. That means that one predicts the Average whenever one predicts a Y value, and the line for the Regression of X on Y is vertical. Again, the Average will be predicted whenever a prediction of X is made.

Study the previous statements very carefully! I know they're tricky, but if you can figure them out you'll know how Regression works. Of course, it's called Regression because the values go back, or regress toward the Average, just as Galton said. Also, you can prove that Formula 8-1, or any of the variations on it, will make the best possible prediction under the circumstances, that is the prediction that will make the least error. Proof of this is rather technical, but it can be found in advanced statistics books.

Now, if the Correlations were gradually reduced from 0.6 to the values 0.4, 0.2, and 0.0, the line for the Regression of Y on X would change as shown in Fig. 8-12. Notice how the Regression

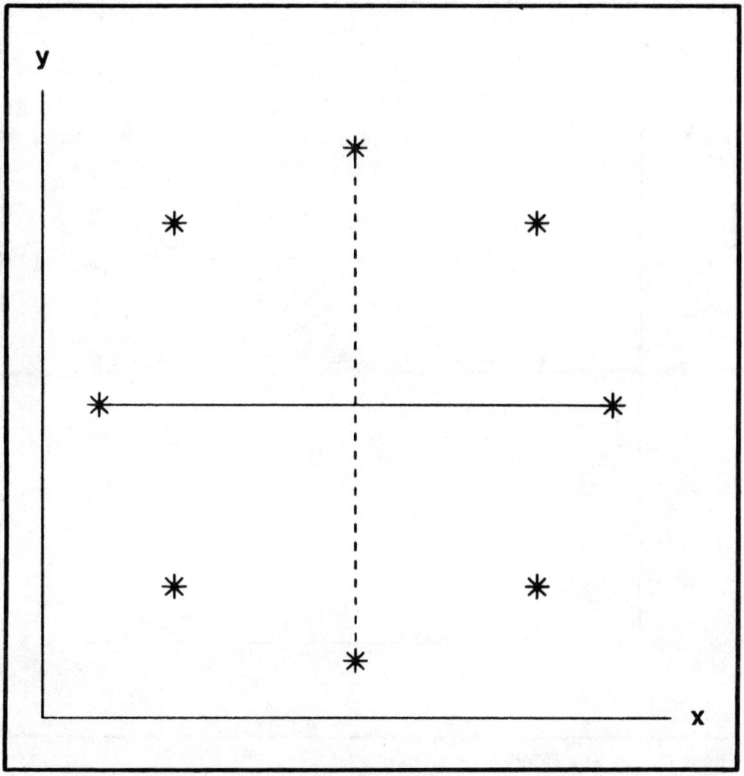

Fig. 8-11. Regression lines in zero Correlation.

Line gradually moves down toward the horizontal, to the Average. Figure 8-13 adds the changes in the line for the Regression of X on Y. That slowly approaches the vertical line, the Average, as the r approaches zero. Thus, both lines regress toward the Average. In the absence of any relationship between the X and Y variables, you can always at least predict the Average.

MORE EXAMPLES

A few more examples may be of use in helping you to understand Regression Lines. Look at Fig. 8-14. This is another perfect geometric figure—a parabola. Recall from the last chapter that I said that Correlation between the X and Y variables in the parabola was zero. Thus, we have another precisely zero Correlation, just as with the circle in Fig. 8-11. Notice again, that the two Regression Lines are horizontal and vertical, and at right angles. In the absence of any relationship, one simply predicts the Average.

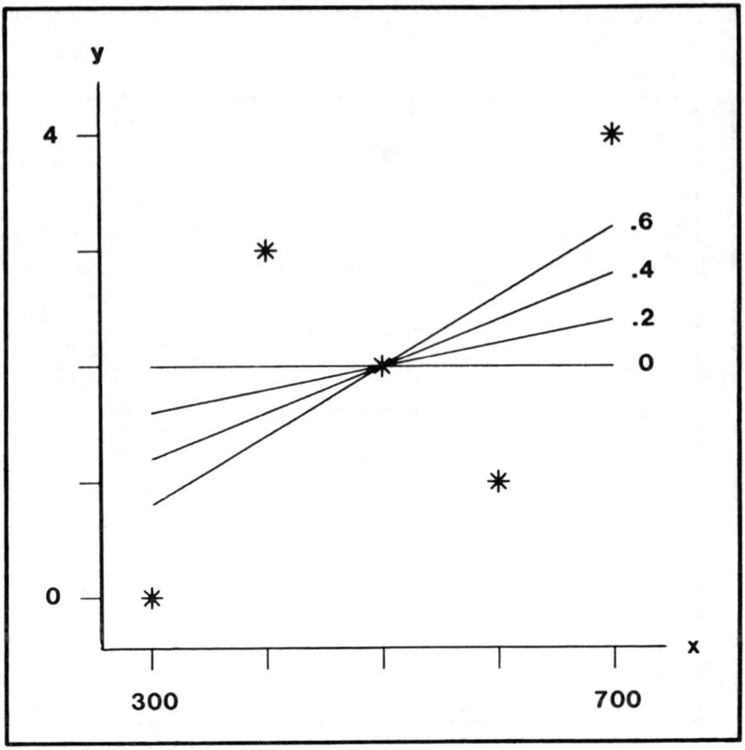

Fig. 8-12. Several Regression lines plotted.

Figure 8-15 shows a very strong relationship—it's the Correlation of +0.9938 obtained on the two electrical meters of the last chapter. Notice here that the two Regression Lines are virtually on top of each other. That's always the situation when the Correlation is very nearly perfect. The two prediction lines become essentially one.

Figure 8-16 plots some fictitious data. Here the Correlation is negative (−0.8632). This is a rather high Correlation, but notice that the two Regression Lines are beginning to be rather far apart, even with this high r.

Figure 8-17 plots the Wechsler Intelligence Test Information-Full Scale data from the last chapter. The r here is +0.4942, but the points are spread out a good deal, and so are the Regression Lines. Clearly, prediction with an r of this magnitude is not going to be very precise.

Figure 8-18 shows the Correlation of the Wechsler Verbal and Full Scale IQs from the last chapter. This r is higher than the last

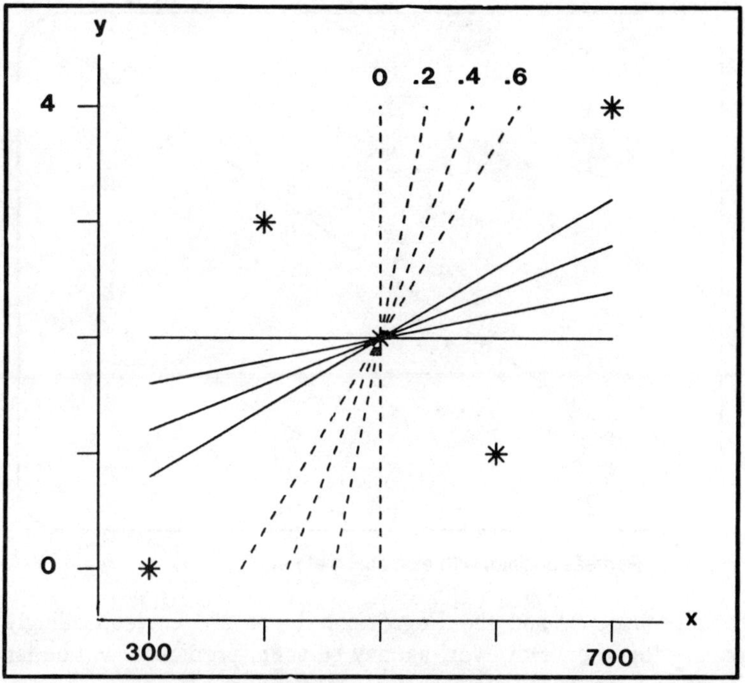

Fig. 8-13. Several Regression lines for the Regression of both Y on X and X on Y.

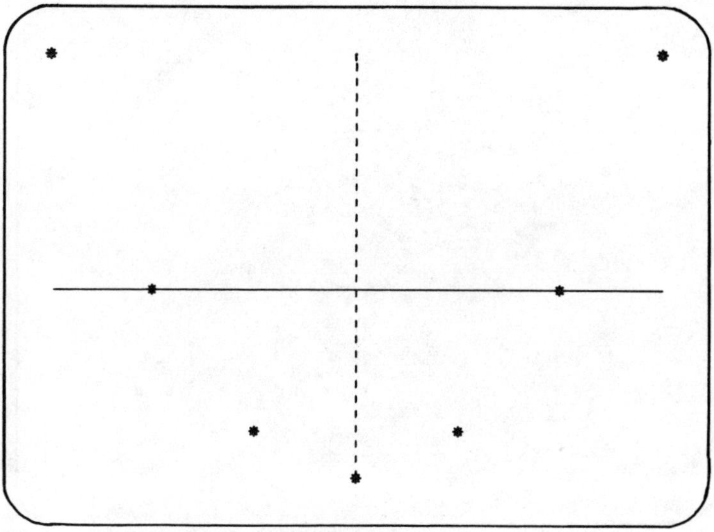
Fig. 8-14. Regression in a parabola.

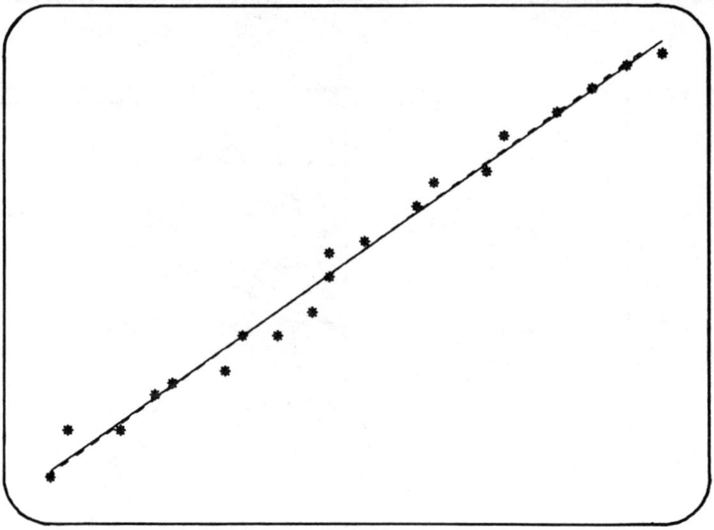

Fig. 8-15. Regression lines with electrical meters.

one (+0.8587), and the Regression Lines are correspondingly closer together. However, as may be seen, prediction will be far from perfect, even with this high Correlation.

The effect of even a slight deviation from an r of zero is shown in Fig. 8-19. Recall from the last chapter that with this data I was

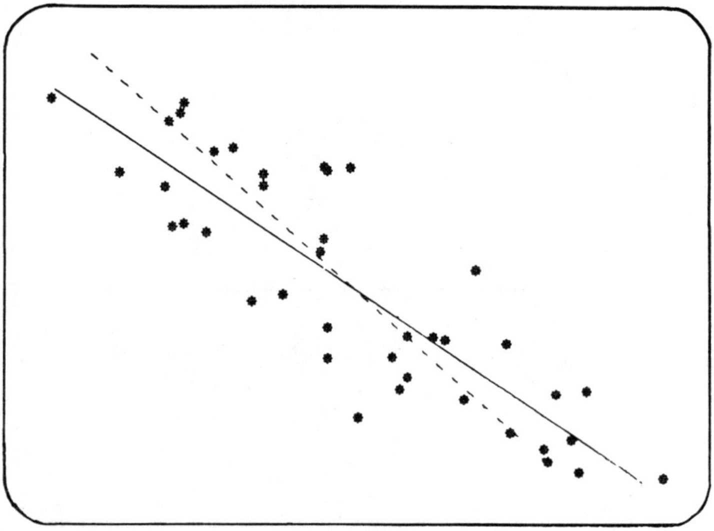

Fig. 8-16. Regression with a strong negative r.

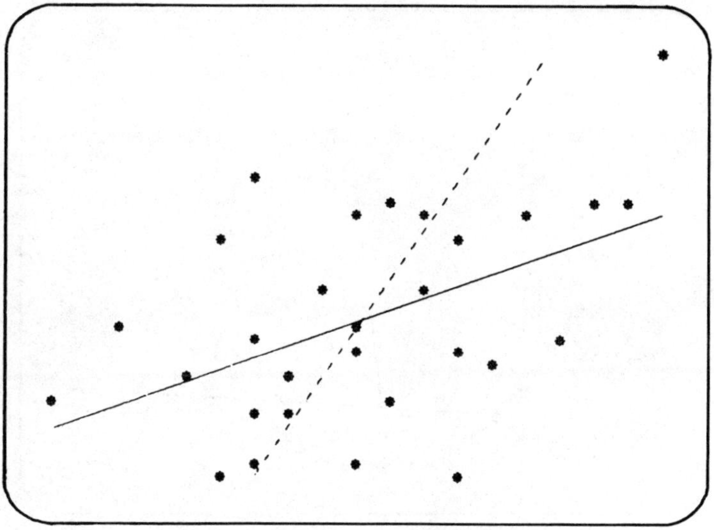

Fig. 8-17. Information full-scale IQ Regression lines.

trying to achieve a parabola (see Fig. 7-14). However, I didn't do it quite properly, and the r turned out to be +0.0833, not zero, as I had planned. Notice the peculiar tilt to the Regression Lines. Deviations from zero r's are easily seen in such situations.

Figure 8-20 shows the personality test Warmth-Self-

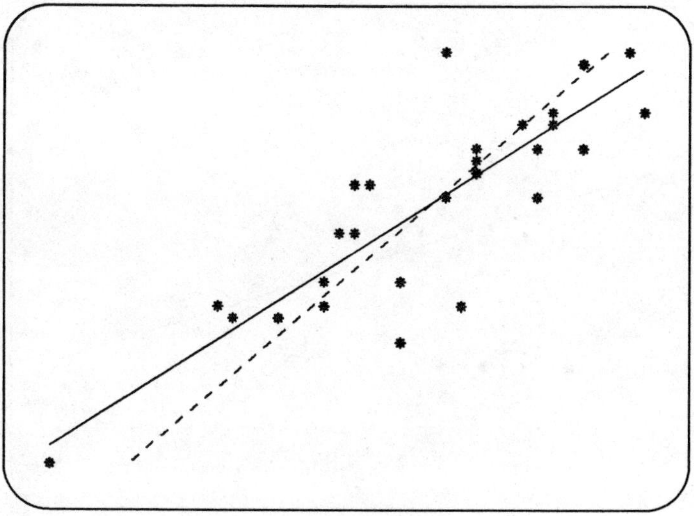

Fig. 8-18. Verbal full-scale IQ Regression lines.

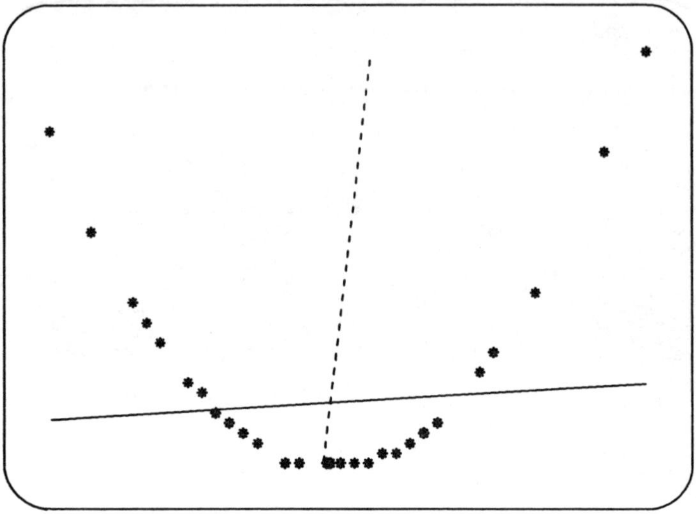

Fig. 8-19. Regression lines with r slightly greater than zero.

Sufficiency scores from the last chapter. Recall that the r was −0.5704. That's a high r for my field, but notice that the two Regression Lines diverge rather markedly.

Figure 8-21 is the one in which the r was so difficult to grasp from the last chapter; the r here is only +0.3663. Notice how far the

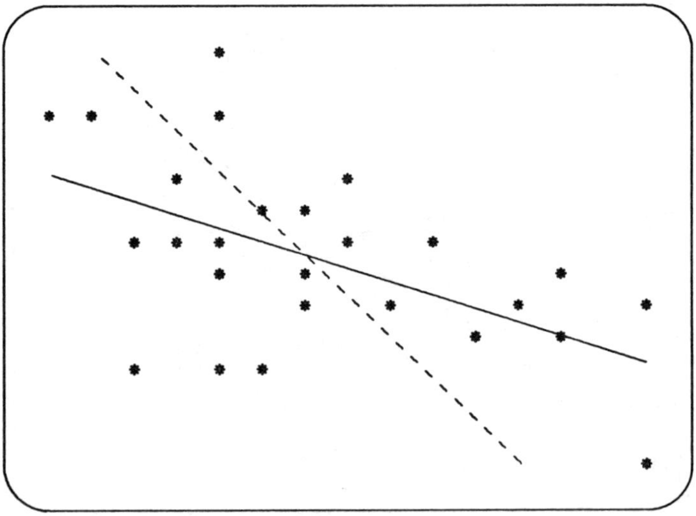

Fig. 8-20. Regression lines with personality test data.

(Y′ = −0.2044X + 73.5.
predict for an X of 60 (61.

4. Scores on a personality
be compared with scores
equation for predicting Y

X 22
Y 25

(Y′ = 1.66780X + 8.42334
lead to the prediction of w

5. Wechsler-type IQ sc
Standard-Binet-type IQ sc
testing *reliability*). Find th
dicting S-B scores from W

X 120
Y 110

(r = +1.00. Y′ = 1.00X −
Wechsler-type score of 11
Notice that the Correlatio
different. Why should this

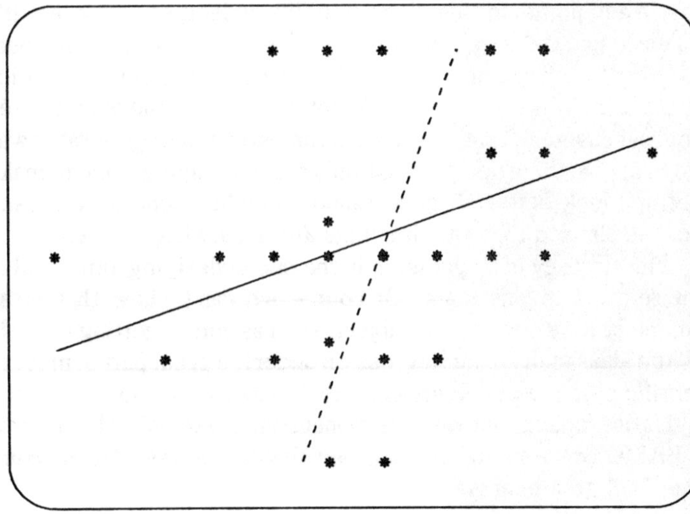

Fig. 8-21. Regression with a moderate positive r.

Regression Lines diverge in this example. With no relationship at all, they would be at a 90 degree angle.

Finally, Sir Francis Galton's data on their relatonship of parents' height to that of their children makes a fitting conclusion to our examples of this chapter, and is shown as Fig. 8-22. There are

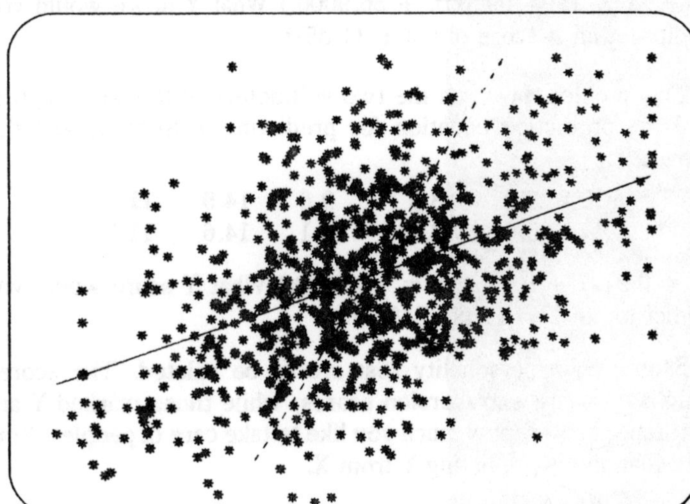

Fig. 8-22. Regression lines in Galton's data.

many, many points in thi
difficult to find the Regre
of all Regression example
that the stars which are p
in the last chapter. That's
used to spread them out fr
the graph look better. A
repeat itself, and thus th

The strategy of Regr
what remains in this bo
elementary treatment of t
fact, the ideas underlying
scientific procedures. Re

Notice, again, that w
old BASIC concepts for
mands will go a long way

PROBLEMS

1. SAT scores (X) are co
on the Wechsler IQ test
predicting Y from X and

X	500
Y	120

(r=+.9377, Y′ = .09900
predict for an X score of

2. This problem involve
Find the prediction equ
Correlation.

X	5.0
Y	5.4

(Y′ = 0.9727 + 0.6454.
predict for an X of 11 (1

3. Scores on a persona
designated X are extrave
nurturance scores (how n
the equation for predicti

X
Y

9
Inferences About Population Averages

We now move into a more difficult area of statistics. Up to this point in the book, we have been content simply with describing things, and as a result the topics we've covered up to this point are called *descriptive statistics*, for they only describe data. Most people can understand descriptive statistics without difficulty.

Now we're going to begin to try to make guesses about things that we do not, or cannot know. We may take a small sample of 30 people and try to determine characteristics of an entire population from the small sample. Statisticians say that they want to make *inferences* about the population from a sample, and hence they call the procedures *inferential statistics*.

In inferential statistics, we're trying to get something for nothing. We're trying to infer population values from sample values, and that's not easy. Consequently, the procedures are more complex than the simple descriptive statistics we've been using to date, and you may feel less comfortable with them. But they're important, and something of a challenge.

A HYPOTHETICAL PROBLEM

Let's use the data from Chapter 3, Program 3-5, as an example. As you recall, it was IQ test data gathered on 5 people, and looked like this:

	IQ
	120
	110
	100
	90
	80

Average:	100
Standard Deviation:	14.1421

Thus, for the 5 people we have an Average of 100 and a Standard Deviation of 14.1421. We'd now like to know what the mean would be for the whole population of the United States, or perhaps even for the whole world.

To do this we must assume that our little group of 5 is really representative of the population. Making sure that this is the case is a special field in itself, one which we haven't room to cover here. One usually strives for a *random* sample—one in which every person in the population has as much chance of being chosen as every other person. Setting up such samples is the province of people who take polls, and as you know, they have become very good at it. We'll just assume that we have somehow obtained a good, random sample.

With the sample values given above (Average = 100, Standard Deviation = 14.1421), what's the population Average? This will take a bit of thought. (See Fig. 9-1). Can the population Average really be 100? Of course, that's possible, but our sample is only 5 people, and it doesn't seem very likely that such a small sample would tell us exactly what the population Average or population Mean is. For that matter, it doesn't seem likely that a sample of 100 would give us the population Average very precisely, or, for that matter, a sample of 1000.

Still, the best guess that we have for the population Average is 100, and we'll have to live with it. However, it would be nice to be able to express our uncertainty about the value of 100 somehow. Incidentally, the population Average is always called μ or mu in statistics books, but we'll try to avoid that usage here.

A SOLUTION

The simplest way out of the problem is that of specifying a

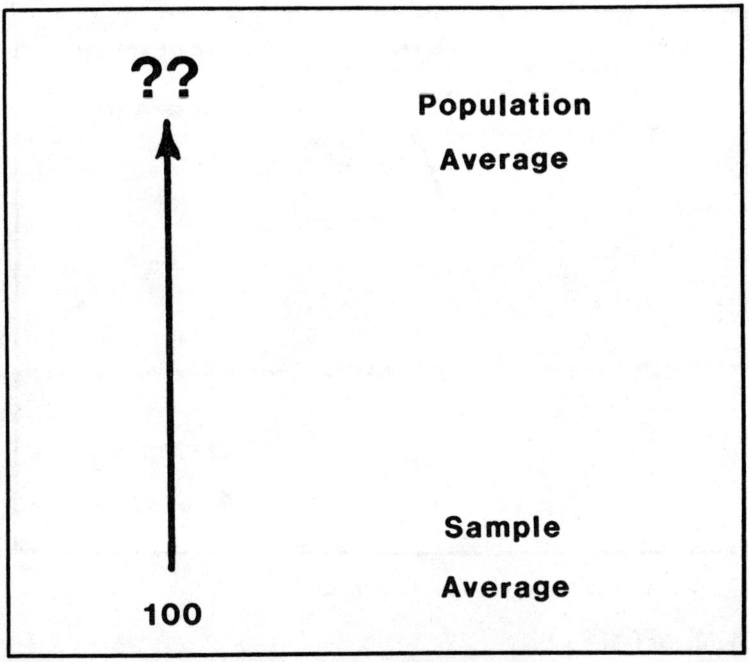

Fig. 9-1. Guessing the population Average from a sample.

range of values into which the population Mean might fall. That is, if our sample Mean is 100, we could say that the population Mean is likely to be found between, say, 86 and 114, as shown in Fig. 9-2. In other words, with a sample Mean of 100, the strategy becomes one of giving a region into which the population Average will fall. That seems eminently reasonable. We can't possibly know where the population Average will go, and so we show our uncertainty by saying that it is between 86 and 114.

It's a little like the situation in which a friend asks you what a woman's age is. Generally, you don't say "She's 30," but rather, something like "She's probably somewhere between 25 and 35." Your uncertainty is expressed by giving an area of values into which the number may fall.

"But," you say, "how do you figure out this range of values? The way in which it may be done certainly isn't obvious." Of course, it isn't. That's why it's the subject of this chapter!

Incidentally, the material that follows is very tricky indeed. This chapter and the next one are perhaps the most difficult ones in the book. You may have to read them over many times before the material sinks in.

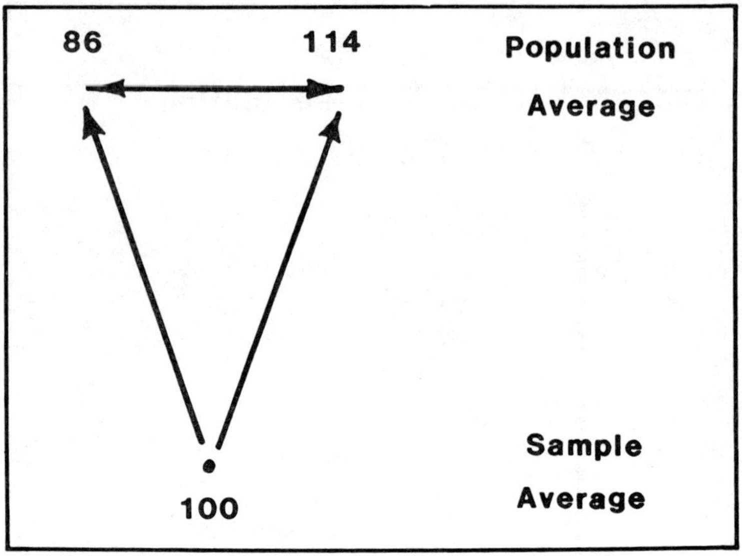

Fig. 9-2. A range for the population Average.

SOME DETAILS

It's obvious that our best guess about the population Average is the sample Average. No one has any trouble accepting that, but to talk about the range into which the population Average might fall, we're going to have to have some sort of Standard Deviation. The Standard Deviation, as you'll recall from Chapter 3, is the way statisticians prefer to talk about a range of values. The next few paragraphs get a bit complex.

If we take samples of size five, just as we did with the values 80, 90, 100, 110, 120 at the beginning of the chapter, the Means of the samples will look something like this (we'll use even numbers to make it easy for you):

Averages of
Samples of Size
5

110
105
100
95
90

That's fictitious data, again, but it's very close to what would happen. This is a bit tricky: in the data just above we have Averages of a lot of samples like those at the beginning of the chapter, not scores from individual people. When you take a sample, even one as small as size 5, it's going to be more stable than an individual person's score. All the things that make for errors tend to cancel out when you have more than one person. If one person in the sample of five taking the IQ test is a little too sleepy, another person may be more alert than usual, and so on. If one person in the sample of five has excellent eyesight, another person may have poor eyesight, and all of these things affect intelligence. Consequently, the scores won't spread out as far for the samples as they did for the individual people, and, indeed, they'll look like Fig. 9-3 (plotted with my EPSON printer). Notice from Fig. 9-3 that the data points for the sample Means still fit the Normal curve, just as the individual scores did, but now they're compressed a great deal, reflecting the fact that samples are more stable than the individual people of the wider curve.

What if we took a lot of samples of size 100? Well, the scores

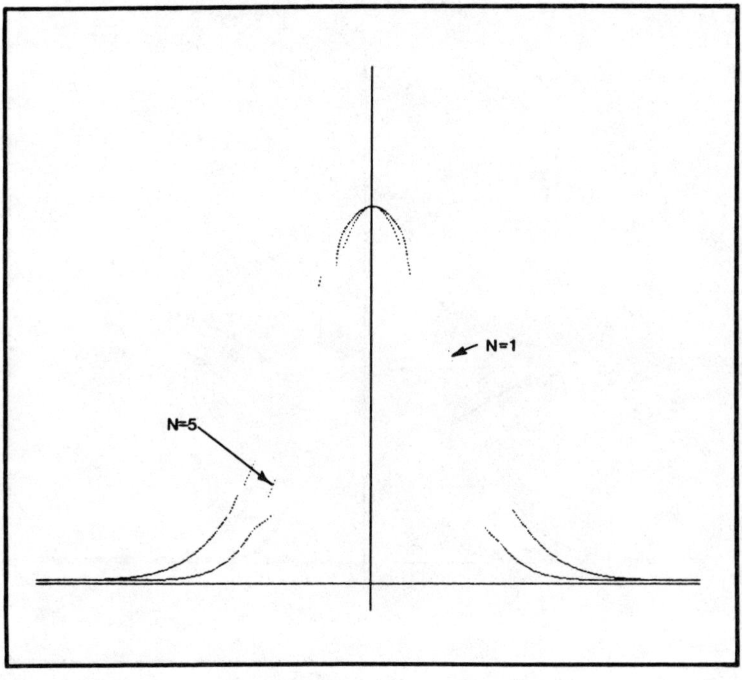

Fig. 9-3. Lower SD in a sample of five.

would be even more stable, something like the ones shown below:

Averages of Samples of Size 100

102
101
100
99
98

Notice that the scores are even more stable than they were with samples of size 5. The situation is plotted as Fig. 9-4. Now, the Averages of the samples of size 100 are very stable, indeed. They have a range of only 4 points, whereas the range of scores for the individual people was 40 points (if you don't recall the way to

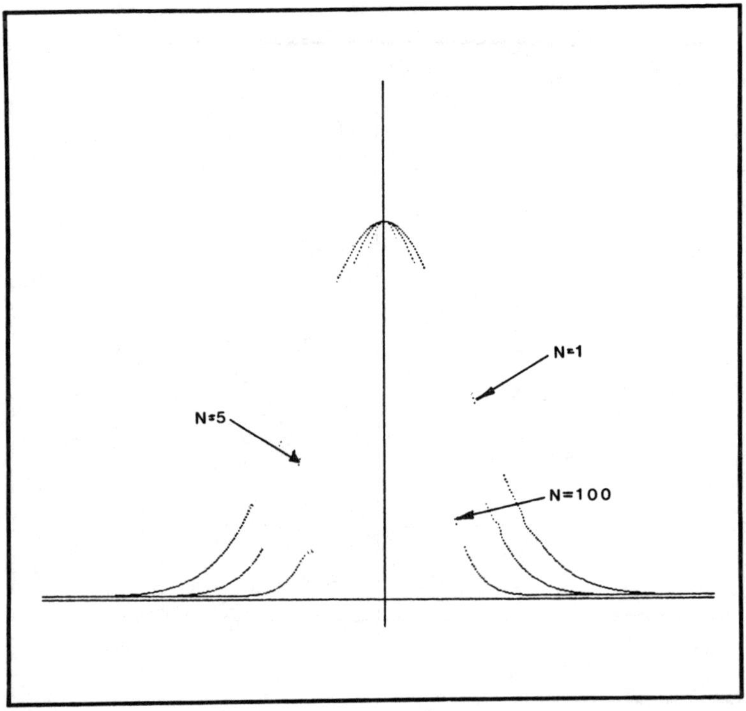

Fig. 9-4. Still lower SD in a sample of 100.

calculate it, look up the Range in Chapter 3). The Standard Deviations for the various sample sizes look as follows:

Sample Size	Standard Deviation
1	14.14
5	7.07
100	1.41

Note that these are Standard Deviations of the samples, not of the individual scores; that's tricky. In short, the bigger the sample size, the smaller the Standard Deviation of the Means of the samples.

However, the data above are only approximate, to give you an idea of what happens in situations like this. How does one really calculate the relationship between sample size and the Standard Deviations of the means? It's fairly easy to do, as we'll see.

THE STANDARD ERROR OF THE MEAN

The Standard Deviation of the Averages of samples of a given size has a name: it's called the *Standard Error of the Mean*. It's too bad that they didn't call it the "Standard Deviation of the Means," for the term would have made a lot more sense to most people. The formula for the Standard Error of the Mean, the Standard Deviation appropriate for means of samples, is

$$s_{\bar{X}} = \frac{s}{\sqrt{N}} \qquad \textbf{9-1}$$

"But wait!", you say. "What's the term 's'? We haven't talked about 's' before." You're right. But before we can talk about the Standard Error of the Mean, we must talk about "s" for a time.

TYPES OF STANDARD DEVIATIONS: A DIGRESSION

Whenever statisticians talk about making inferences, they use a special form of the Standard Deviation, denoted by the small letter s. The only difference between the plain old Standard Deviation, SD, and the new one, s, lies in the fact that in calculating s you do the last division using (N-1) rather than N. Thus the formula for the old Standard Deviation (Formula 3-5) followed by the formula for s look as follows:

$$SD = \sqrt{\frac{\Sigma X^2 - \frac{(\Sigma X)^2}{N}}{N}}$$

$$s = \sqrt{\frac{\Sigma X^2 - \frac{(\Sigma X)^2}{N}}{N-1}}$$

In short, if you are talking about making inferences about population values from sample values, you use s, with the N-1 in the denominator. If you just want an index of the spread of scores in a group of people, then SD is used. Of course, in a sample of any size the two things are going to be almost identical. After all, there's only a one point difference in the denominator. It's possible to prove with complex mathematics that s works better in predicting population values than does SD.

A computer program for calculating s and SD is given as Program 9-1. You'll notice that it's almost identical with Program 3-6, the old Standard Deviation program.

```
10 REM PROGRAM 9.1. "S" PROGRAM
20 READ X                      :REM GET DATA
30 IF X=-1 THEN 80             :REM TEST FOR END
40 S=S+X                       :REM SUM NUMBERS
50 Q=Q+X*X                     :REM SUM SQUARED NOS
60 N=N+1                       :REM GET COUNT
70 GOTO 20                     :REM CONTINUE
80 A=S/N                       :REM GET AVERAGE
90 PRINT "AVERAGE IS",A        :REM PRINT IT
100 V=Q - S*S/N                :REM TOP OF FORMULA
110 D =SQR(V/N)                :REM GET SD
120 SS=SQR(V/(N-1))            :REM GET S
130 PRINT "SD=",D              :REM PRINT D
140 PRINT "S =",SS             :REM PRINT S
150 DATA 80,90,100,110,120,-1 :REM DATA TO USE
```

(See Fig. 9-5). Run the program. It will tell you that SD is 14.1421, and s is 15.8114. As mentioned, this program is just like Program 3-6, with the addition of the lines for calculating and printing of the Average and s, which is called SS in the program so that it won't be confused with S. Calculation of the new statistic is done in line 120. Notice in that line (N-1) is on the bottom, rather than N, as in line 110, where SD is calculated. N-1 has to be in parentheses, as (N-1),

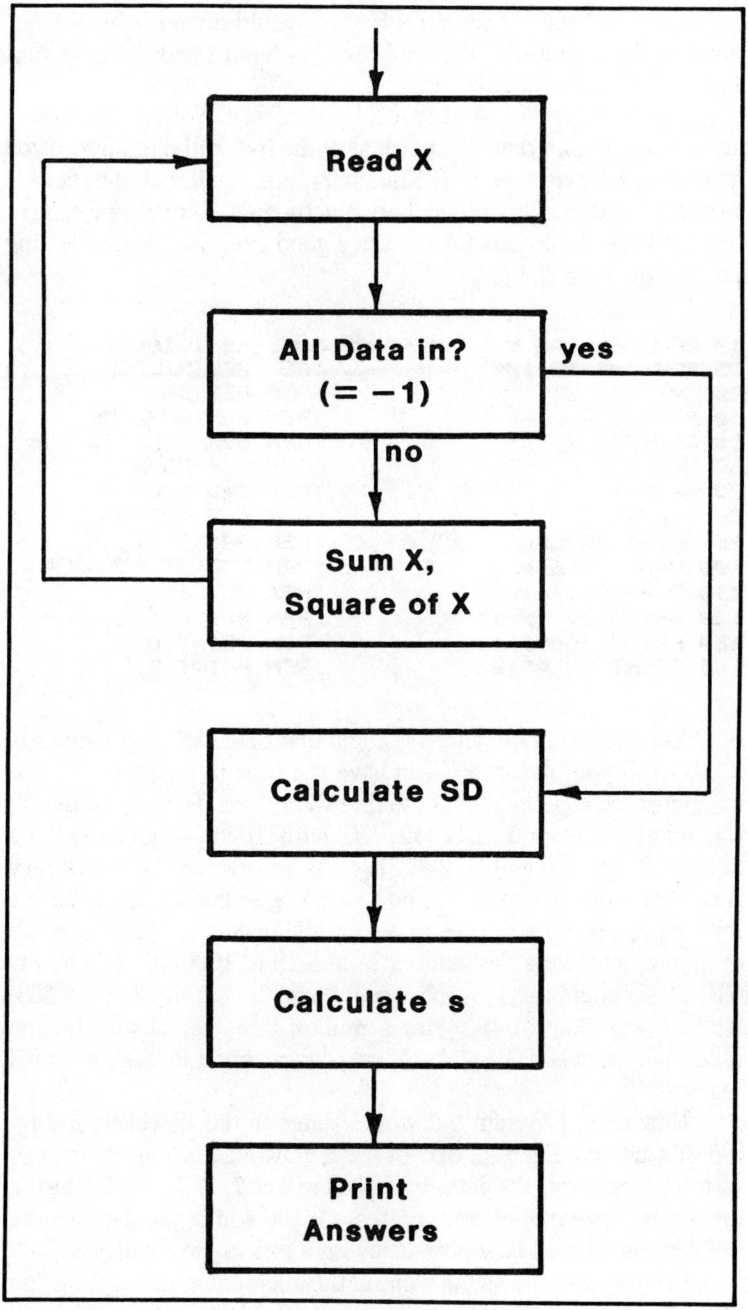

Fig. 9-5. Flowchart for the s Standard Deviation program.

for otherwise your computer's BASIC would divide V by N first, since it does multiplication and division before addition and subtraction. Details, details, details!

Run Program 9-1 with several sets of data. You'll find that s is always a bit higher than SD, because of the N-1, rather than N. If you have the patience, type in 100 numbers, and you'll find that there is very little difference, indeed, between the two. Actually we can get the computer to do this for us. It's a good exercise. The following program will do the job:

```
10 REM PROGRAM 9.2. SD'S OF NUMBERS 1-100
15 FOR I=1 TO 100              :REM BEGIN LOOP
20 X=I                         :REM GET DATA
40 S=S+X                       :REM SUM NUMBERS
50 Q=Q+X*X                     :REM SUM SQUARED NOS
60 N=N+1                       :REM GET COUNT
70 NEXT I                      :REM END LOOP
80 A=S/N                       :REM GET AVERAGE
90 PRINT "AVERAGE IS",A        :REM PRINT IT
100 V=Q - S*S/N                :REM TOP OF FORMULA
110 D =SQR(V/N)                :REM GET SD
120 SS=SQR(V/(N-1))            :REM GET S
130 PRINT "SD=",D              :REM PRINT D
140 PRINT "S =",SS             :REM PRINT S
```

Run this program, and you'll find that SD is 28.8661 while s is 29.0115. If you like, and if you have the time to wait a bit for the computer, change the number in line 15 to 1000. Then it will use all the numbers from 1 to 1000. SD with 1000 numbers will be 288.6750, while S will be 288.8194. As you can see, there is very little difference between SD and s with large samples. If you have a lot of time, and nothing else to do, try all the numbers from one to a million by changing the number in line 15 to 1000000. It took my VIC-20 six hours to do the job, and it found the SD was 288674.353, while s was 288674.45. With a million samples, obviously, the difference between SD and s is very, very small indeed, as you'd expect.

How does Program 9-2 work? Refer to the flowchart in Fig. 9-6. It's just like Fig. 9-5, except that a FOR-NEXT loop, from lines 15 to 70, generates the data. Variable I will be 1, 2, 3, . . ., 100 as the loop progresses. Of course, no test for the end of the data will be needed, since we know how many data points are involved.

Program 9-2 shows the value of the microcomputer to someone studying statistics. Think of how long it would take you to get s on a million numbers by hand!

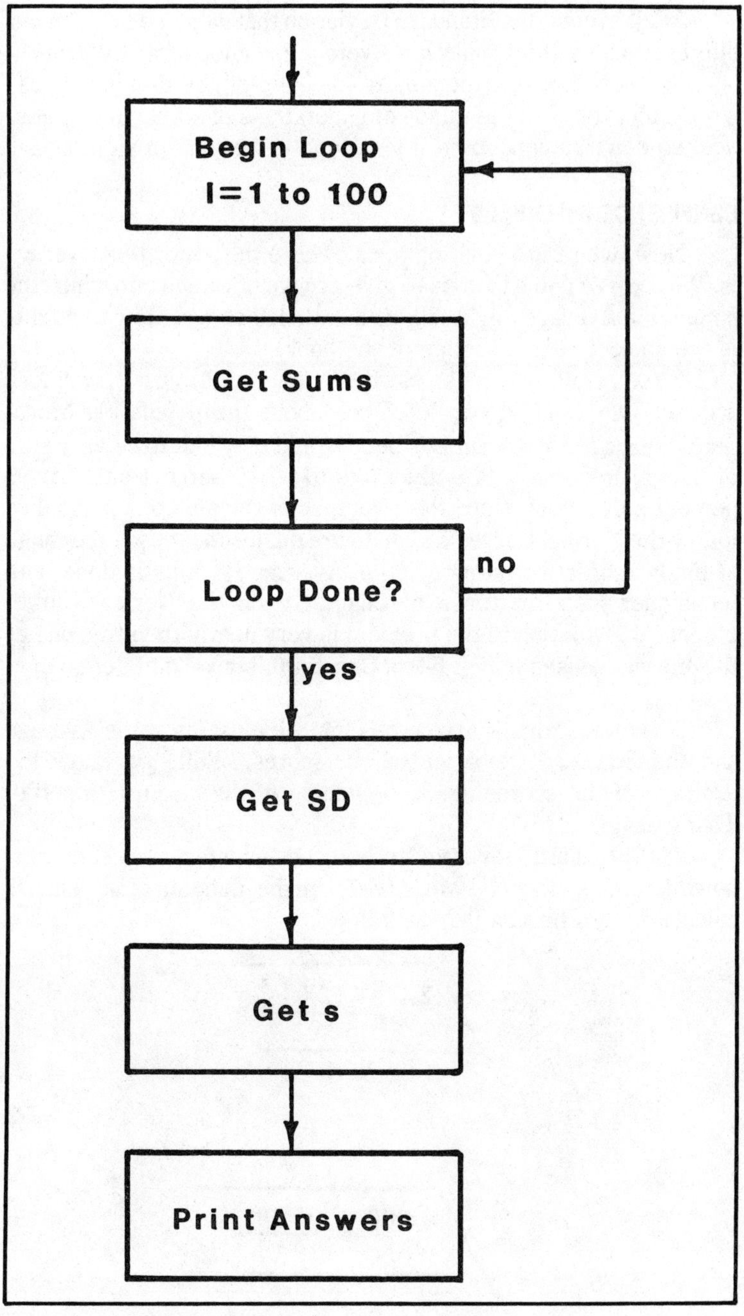

Fig. 9-6. Flowchart for the program for s and SD of 1-100.

At all events, the Standard Deviation that we need to make our inference about the population Average is called s, and it's calculated with N-1 in the denominator. This slight distinction has wreaked havoc with the minds of generations of students of statistics, so don't be concerned if you don't understand it right away.

CONFIDENCE INTERVALS

Now, we're in a position to say where the population Average is. What we're going to do is to give a range of values into which the population Average might fall, as mentioned above. How to do this is the subject of this section of the book.

First of all, we must clarify in our minds just how much accuracy we want in our inference about the population Mean. Statisticians are very cautious people, and they like to be very sure of things. In point of fact, they like to be 95 percent sure, or 99 percent sure. Now, since the Averages of samples of a given size follow the Normal Curve, we can figure out just how wide the range of likely scores for the population Average is. This is done with techniques just like those of Chapter 6 on the Normal Curve. Indeed, the solution to the problem is very nearly the same one as the one we devised in Fig. 6-9 of Chapter 6, but we'd better go over it again.

To work Normal Curve problems, we must know the Average and the Standard Deviation of our scores. Well, we know the Average of the scores at the beginning of the chapter; it's 100. That's easy.

The Standard Deviation to be used for inferential statistics, however, is s, the one with (N-1) in the denominator. This is calculated by Formula 9-2, as follows:

$$s = \sqrt{\frac{\Sigma X^2 - \frac{(\Sigma X)^2}{N}}{N-1}}$$

9-2

$$= \sqrt{\frac{51000 - \frac{(500)^2}{5}}{4}}$$

$$= 15.8114$$

And that's s, the Standard Deviation for inferences. However, that isn't quite the figure we want. We want the Standard Deviation for Averages, which is given by Formula 9-1. Putting in the figures for our problems we get the following:

$$s_{\overline{X}} = \frac{s}{\sqrt{N}} = \frac{15.8114}{\sqrt{5}} = 7.0711$$

Thus, the Standard Error of the Mean, as it's called formally, is 7.0711 in our case. That's the Standard Deviation for the Means of samples of size 5, and will now be used in our Normal Curve problem.

Figure 9-7 is a repetition of Fig. 6-9 from Chapter 6. It shows the z-scores that we need to take in 95 percent of the population. Remember, the z's needed are plus and minus 1.96. Now, by using Formula 4-2, or a variation of it, we can find the points that we need as follows:

Confidence Limits are $\overline{X} \pm (1.96)(s_{\overline{X}})$ **9-3**

or $100 \pm (1.96)(7.0711)$
or 86.1407 to 113.8593

In short, if the sample Mean is 100, and the sample Standard Error of the Mean 7.0711, we can be 95 percent sure that the population Mean will fall between 86.1407 and 113.8593.

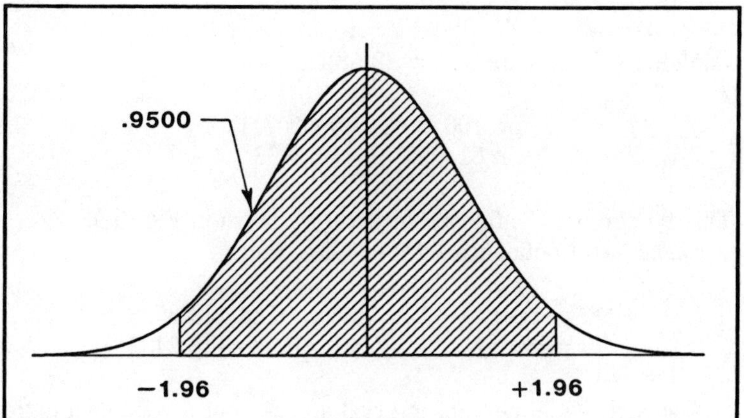

Fig. 9-7. z-scores for 95 percent of the population.

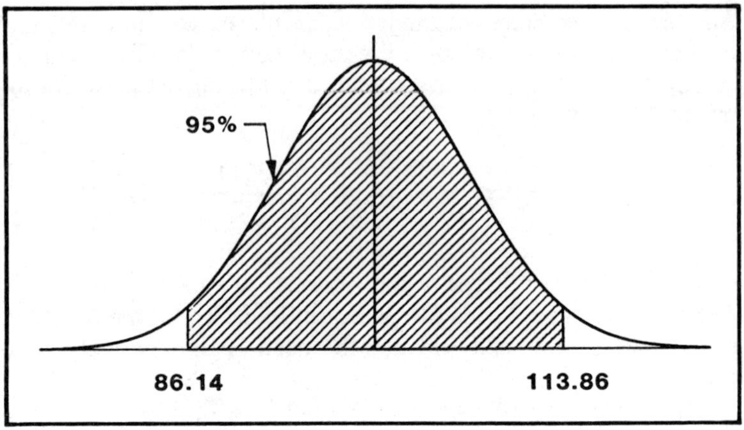

Fig. 9-8. The 95 percent Confidence Interval.

That's a pretty wide range, but it's a lot better than our previous situation. Before we used Formula 9-3, we had no idea where the population Mean might be. It could have been anywhere from minus to plus infinity. Now we have it nailed down to a range of about 28 points. That's quite an improvement! The 95 percent Confidence Interval, as it's called, is shown in Fig. 9-18.

Of course, we might want to be a bit surer about things—say 99 percent sure. Recall from Fig. 6-10, that the values needed to take in 99 percent of the Normal Curve are plus and minus 2.58. We can calculate the 99 percent Confidence Interval points with Formula 9-3, as follows:

Confidence Limits are $\overline{X} \pm (2.58)(s_{\overline{X}})$

or $100 \pm (2.58)\ (7.0711)$
or 81.7566 to 118.2433

This 99 percent Confidence Interval is shown in Fig. 9-9.

The two Confidence Intervals are, then:

95 percent	86.1407 to 113.8593
99 percent	81.7566 to 118.2433

Notice that the 99 percent interval spreads out a good deal more than the 95 percent interval. That makes sense, for, after all, one should have to take in more area to be surer about things. Also,

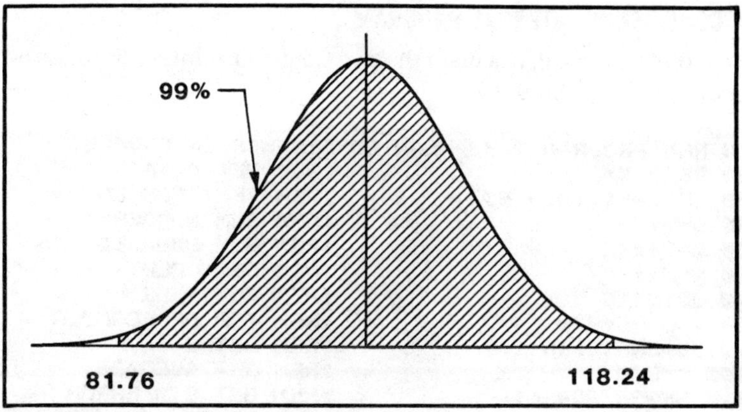

Fig. 9-9. The 99 percent Confidence Interval.

notice that we're almost certain that the population Average is between about 82 and 118. There's a one percent chance for mistakes. That's certainly good enough for most situations.

Thus, our solution to the problem of the location of the population Average is shown in Fig. 9-10.

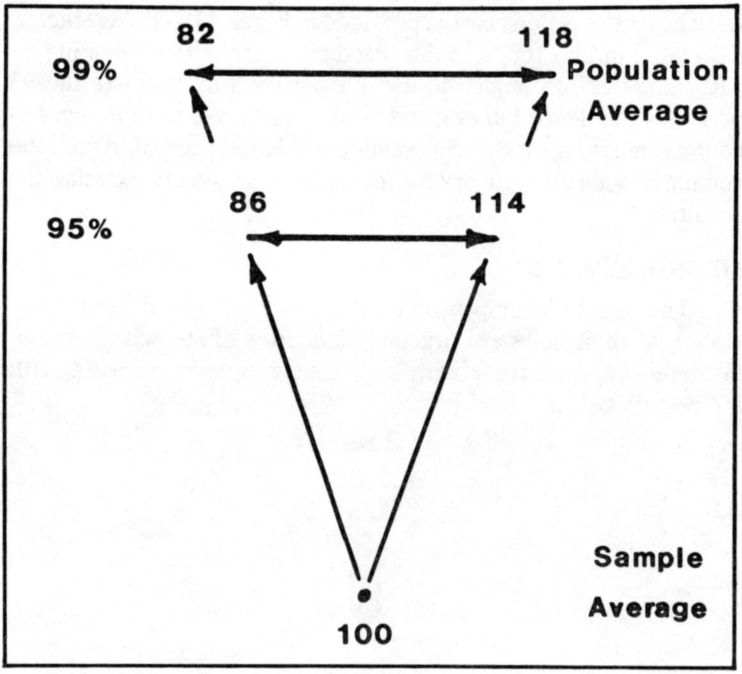

Fig. 9-10. Both Confidence Intervals.

A CONFIDENCE INTERVAL PROGRAM

A program will calculate the two Confidence Intervals for us is shown as Program 9-3.

```
10 REM PROGRAM 9.3. MEAN CONFIDENCE INTERVALS
20 READ X                          :REM GET DATA
30 IF X=-1 THEN 80                 :REM TEST FOR END
40 S=S+X                           :REM SUM NUMBERS
50 Q=Q+X*X                         :REM SUM SQUARED NOS
60 N=N+1                           :REM GET COUNT
70 GOTO 20                         :REM CONTINUE
80 V=Q -S*S/N                      :REM TOP OF FORMULA
90 SS=SQR(V/(N-1))                 :REM GET S
100 A=S/N                          :REM GET AVERAGE
110 SE=SS/SQR(N)                   :REM GET STD ERROR
120 L5=A-1.96*SE                   :REM LOWER 95% LIMIT
130 U5=A+1.96*SE                   :REM UPPER 95% LIMIT
140 PRINT "95% INTERVAL",L5,U5   :REM PRINT 95%
150 L9=A-2.58*SE                   :REM LOWER 99% LIMIT
160 U9=A+2.58*SE                   :REM UPPER 99% LIMIT
170 PRINT "99% INTERVAL",L9,U9   :REM PRINT 99%
180 DATA 80,90,100,110,120,-1   :REM DATA TO USE
```

As you can see, this program is almost identical with Program 9-1, the s program. A flowchart is provided in Fig. 9-11. The Average is calculated in line 100, and the Standard Error of the Mean in line 110. Lines 120 through 140 use Formula 9-3 to calculate the 95 percent Confidence Interval and print it, and lines 150 through 170 do the same thing for the 99 percent Confidence Interval. Again, the computer does all the work for us, and we can simply examine the results!

ANOTHER EXAMPLE

The Ajax Bolt company makes ¼-20 bolts—lots of them. They state that their bolts are precisely a quarter of an inch (0.25″) in diameter. We don't trust them, and take a sample of five bolts, with the results below:

	Size
	.253
	.258
	.262
	.265
	.270

Average	.262
s	.007

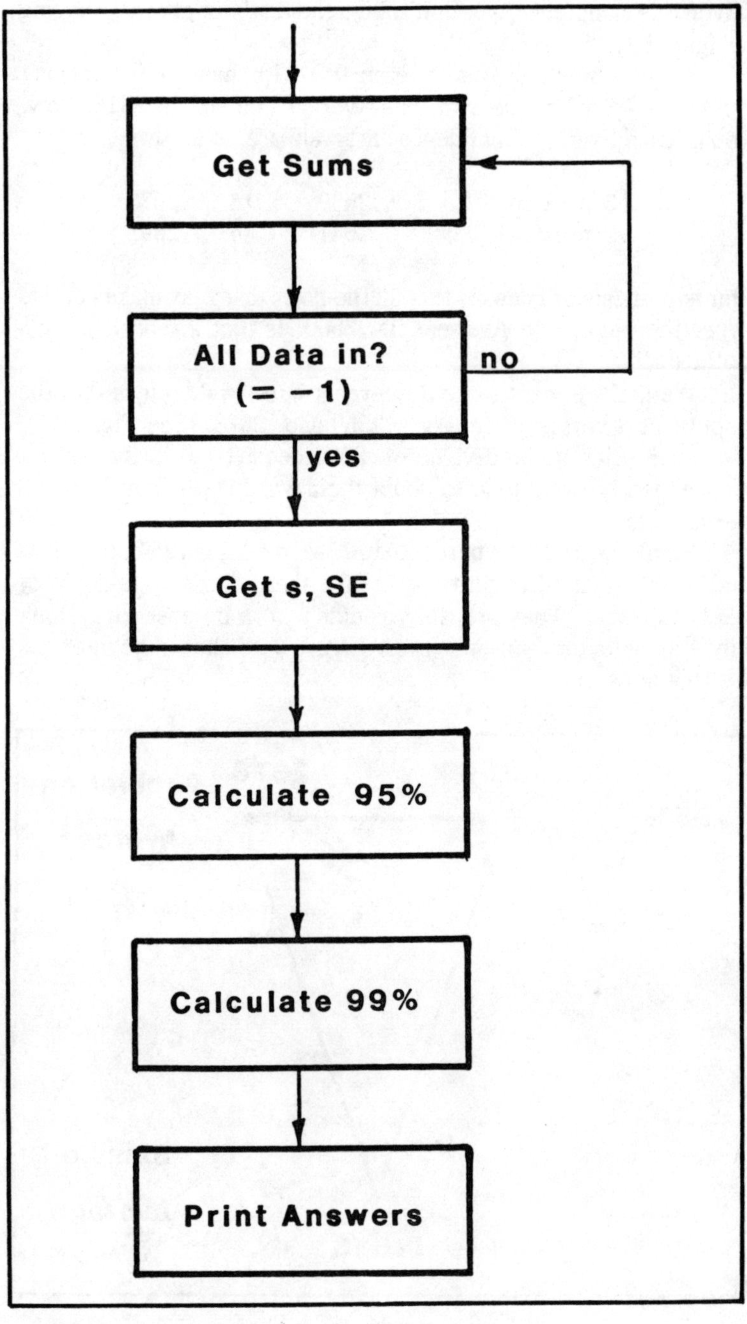

Fig. 9-11. Flowchart for the Confidence Interval program.

Given this sample, is Ajax Bolt telling the truth, or are they fudging things a bit?

To find out, simply run Program 9-3, with the data above. Don't forget the −1 to end the data. The Average bolt size is .2616. Now, you'll learn that the Confidence Intervals are as follows:

95 percent	.2559	to	.2673
99 percent	.2541	to	.2691

Our hypothesis or guess is that all the bolts that Ajax makes of this type, the population Average, is .2500. Is that a reasonable hypothesis?

Well, if we want to be 95 percent sure, we'd guess that the population Average is between .2559 and .2673. (See Fig. 9-12). However, Ajax's stated value of .2500 doesn't fall in that range. Consequently, we'd tend to doubt their word if we want to be 95 percent sure.

Similarly, to be 99 percent sure, we'd take in .2541 to .2691, and .2500 doesn't fall in that interval either. Ajax's bolts are very likely too large. They probably wouldn't work in most nuts. Don't buy Ajax bolts for this application. Figure 9-13 shows the intervals for this situation.

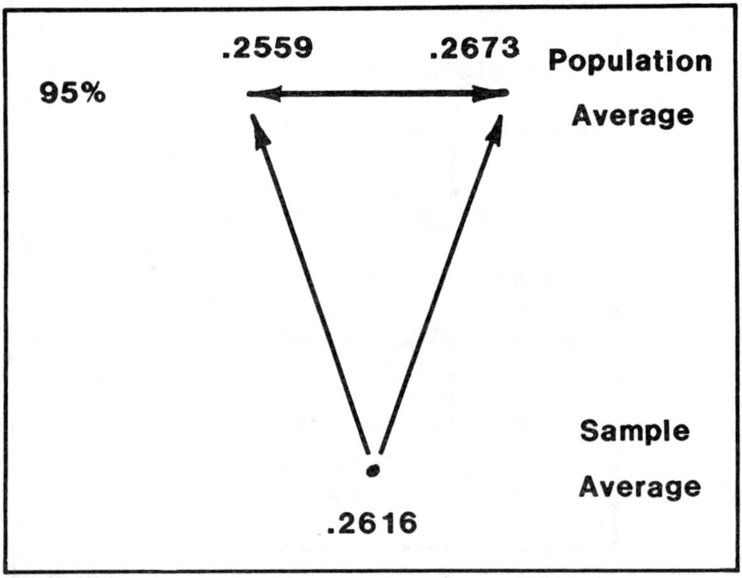

Fig. 9-12. Ajax Bolt's 95 percent Confidence Interval.

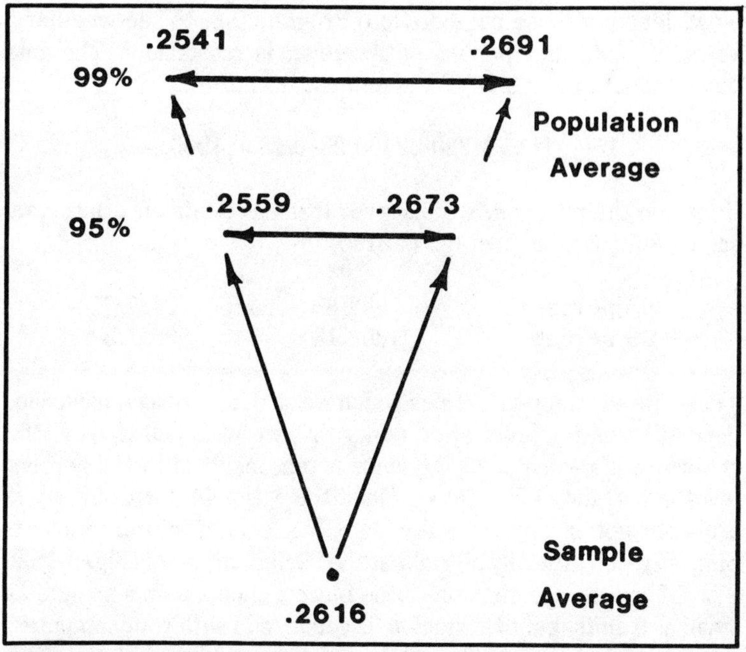

Fig. 9-13. Ajax Bolt's 99 percent Confidence Interval.

Of course, you can work out the problem by hand fairly easily, but notice how easily and simply your microcomputer does even this elementary problem for you. It saves a lot of time!

ANOTHER EXAMPLE

Let's say that the government has posted a new regulation saying that all cars must weigh 2000 pounds. We'd like to find out whether the cars from the Major Motor Company do in fact meet this criterion. We take a sample of five cars, with the following results:

	Weight
	2200
	2400
	2500
	1800
	1900

Average	2160
s	305

Now, let's put those numbers into Program 9-3, to see whether a value of 2000, as a population Average, is reasonable. The data points should go in a DATA statement, as follows:

```
180  DATA 2200,2400,2500,1800,1900,-1
```

If we run the program, we discover that the Confidence Intervals are as follows (the Average is 2160):

95 percent	1892.69	to	2427.31
99 percent	1808.13	to	2511.87

These are the ranges of values which we shall regard as reasonable, depending on the level of certainty that we want. Obviously, the government's value of 2000 pounds is reasonable at the 95 percent level of certainty (Fig. 9-14). Thus, the Major Motor Company is probably not in any difficulty with Uncle Sam at the moment. Similarly, it's probably all right at the 99 percent level (Fig. 9-15).

Of course, we shouldn't base our conclusions on a sample as small as 5 units, but the principle is sound, and with your computer, you can analyze almost any number of pieces of data with the same ease that you can do this small problem!

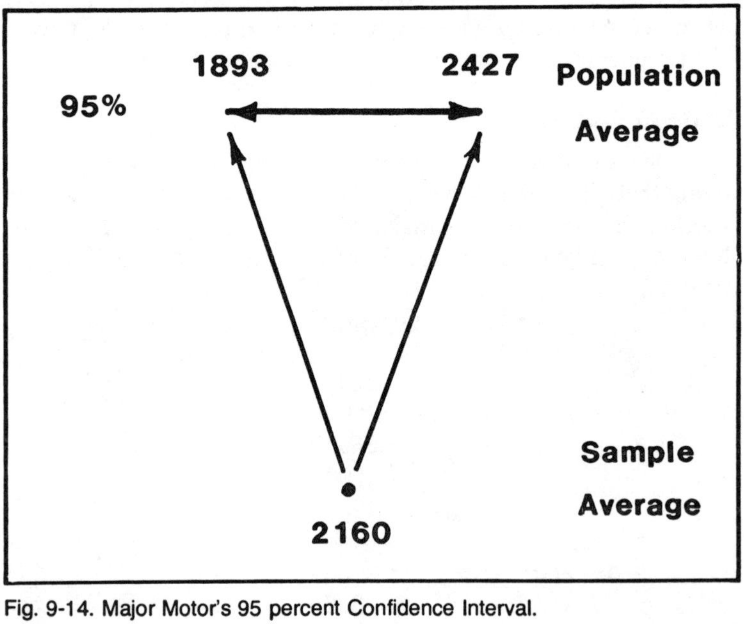

Fig. 9-14. Major Motor's 95 percent Confidence Interval.

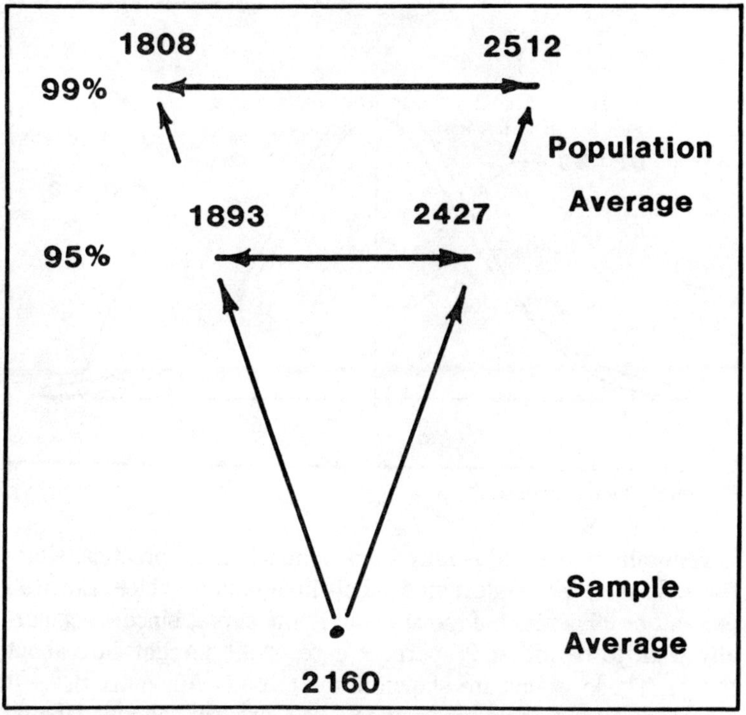

Fig. 9-15. Major Motor's 99 percent Confidence Interval.

A FURTHER COMPLICATION

In general, as you have probably gathered by this time, it is relatively safe to assume that most things fit the Normal Curve. This is generally the case with sample Means, also, as long as the samples are relatively large. How large must a sample be to be *relatively large* is open to question, but most texts consider samples larger than thirty or so to be *large*.

What happens in the situation in which one does not have a sample of at least 30? Things become much more complicated. The Normal Curve tables no longer quite apply to very small samples; a special table is needed. Indeed, one table is needed for each sample size. This is because the form of the distribution tends to get flatter and flatter as sample sizes become smaller. A precise drawing of this effect follows as Fig. 9-16. There is a different curve for each sample size! Obviously, we can't print a different table like the Normal Curve table for each different sized sample. If there were 100 possible sample sizes, we'd need 100 tables. Fortunately, one

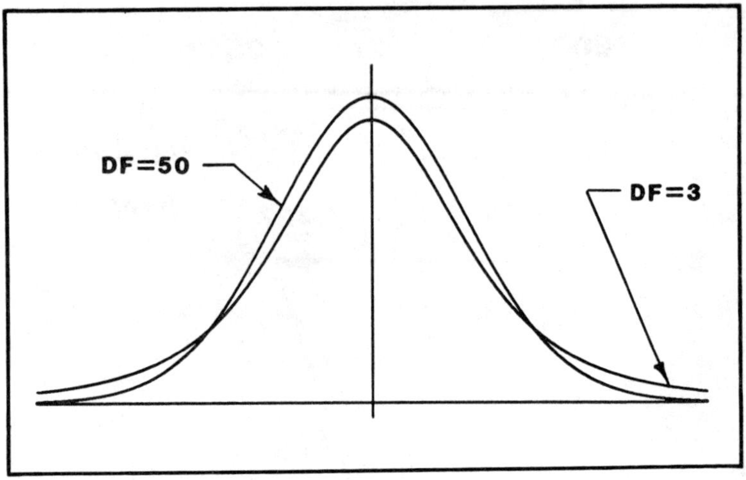

Fig. 9-16. The t-Distribution.

is generally interested in only a few values in most practical situations. That is, we're interested mainly in the values which take in 95 percent or 99 percent of the area under the curve, since we generally want to be either 95 percent sure or 99 percent sure about things. These values are shown in the table in Appendix B.

Turn to that table, and notice that there are columns labeled .95 and .99. These correspond to the desire to be 95 percent and 99 percent sure, respectively. This part of the table is reasonably easy to understand. As the sample size becomes very large the values become 1.96 and 2.58, precisely the values needed to take in 95 percent or 99 percent of the Normal Curve area, although they're not shown in our limited table.

DEGREES OF FREEDOM

The thing that is the most confusing about the table in Appendix B is the column marked DF, which stands for Degrees of Freedom. If that term looks confusing to you, don't worry, for it makes little direct sense, in the context of statistics. It is a carry-over from dynamics, the science of moving bodies. Karl Pearson, the father of statistics as we know it, taught dynamics for a living, and he naturally tended to use terms with which he was familiar.

The term *Degrees of Freedom* in dynamics has to do with the number of ways in which something is free to move. Essentially, for every moveable joint, one gains at least one Degree of Freedom. The front suspension of a car has a great many ways in which it may move (in excess of 25); all of them must be considered in making a car that handles well.

However, the term does not make sense in the context of statistics, but it is a fixed part of the literature, and one must learn to live with it. You can get some idea of this difficult concept from the following example. The following numbers of course contain three pieces of information:

	1
	5
	6

No one would argue about that. Now, if we add the Average you might be tempted to think that there are now four pieces of information.

	1
	5
	6

Average	4

There are three scores plus the mean, making four, but there are still only three independent pieces of information. To show this, assume that the following numbers are known:

	1
	5
	?

Average	4

With this set of numbers what does the unknown value have to be? Well, if the Mean is 4, then the sum of the numbers must be 12 (12/3=4). If the sum of the numbers must be 12, and we have data points of 1 and 5 (=6), then "?" must be 6. There is no escaping it; we have only three pieces of information in that set of numbers. Pearson, because of his familiarity with dynamics, chose to say that there are only three Degrees of Freedom.

The point is, generally, that every time a mean is calculated, a Degree of Freedom is lost. Of course, things are not all that simple, and there are violent arguments at times among professional statisticians concerning just how many Degrees of Freedom there are in any given problem. However, in this chapter, dealing with inferences about population Averages from sample Averages, a Degree of Freedom is lost whenever the Mean is calculated. Thus, we'll look up values in the DF column of Appendix B with (N-1) Degrees of Freedom. For example, with 30 Degrees of Freedom we need a value of 2.04 to be 95 percent sure, and one of 2.75 to be 99 percent sure. In short, small samples, Means, and many other things are not quite normally distributed, but are distributed instead according to the *t-distribution* in Appendix B. Notice that the effect of the t-distribution is simply that of making the requirements more stringent as the sample size grows smaller. That makes sense; small samples should't be as stable as large samples.

CONFIDENCE INTERVALS WITH t

In short, to calculate Confidence Intervals for small samples, you use the values from the t table, rather than the Normal Curve values of 1.96 and 2.58. We'll have to change Program 9.3 a bit to allow this:

```
10 REM PROGRAM 9.4. T CONFIDENCE INTERVALS
20 READ X                          :REM GET DATA
30 IF X=-1 THEN 80                 :REM TEST FOR END
40 S=S+X                           :REM SUM NUMBERS
50 Q=Q+X*X                         :REM SUM SQURED NOS
60 N=N+1                           :REM GET COUNT
70 GOTO 20                         :REM CONTINUE
80 V=Q -S*S/N                      :REM TOP OF FORMULA
90 SS=SQR(V/(N-1))                 :REM GET S
100 SE=SS/SQR(N)                   :REM GET STD ERROR
110 A=S/N                          :REM GET AVERAGE
120 DF=N-1                         :REM GET DF
130 PRINT "T-VALUE FOR DF=",DF :REM GET T
140 INPUT TT                       :REM GET T
150 L=A-TT*SE                      :REM LOWER 95% LIMIT
160 U=A+TT*SE                      :REM UPPER 95% LIMIT
170 PRINT "CONF INTERVAL",L,U  :REM PRINT INT
180 DATA 80,90,100,110,120,-1    :REM DATA TO USE
```

If you run this program, you'll have to supply the t-table values (Appendix B) for the 95 percent or 99 percent level. In our case, with 4 Degrees of Freedom, we need a 95 percent t-value of 2.78. The new Confidence Interval with t-values is

80.3424 to 119.6575

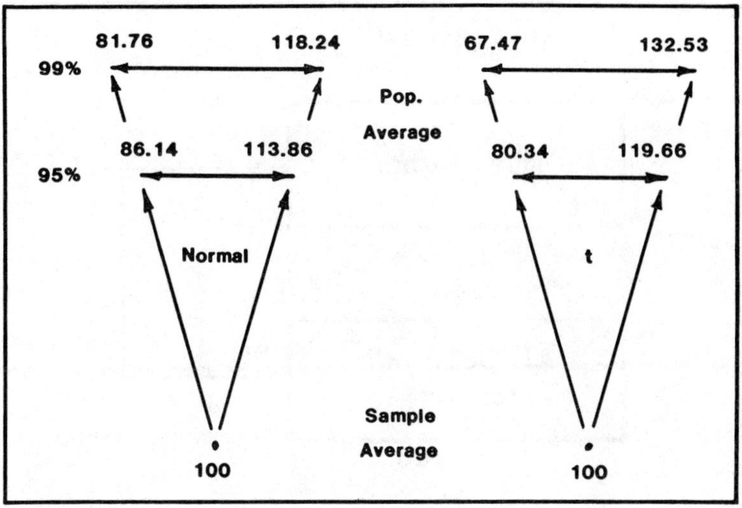

Fig. 9-17. t Confidence Intervals.

and the old one, using Program 9-3 and Normal Curve values was

86.1407 to 113.8593

The values are plotted as Fig. 9-17. Notice that the new values spread out a great deal more than the ones with the Normal Curve values. Obviously, small samples are less stable, and the statistic shows it. The program works pretty much like the last one, and I'll bet you can figure it out on your own. A flow chart is provided in Fig. 9-18. Notice, by the way, that this program will do the Normal Curve values if you supply 1.96 or 2.58 when it asks for t values.

NORMAL CURVE AND t EXAMPLES

Let's look at a few more examples. Here are some examination scores from one of my Introductory Psychology classes:

	Score
	47
	49
	36
	36
	39
	37

Average	40.67
s	5.82

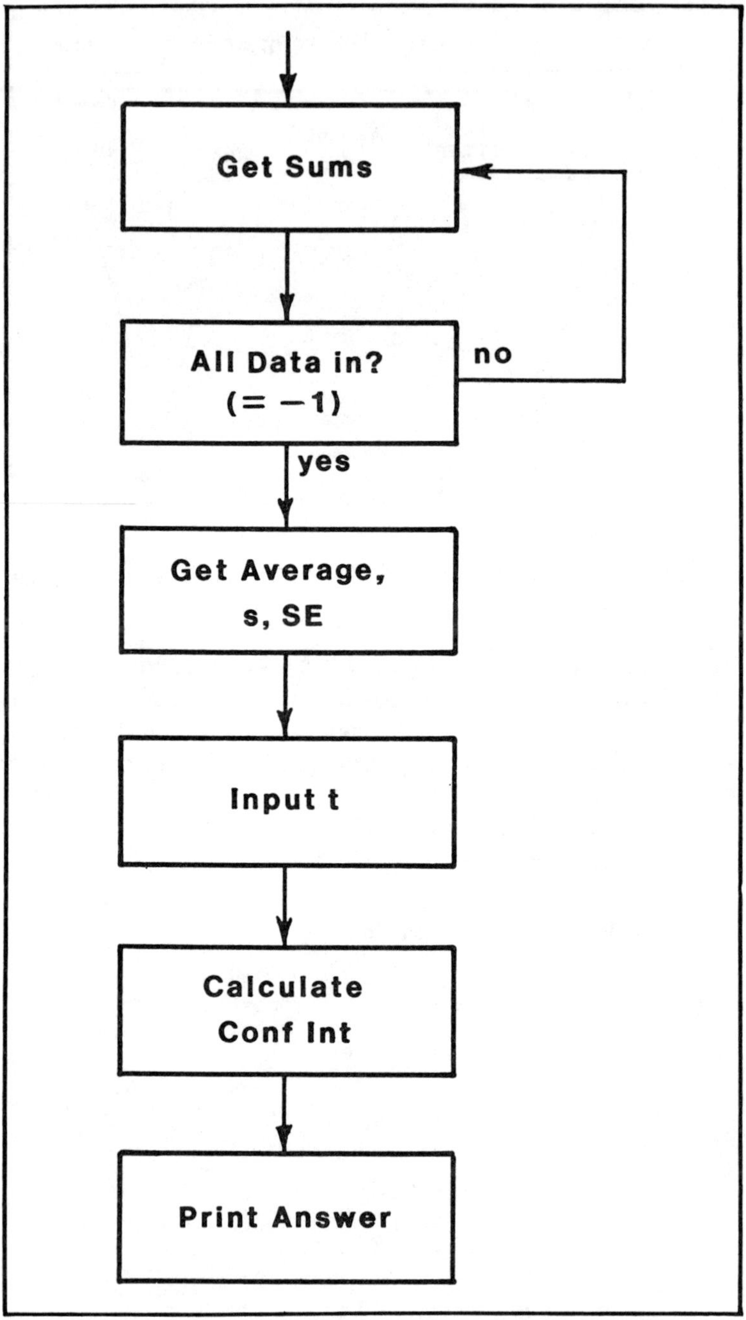

Fig. 9-18. Flowchart for the t Confidence Interval program.

This is an interesting example, for we know the Average (44.83) for the whole class of 100. It will be interesting to see whether the system works. (Actually, a more complex formula is needed with a population as small as 100, but the results will be instructive.)

Putting the data into Programs 9-3 and 9-4 (with 5 DF we need t-values of 2.57 and 4.03 from Appendix B), we find that the Confidence Intervals are:

Normal

36.0101 to 45.3232	**95 percent**
34.5371 to 46.7962	**99 percent**

t

34.5609 to 46.7725	**95 percent**
31.0922 to 50.2412	**99 percent**

These values are shown in Fig. 9-19. Notice that the Confidence Intervals using the t-values are a good deal wider than those using Normal Curve values. Notice, also, that the Average for the entire class, 44.83, falls into all the intervals. Thus, our system does a decent job of predicting the location of the population Average.

What would happen if a very small sample were used? The data below are from the same class, but this time a sample of only three students was taken.

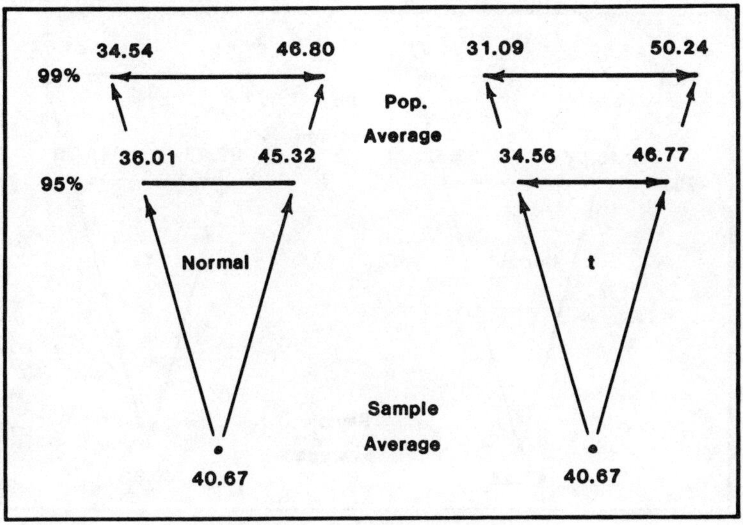

Fig. 9-19. Introduction psychology Confidence Intervals.

	Score
	52
	45
	51

Average	49.33
s	3.79

Here we have only two DFs, which require t-values of 4.30 and 9.92 at the 95 percent and 99 percent levels, respectively. Putting all of this information into Programs 9-3 and 9-4, we obtain the following Confidence Intervals:

Normal	
45.0492 to 53.6175	**95 percent**
43.6940 to 54.9727	**99 percent**
t	
39.9344 to 58.7323	**95 percent**
27.6501 to 71.0165	**99 percent**

These intervals are shown in Fig. 9-20. Values from a sample as tiny as this one are very unstable. Notice that the Normal Curve, 95

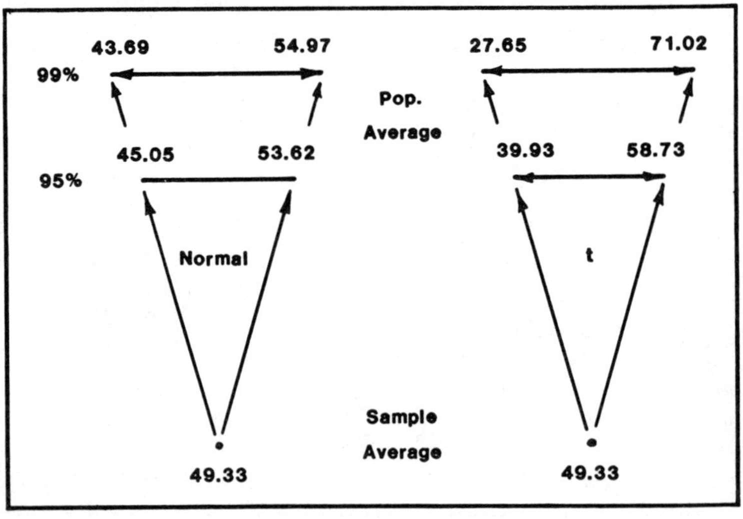

Fig. 9-20. Confidence Intervals from a small sample.

percent interval does not take in the Average for the whole class, 44.83 (it's the same class as last time!). However, all of the other Confidence Intervals do, and notice just how much farther the t-value Confidence Intervals spread out in this case. The table in Appendix B is saying, "You have a very tiny sample, so I'm going to make your Confidence Intervals much larger."

Here's a similar example, using some actual SAT (Scholastic Aptitude Test) data:

	Score
	460
	350
	530
	520
	700
	670
	620
	470
	620
	480

Average	542.0
s	108.91

Here we have 9 DF, and from Appendix B we discover that values of 2.26 and 3.25 are needed at the 95 percent and 99 percent levels, respectively. Programs 9-3 and 9-4 give us the following Confidence Intervals:

Normal		
	474.495 to 609.506	**95 percent**
	453.141 to 630.859	**99 percent**
t		
	464.162 to 619.838	**95 percent**
	430.065 to 653.935	**99 percent**

These are shown in Fig. 9-21. The actual Average is 552.7, for the entire sample of 50. Notice that all of these Confidence Intervals contain 552.7. Thus the system works, again. By predicting from a small sample, we're able to guess the range in which the population Mean will fall fairly well.

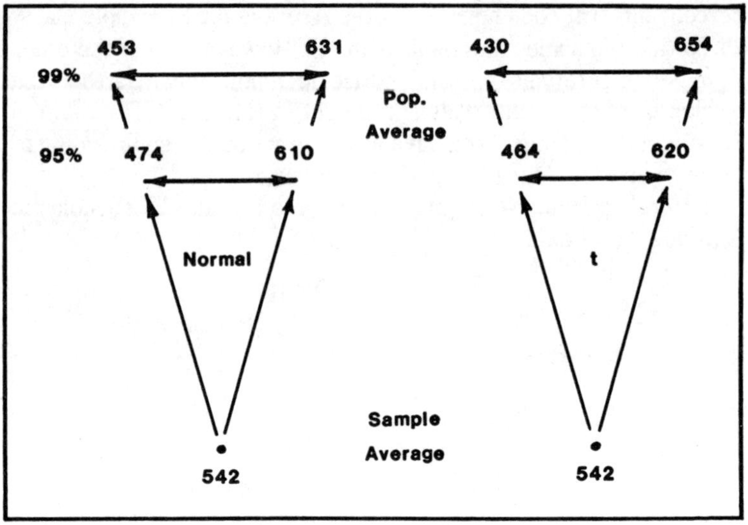

Fig. 9-21. Confidence Intervals from SAT scores.

Again, let's try a smaller sample. Here we'll take a sample of only size 5 from the same data as the previous one (in this case it's the first five numbers). The Confidence Intervals should be much larger.

	Score
	460
	350
	530
	520
	700

Average	512
s	127.16

Here, with only 5 people, we have 4 DF. Appendix B shows that we need values of 2.78 and 4.60 in this situation. Programs 9-3 and 9-4 give the following information:

Normal

400.538 to 623.462	**95 percent**
365.280 to 658.720	**99 percent**

t

353.906 to 670.094	**95 percent**
250.406 to 773.594	**99 percent**

These Confidence Intervals are plotted in Fig. 9-22. Notice that, indeed, the intervals are much wider than the last ones, and in every case, the population Mean of 552.7 is included in the intervals.

Thus, our system works. We can obtain a pretty good idea of the location of the population Mean from our samples. Of course, our populations aren't very large in these examples, and a somewhat more elegant formula should be used to obtain the Standard Deviation with less than very large populations. Still, the concepts hold, and we are in a good position to make guesses about Averages of Populations.

AN AUTOMATIC CONFIDENCE INTERVAL PROGRAM

When running Program 9-4, it is an annoyance to have to look up the t-values in Appendix B. It would be far easier to have our little microcomputer do the job for us; after all that's what our little computers are for. Consequently, we need a program to give us the needed t-values when supplied with the proper Degrees of Free-

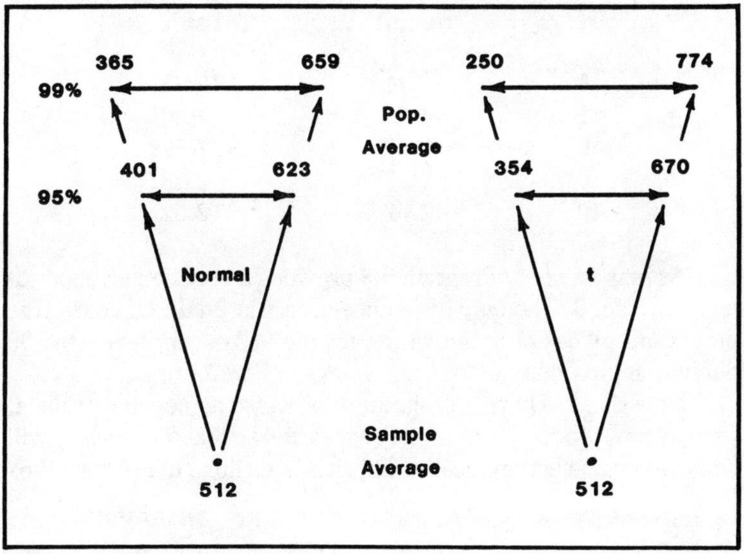

Fig. 9-22. SAT Data, but from a smaller sample.

dom. Program 9-5 will do the job for us:

```
10 REM PROGRAM 9.5. BACKWARDS T PROGRAM
20 INPUT "DF=";DF                :REM GET DF
30 IF DF<>1 THEN 50              :REM JUMP IF NOT 1
40 TT=12.71:GOTO 110             :REM HANDLE 1 DF
50 TT=1.959964:DF=1/DF           :REM SET CONSTANTS
60 TT=TT+2.372271279*DF          :REM CALCULATE T
70 TT=TT+2.822498705*DF*DF       :REM  IN NEXT LINES
80 TT=TT+2.555849791*DF^3
90 TT=TT+1.589534146*DF^4
100 TT=TT+.7328982604*DF^5
110 PRINT TT                     :REM PRINT ANSWER
120 GOTO 20                      :REM CONTINUE
```

This program is based on a paper by one of the fathers of modern statistics, Sir Ronald Fisher (Fisher and Cornish, 1960). Unfortunately, Fisher omitted several of the needed coefficients in his paper, and did not provide sufficient accuracy in others, so that the values all had to be recalculated. It's not often that one finds an error in a paper by Sir Ronald Fisher!

At any rate, if one gives the program the value of a DF, it will return the t-value needed for significance at the 95 percent level; the idea is very similar to that of Program 6-3, for the Normal Curve. If we try various DF values, we obtain t-values as follows:

DF	Program Result	Value from Table
1	12.71	12.71
2	4.29	4.30
5	2.57	2.57
10	2.23	2.23
40	2.02	2.02

As may be seen, Program 9-5 provides a very decent approximation, indeed. The only difference occurs at 2 DF. Of course the program provides only the values for the 95 percent level (the 99 percent approximation does not work very well, alas).

Now that we have a program to provide the needed t-values, we may now proceed to change Program 9-4 so that the t-values will be incorporated automatically. This has been done in Program 9-6:

```
10 REM PROGRAM 9.6 AUTOMATIC T CONF INTERVAL
20 READ X                        :REM GET DATA
30 IF X=-1 THEN 80               :REM TEST FOR END
```

```
40 S=S+X                          :REM SUM NUMBERS
50 Q=Q+X*X                        :REM SUM SQUARED NOS
60 N=N+1                          :REM GET COUNT
70 GOTO 20                        :REM CONTINUE
80 V=Q -S*S/N                     :REM TOP OF FORMULA
90 SS=SQR(V/(N-1))                :REM GET S
100 SE=SS/SQR(N)                  :REM GET STD ERROR
110 A=S/N                         :REM GET AVERAGE
120 DF=N-1                        :REM GET DF
130 IF DF<>1 THEN 150             :REM JUMP IF NOT 1
140 TT=12.71:GOTO 210             :REM HANDLE 1 DF
150 TT=1.959964:DF=1/DF           :REM SET CONSTANTS
160 TT=TT+2.372271279*DF          :REM CALCULATE T
170 TT=TT+2.822498705*DF*DF :REM   IN NEXT LINES
180 TT=TT+2.555849791*DF^3
190 TT=TT+1.589534146*DF^4
200 TT=TT+.7328982604*DF^5
210 L=A-TT*SE                     :REM LOWER 95% LIMIT
220 U=A+TT*SE                     :REM UPPER 95% LIMIT
230 PRINT "CONF INTERVAL",L,U  :REM PRINT INT
240 DATA 80,90,100,110,120,-1 :REM DATA TO USE
```

A flowchart for this program is provided in Fig. 9-23.

Notice the program is nearly the same as Program 9-4. However, when the value for t is needed, in line 130, the needed lines from Program 9-5 are inserted, and the t-value is calculated in lines 130 to 200. Program 9-4 picks up again in line 210, just as before. In the next chapter, we'll show how this change might be made in a simpler fashion with subroutine (GOSUB) statements, but Program 9-5 is simple enough so that we can simply incorporate it as a unit in our program.

If we run this program, we discover that the 95 percent confidence limits are 80.369 to 119.631. These may be compared with the values obtained with Program 9-4, using numbers from Appendix B.

Appendix B	80.3424 to 119.6575
Program 9-6	80.369 to 119.631

Notice that the values are very, very similar. Program 9-6, however, is much simpler to use.

We have completed another chapter, one which is included in the book primarily as an introduction to the next one, on differences between Averages. Notice that there are no new BASIC concepts in this chapter; indeed, there will be few new ones in the remainder of the book.

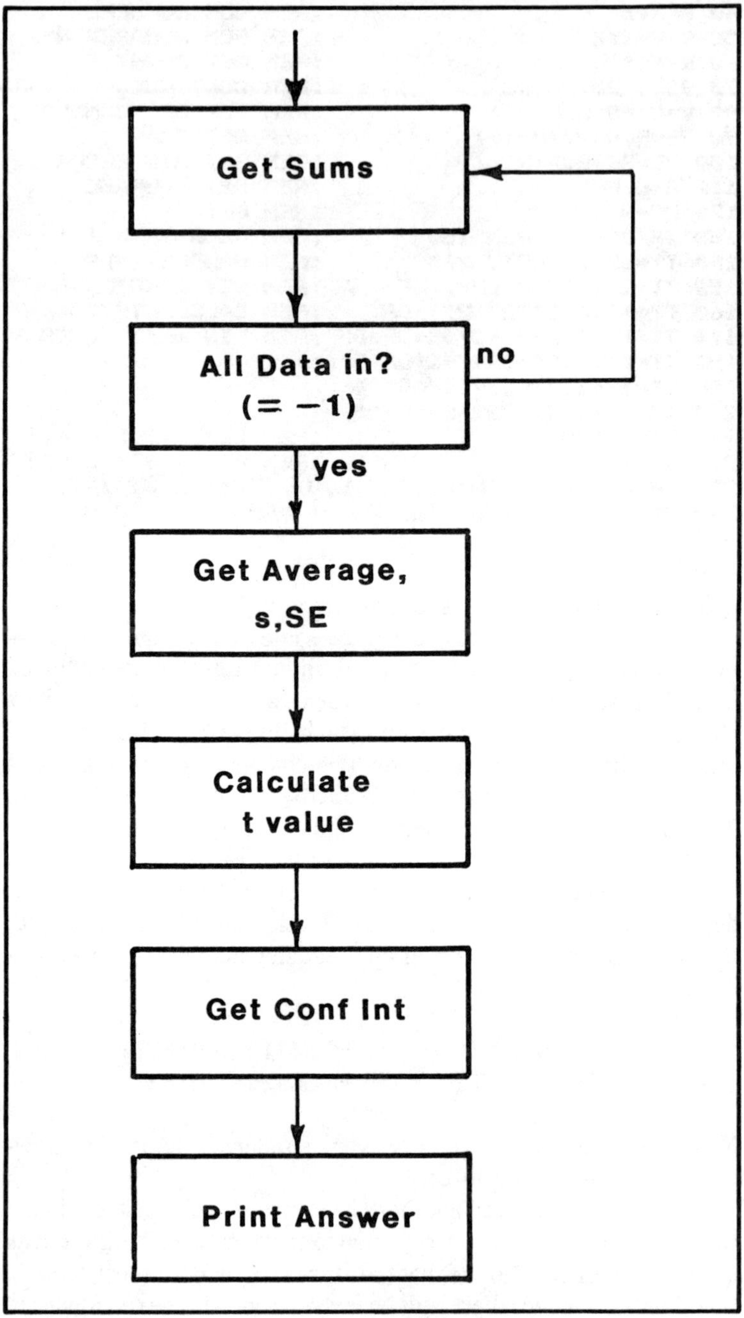

Fig. 9-23. Flowchart for Automatic t Confidence Intervals.

PROBLEMS

Work these problems by both the Normal Curve and the t-value approach. Notice that in all cases the sample sizes are really too small. In real life, one should use larger samples, but these make good problems. Don't forget the −1 to end the data!

1. A supermarket chain wants to obtain some one pound blocks of cheese from a supplier. They take six samples of the product, and obtain the following weights: 0.95, 0.96, 1.01, 0.90, 0.92, 0.88. Is the one pound claim reasonable? (Perhaps not. For example, the 95 percent Normal Curve interval is .8993 to .9741, and one pound doesn't fall in that interval. However, the 99 percent t-value interval is .8597 to 1.0136, but that's the only one of the four intervals in which one pound looks reasonable.)

2. The Varoom Motorcycle Company claims that its bikes have a top speed of 120 miles per hour. A mechanics magazine does a survey, and gets the following values for the top speed: 110, 120, 105, and 100 mph. Is the company reasonable in its claims? (Perhaps! The 95 percent Normal Curve interval is 100.382 to 117.118, and 120 isn't in that interval, but the sample size is only 4, and thus the t-values should be used. The 95 percent t-value interval is 95.1728 to 122.3272, and 120 is in that interval.)

3. A bright students' high school claims that its students have an average IQ of 140. A college wants to test this out, and gets these values on a sample of 10 students from the high school: 151, 122, 134, 176, 83, 140, 131, 146, 161, 100. Is the high school telling the truth? (Undoubtedly. The 95 percent Normal Curve interval is 117.275 to 151.52, and the 95 percent t-value interval is 114.654 to 154.146. 140 is comfortably in these intervals, and the others, too.)

4. At the other end of the range, a school for the mentally retarded says that the average IQ of its students is 60. A sample of 7 students gives these IQ scores: 103, 94, 82, 77, 63, 50, 78. Is the statement correct? (It depends on how you look at it. With the Normal Curve values, the lower end of the 95 percent interval is 64.9363 and the lower end of the 99 percent interval is 60.7587. Thus, with the Normal Curve, you'd probably disagree. But with the t-values, the lower ends of 95 percent and 99 percent intervals are 61.6346 and 53.1447. Thus, at the 99 percent t-value level, the claim seems reasonable.)

5. The Zippy Printer Company makes printers for microcomputers. They claim that their printer prints 1,000 characters per second. A computer magazine tests 8 printers, with the following speeds as a result: 1100, 1203, 1401, 1050, 1192, 1234, 1460, 1302.

Should you buy a Zippy Printer? (Use your judgment. The 95 percent Normal curve interval is 1145.62 to 1339.88, and the 99 percent t-value interval is 1069.31 to 1416.19. Consequently, it's not very likely that the printers do exactly 1,000 characters per second, but then again, there's nothing wrong with a printer with an average speed of 1242.75 characters per second! One shouldn't get overly lost in the statistics!)

I THINK GIRLS ARE SMARTER THAN BOYS!

10 Testing for Differences Between Groups

The single Averages about which we were making guesses in the last chapter are interesting, but usually our principal interest is in differences between Means of groups. Again and again, we want to know whether men are smarter than women, whether dogs are smarter than cats, whether Chevrolets are faster than Fords, whether summers are warmer in Michigan or Minnesota. The procedures that we'll use for this purpose are directly analogous to the ones in the last chapter, but statistical thinking in this area is very, very tricky. To keep from getting too technical, we won't go into details very deeply, and consequently you may not be able to see the intimate connection between this chapter and the last one. Fortunately, there are ways of explaining the procedures, which while not mathematically rigorous, will give you a decent understanding of the process.

AN EXAMPLE

Let's try some fictitious data. Assume that we want to find out whether women are smarter than men, or vice versa. (Fortunately for the relationship between the sexes, a great deal of psychological research has shown that both men and women are about equally bright.)

We give some people a good intelligence test, something like the Wechsler test. The results are as follows:

	Group 1 Males	Group 2 Females
	102	105
	101	104
	100	103
	99	102
	98	101
	--------	--------
Average	100	103
SD	1.4142	1.4142

The question to be answered, then, is that of whether females are smarter than males.

"That's silly!" you say. "On the basis of that set of data, obviously males aren't as smart as females. The males have an Average of 100, and the females an Average of 103. There's no doubt about it; females are the smarter of the two groups."

Ah, but it's not so simple. First of all, notice that we have very, very small groups. It is very difficult to make reliable decisions on the basis of a sample of only ten people. The use of such a small number of people to make conclusions about the population as a whole seems very hazardous. Second, all present-day intelligence tests are far from perfect. No doubt there's some error in the measuring process. A difference in Averages of 3 points between males and females could easily be due to error, or to pure chance. Further, by looking at Fig. 10-1, we notice that the groups overlap a good deal. (The plots in this chapter of actual data, in keeping with our use of the microcomputer, were done on a Radio Shack #26-1190A plotter, and my TRS-80, and are precisely to scale. A Normal Curve is plotted to show the Standard Deviations, although with such small samples the Normal Curve may not apply very well.) Thus, we need some dependable way to decide whether two group Means are really different or not, and hence this chapter.

A SIMPLE EXPLANATION

As mentioned above, more advanced statistics books have a formal way of explaining the way the decision is made about a real difference that is very tricky, but I can explain it fairly easily in simplified terms that you'll be able to understand. In the process I'll build up the formula for the t-test, as it is called.

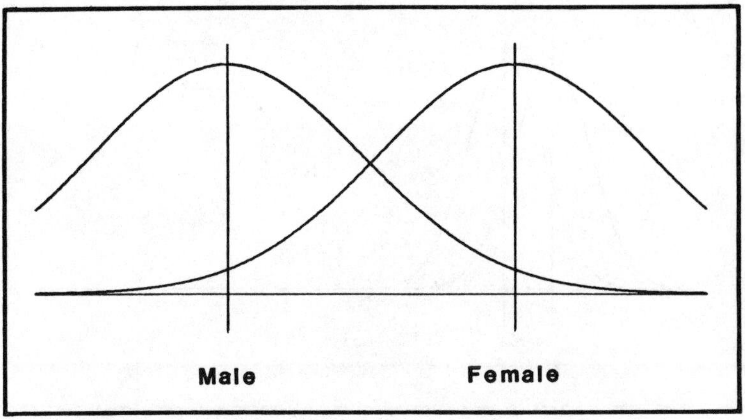

Fig. 10-1. A plot of fictitious male-female IQs.

Assume that we want to discover whether males are smarter than females.

The object of most statistical tests is to come out with one single number that gives the information desired. We've developed such numbers with the Average, the Standard Deviation, the Correlation, and so on. Now we want a number to tell us about the difference between group Averages. Statisticians call the single number that tells us about differences between group Averages t.

Let's take several fictitious examples. First of all, say that samples are taken, and it is found that males have an average IQ of, say, 100, and females an average IQ of 101. Further, assume that the Standard Deviation for both groups is rather small. This situation is shown in Fig. 10-2 (the next series of figures are only for purposes of illustration and are not to scale). If we look at Fig. 10-2, and use our common sense, is it likely that there is a real, non-chance difference between the two groups?

It's hard to say, for the figure is compressed together at the left-hand side of the drawing, and there is a great deal of overlap between the groups. However, since it's hard to make a decision, it seems likely that the difference between the two groups can't be very strong.

Another example will help: assume that we know that males have an Average IQ of 100, and females one of, say, 125, with the Standard Deviations remaining the same. This is shown in Fig. 10-3. Is it likely that the two groups really differ in intelligence? (Keep in mind that this is all fictitious data!)

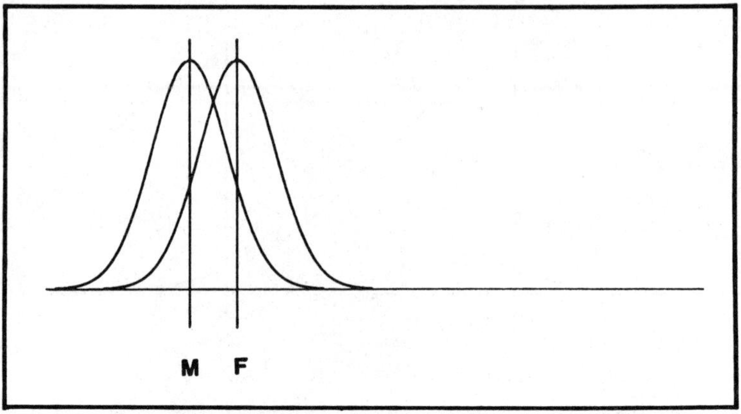

Fig. 10-2. Male-female Means close together.

That example is interesting. Now, to the eye, it looks a lot more likely that there is a real difference between males and females. Also, notice that the difference between the Averages of the two groups has gone from one (101-100) to 25 (125-100). It looks very much as though the farther apart the Averages of the groups are, the more likely it is that the groups really differ. Certainly, it seems much more likely that there is a real difference in Fig. 10-3 than was the case for Fig. 10-2. The groups overlap only slightly.

Figure 10-4 clinches the matter. Here the fictitious difference between the males and females is, say, 50 points, with the Averages at 100 and 150. There is no question of overlap between the groups here. Obviously, there is no doubt, to the eye, that there is a real difference between the groups.

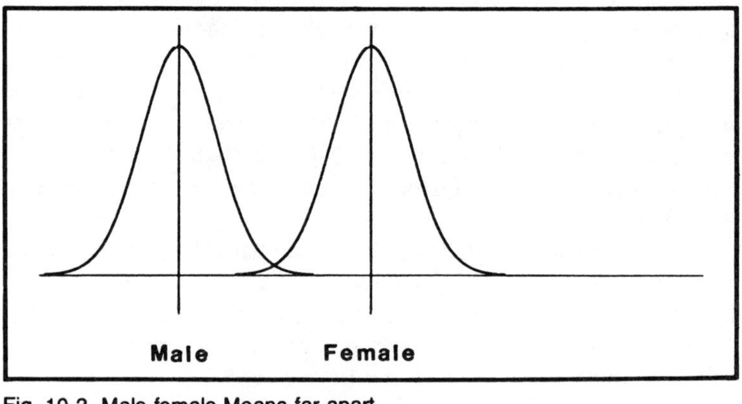

Fig. 10-3. Male-female Means far apart.

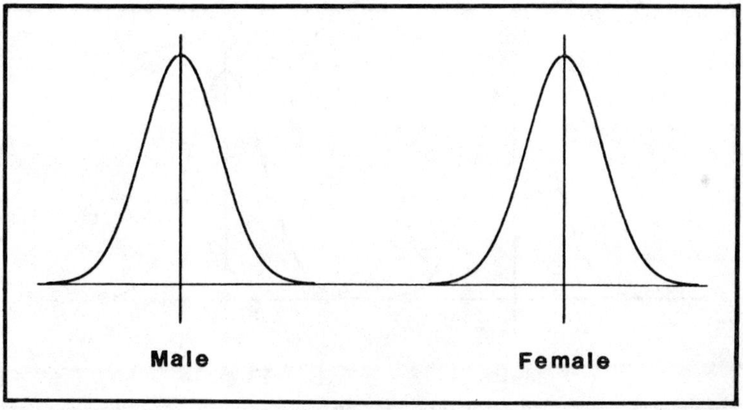

Fig. 10-4. Male-female Means very far apart.

Now, if we can figure out how our eye and brain know that there is a real difference between the groups, we have a start on our formula. The difference between Fig. 10-2, 10-3, and 10-4 is simply that the difference between the Averages grows larger and larger. In short, it seems reasonable that the greater the difference between the Averages, the more likely there is to be a real difference between the groups. We want to set up a formula, in which the variable t will get larger when there is more reason to think that there is a real difference between the Averages will go on the top, as shown:

$$t = \frac{\overline{X}_1 - \overline{X}_2}{?} \qquad \textbf{10-1}$$

The ? means that we don't know what to put on the bottom yet.

Thus, in Fig. 10-2, we have a lot of doubt about whether the groups are different, and the values are only one point apart. Therefore our t-value should be small, and it is. In Fig. 10-5, we have no doubt whatever about whether the groups are different, and the Averages differ by 50 points. Therefore, it makes sense that the difference between the Averages should go on the top of the formula, and t will be directly proportional to this difference. Study Figs. 10-2 through 10-4 until you understand this.

THE BOTTOM OF THE FORMULA

Now, what about the ? in the Formula 10-1, above? Again, we'll

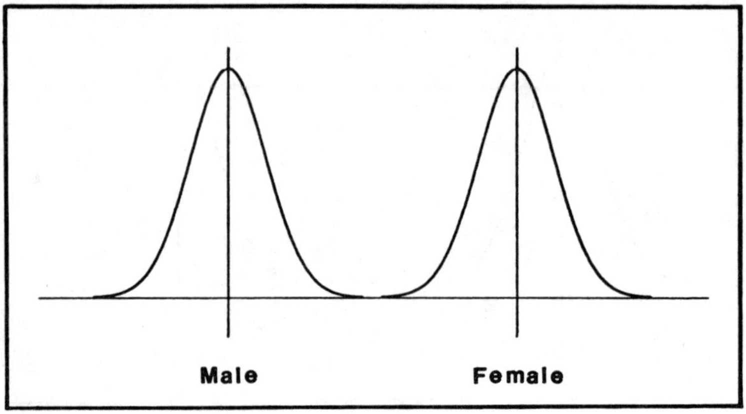

Fig. 10-5. Male-female data with small SD.

assume that we're working with males and females, and let's assume that the real difference between the Averages will always be something like 25 points, so that males have an Average IQ of 100, and females an Average of 125. Thus, the scores would correspond, roughly, to Fig. 10-3.

Let's experiment with the Standard Deviations a bit. Figure 10-5 will start the series. Again, there is little doubt that there is a real difference between the groups.

In Fig. 10-6, things become a bit fuzzier. It looks very much as though there could be a real difference between the groups, but there's a lot of overlap. The Standard Deviations of the groups have become larger, so that the groups spread out more. Perhaps the

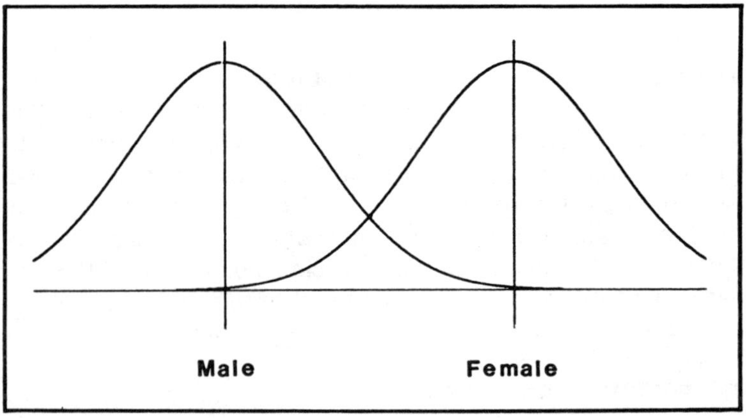

Fig. 10-6. Male-female data with larger SD.

more the groups spread out, the less likely there is to be a real difference.

Figure 10-7 completes the progression. In this figure, the Standard Deviations are so large that it seems very unlikely indeed that there is a real difference between the groups. There seems to be what approaches complete overlap.

Thus, from Figs. 10-5 through 10-7, we find that the more the scores spread out, the less likely there is to be a real difference between the groups.

To add this concept to Formula 10-1, notice that we want the t to be larger if there is more likely to be a real difference between the groups. However, the more the scores spread out, the less likely is there to be a real difference. The t should be inversely proportional to the measure of spread, as a mathematician would say. Hence, the measure of spread should go on the bottom of the formula, replacing the question mark, as follows:

$$t = \frac{\overline{X}_1 - \overline{X}_2}{\text{spread}} \qquad \textbf{10-1}$$

"I'll bet the measure of spread is the Standard Deviation, or the Standard Error of the Mean," you say. "After all, those are the only ones we know."

You're right, in a way, but keep in mind that we have two groups, two Means instead of one. Consequently, we need some sort of measure of spread that takes into account the fact that there

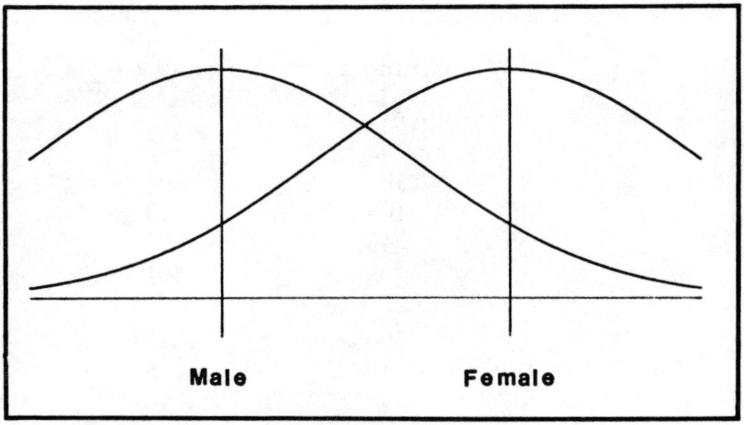

Fig. 10-7. Male-female data with large SD.

are two groups. Indeed, the measure of spread reflects this in its name, for it's called the Standard Error of the Difference between the Means. The formula looks like this:

$$s_{\overline{X}_1-\overline{X}_2} = \sqrt{\left(\frac{\Sigma x_1^{\ 2} + \Sigma x_2^{\ 2}}{(N_1-1)+(N_2-1)}\right)\left(\frac{1}{N_1} + \frac{1}{N_2}\right)} \qquad \textbf{10-2}$$

Now that is a frightening formula! But it isn't as bad as it looks. The sums of little x's squared, on the top (Σx^2), are the symbols that were the top of the Standard Deviation formula (Formula 3-5), and also in line 80 of Program 3-7. They're also the terms that were found again in the Correlation formula (Formulas 7-2 and 7-3). Thus, we obtain the sums of little x squared for both groups, and then add them up. That's the hard part of the formula. The remainder is simply adding up various combinations of the numbers in each group.

$(N_1-1) + (N_2-1)$, on the bottom should have a familiar look. Remember from the last chapter that the Degrees of Freedom when working with a single Mean were (N-1). Well, when checking for differences between two groups, we have two Averages, and hence will lose two Degrees of Freedom, one for each Mean. We can write this expression as $N_1 + N_2 - 2$, and that quantity will be the Degrees of Freedom for the problem.

The remainder of the term under the square root sign is simply the sum of the reciprocals of the two group sizes. Put non-mathematically, we simply divide the number one (1) by each group size, and add them up. That's the right side of the formula. To work it out, we'll need some more information about the problem:

	Group 1 Males	**Group 1 Females**
	102	105
	101	104
	100	103
	99	102
	98	101
	-------	-------
ΣX	500	515
ΣX^2	50010	53055
Σx^2	10	10
$\overline{X}$	100	103

The sum of little x squared for Group 1 is gotten as follows, by Formula 7-3.

$$\Sigma x^2 = \Sigma X^2 - \frac{(\Sigma X)^2}{N} = \qquad \textbf{7-3}$$

$$= 50010 - \frac{(500)^2}{5} = 10$$

The sum of little x squared for the second group is obtained the same way. Now we're ready to get the Standard Error of the Difference between the Means, using Formula 10-2:

$$s_{\overline{X}_1 - \overline{X}_2} = \sqrt{\left(\frac{10 + 10}{5 + 5 - 2}\right)\left(\frac{1}{5} + \frac{1}{5}\right)}$$

$$= \sqrt{\frac{20}{8} * (0.4)} = 1.00$$

That's almost too good to be true. That awesome looking formula works out to be just the number 1. Of course numbers that are this convenient almost never happen, but it's nice for our purposes.

CALCULATION OF THE t

Now we're in a position to finish up our work. We put the Standard Error of the Difference between the Means in the bottom of Formula 10-1, put in the two Averages, and have the answer.

$$t = \frac{\overline{X}_1 - \overline{X}_2}{s_{\overline{X}_1 - \overline{X}_2}} \qquad \textbf{10-3}$$

$$= \frac{3}{1} = 3$$

Thus, the value of t that we want is 3.00.

"So what?" you say. "What do you do with that value?" Well, we look up in the t-table, which as you recall is Appendix B. The

Degrees of Freedom are equal to $N_1 + N_2 - 2$, or $5 + 5 - 2 = 8$ (remember that this quantity is to be found in the bottom of Formula 10-2). If we look in Appendix B under 8 Degrees of Freedom, we obtain the values 2.31 and 3.36. Notice, by the way, that there is a t-table, and now we're calculating t-values. The similarity (and difference!) between these terms should be carefully noted; it has driven generations of students to distraction.

If our t-value of 3 exceeds 2.31, we can be 95 percent sure that there is a real difference between the two groups. If the t-value of 3 exceeds 3.36, we can be 99 percent sure that there is a real, not chance, difference between the groups. In either case, statisticians say that there is a significant difference between the Means.

Our t of 3 exceeds 2.31 (95 percent), but not 3.36 (99 percent), so we can be 95 percent sure that there is a real difference, but not 99 percent sure. Thus, there is some element of uncertainty about the process. If we're willing to be 95 percent sure, there is indeed a difference. But if we want to be 99 percent sure, we must conclude that there isn't one. Personally, I'd be happy with 95 percent certainty, but some workers prefer more accuracy. At any rate we may be 95 percent certain that women are smarter than men, from the data we've collected. This complex process no doubt has confused you, and we'll have more examples in a few pages.

THE t-TEST WITH THE COMPUTER

Obviously, calculating the t-test, as it is called, with the forfulas on the preceding page is a great deal of work. It would be even more work if we had many, many numbers. One must calculate the Average for each group, the sum of little x squared, Formula 10-2, etc., etc., and that's a great deal of work. Even worse, in computations that complex, it is very, very likely that even the most careful person will make an error.

Well, we have our microcomputers to do the work, and they make very few errors, indeed. Up to this point in the book, you may have thought that your little computer wasn't really needed, for with small problems in the chapters up to this point, there may not have seemed to be a large savings in time or effort. With the t-test, all that has changed, and the following program will go through all the complicated computations of Formula 10-3 almost instantly, and it won't make any errors!

```
10 REM PROGRAM 10.1. T-TEST PROGRAM
15 REM COMPUTATIONS FOR GROUP 1 FOLLOW:
20 READ X                          :REM 1ST GROUP DATA
```

```
30 IF X=-1 THEN 70            :REM TEST FOR END
40 S1=S1+X:Q1=Q1+X*X          :REM GET SUMS
50 N1=N1+1                    :REM GET COUNT
60 GOTO 20                    :REM CONTINUE
70 A1=S1/N1                   :REM GET 1ST AVERAGE
80 V1=Q1-S1*S1/N1             :REM GET LTL X SQRD
110 REM COMPUTATIONS FOR GROUP 2 FOLLOW:
120 READ X                    :REM 2ND GROUP DATA
130 IF X=-1 THEN 170          :REM TEST FOR END
140 S2=S2+X:Q2=Q2+X*X         :REM GET SUMS
150 N2=N2+1                   :REM GET COUNT
160 GOTO 120                  :REM CONTINUE
170 A2=S2/N2                  :REM GET 2ND AVERAGE
180 V2=Q2-S2*S2/N2            :REM GET LTL X SQRD
200 H=A1-A2                   :REM AVERAGE DIFF
210 DF=N1+N2-2                :REM DEGREES OF FRDM
220 SE=SQR(((V1+V2)/DF)*(1/N1+1/N2)) :REM GET SE
230 T=H/SE                    :REM GET T
240 PRINT "T=", T, "WITH DF=", DF :REM PRINT ANS
500 DATA 98,99,100,101,102,-1    :REM 1ST GROUP
510 DATA 101,102,103,104,105,-1 :REM 2ND GROUP
```

This program is a rather long one, but, then, the statistical concepts are getting more and more complex. The program gets Averages, Standard Deviations, differences between Averages, a Standard Error, and so on. It's a very complex business.

How does the program work? If you look closely at it and the flowchart in Fig. 10-8, you'll notice that the first 17 lines are very, very similar to past programs, such as Program 3-7, or, more recently, Program 9-1. For the t-test, we need to do the same operations twice, and the line numbers in the present program are arranged to make comparisons between the two halves as easy as possible.

Lines 10-80 and lines 110-180 carry out the same operations, 10-80 for the first group, and 110-180 for the second group. Lines 20 and 120 get the data, and lines 30 and 130 check for the end of the data. Lines 40 and 140 get the sums and sums of squared numbers for the two groups. Notice that S has now become either S1 or S2 so that we can distinguish between the groups; this convention will be used from now on. Also notice that we've put two operations on one line to each operation, especially if they logically go together! Lines line to each operation, especially if they logically go together! Line 50 and 150 count the size of each group, and lines 60 and 160 loop back to get more data. The data must be arranged, then, so that all the data for Group 1 comes first, then a −1 to signal the end of the Group 1 data, and the data for Group 2, and another −1 to signal the end of all the data.

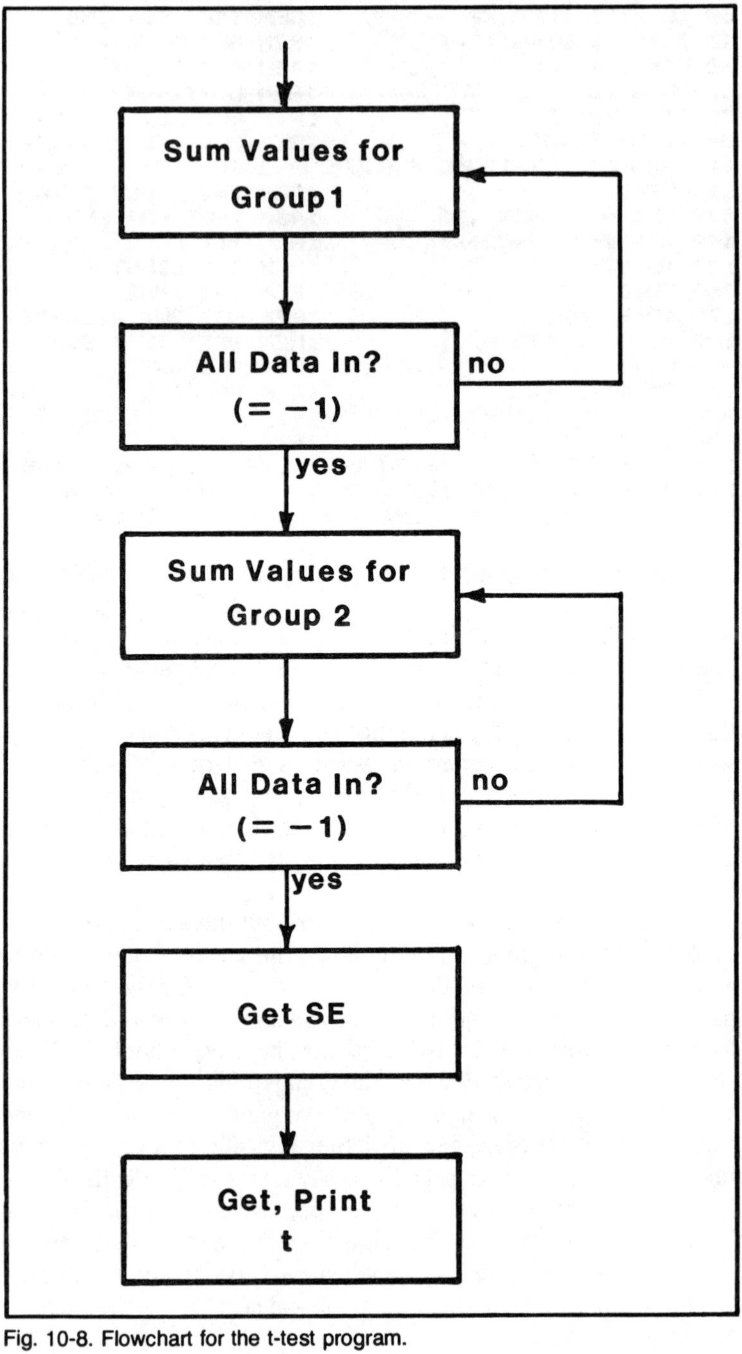

Fig. 10-8. Flowchart for the t-test program.

Lines 70 and 170 calculate the Average for each group, and lines 80 and 180 calculate the sum of little x squared for each group (just as in Formula 7-3). Thus, by line 180, we have all the information we need to compute Formula 10-3.

Line 200 computes the difference between the Averages of the two groups, A1 and A2. Line 210 computes the Degrees of Freedom. We want this for Formula 10-2, and also you'll need it to look up values in the table in Appendix B. Line 220 is the grand finale: it calculates Formula 10-2; SE stands for Standard Error. In Line 220, the sums of little x squared are added together, as in Formula 10-2, then that sum is divided by the Degrees of Freedom, and that product multiplied by the sum of the reciprocals of the sample sizes. The square root is taken to yield SE. It's a tricky business.

Line 230 calculates Formula 10-3, and line 240 prints out the results. If you like, you can add another statement to print out the Averages, Standard Deviations, and so on. You certainly have the ability to do this by now. Thus, you could add the following line

```
205  PRINT "AVERAGES=",A1,A2
```

if you wanted the Means, and so on with the others.

Run the program. You'll find that it gives the answer very rapidly indeed. If you have the time, work it out by hand, and you'll see the advantage of the computer very quickly.

Notice that the t-value produced by the program is sometimes negative, and sometimes positive. The sign indicates the direction of the difference, that is, which group is larger. If you don't like the sign, you can eliminate it by using the ABS function in Line 230, as follows:

```
230  T=ABS(H/SE)      :REM GET ABS T
```

Remember the ABS function from Chapter 3? It simply instructs BASIC to ignore the sign. I personally like the sign, and most large computer statistical packages print it out.

EXAMPLES USING THE COMPUTER

We'll now proceed to work a number of problems so that you can get a feel for the t-test. This ability to work many problems in a short period of time may well be the principal advantage of using a microcomputer.

The Zippy Printer Company is in competition with the Turtle

Printer Company to see who makes the fastest computer printer. A number of samples of the printers from both companies are taken, with results as follows (values are characters per second). Is one company's printer really faster than the other?

	Zippy	**Turtle**
	600	700
	700	800
	800	900
	900	1000
	1000	1100
	1100	1200
	1200	1300
	1300	1400
	1400	1500
	----------	----------
Average	1000	1100
SD	258.199	258.199

Using a diagram much like those found at the beginning of the chapter (but now to scale), the data above looks very much like that shown in Fig. 10-8. Is there a real difference between the companies' products? We could work the problem out by hand, or we could put it into Program 10-1, by entering new DATA statements, as follows:

```
500  DATA 600,700,800,900,1000
510  DATA 1100,1200,1300,1400,-1
520  DATA 700,800,900,1000,1100
530  DATA 1200,1300,1400,1500,-1
```

The data is split up among four lines, so that it will fit on the screen.

Now, is there a real difference between the two product lines, or not? Running Program 10-1, we find that t = .7746. From Appendix B, we find that with 16 Degrees of Freedom we need values of 2.12 or 2.92 to say that there's a real difference. Our value of .7746 is well below either value, and we must conclude that there is no difference between the groups. Thus, while it appears that Turtle printers are somewhat faster than Zippy printers, it just doesn't work out statistically. Of course, a look at Fig. 10-8 should give us a strong hint that the difference is not significant, as statisti-

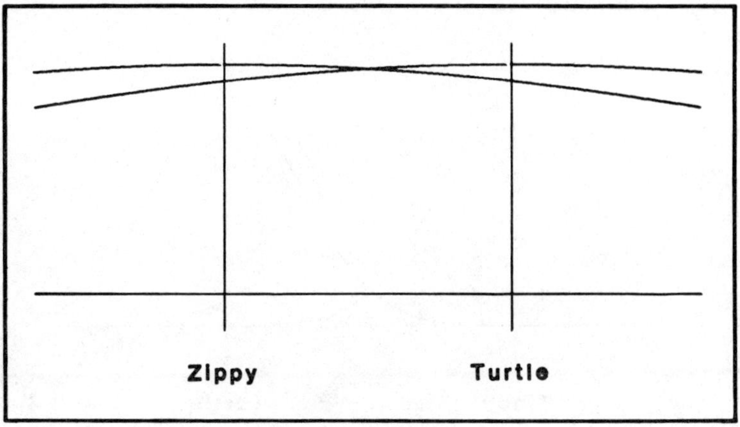

Fig. 10-9. Zippy-Turtle Printers with large SD.

cians say, not really different. The groups, as depicted there, overlap a great deal, but it takes the statistical test to make certain that there is no real difference.

This next example (again fictitious) shows the effect of the Standard Deviation of the data on the results. Again we have Zippy vs. Turtle printers, with the same two averages. However, here the scores do not spread out as far:

	Zippy	**Turtle**
	960	1060
	970	1070
	980	1080
	990	1090
	1000	1100
	1010	1110
	1020	1120
	1030	1130
	1040	1140
	----------	----------
Average	1000	1100
SD	25.820	25.820

Plotted, this data looks a great deal like Fig. 10-10.

Put the data into DATA statements, and run the program. Or, if you're a glutton for punishment, work it out by hand. In either case, you'll find that the t is 7.7460. We still have sixteen Degrees of

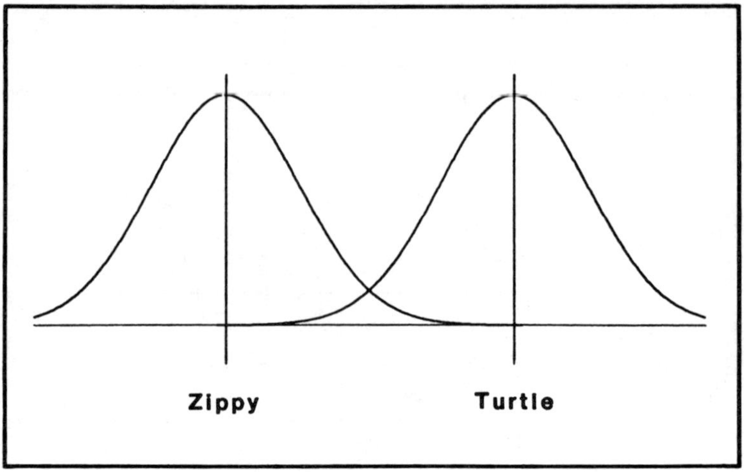

Fig. 10-10. Zippy-Turtle Printers with smaller SD.

Freedom, so that the values needed remain 2.12 and 2.92. Our value of 7.7460 exceeds both of these, and we can thus be 99 percent certain that Turtle printers are indeed faster. The difference is "significant at the 99 percent level," as statisticians say.

Notice the difference between this problem and the last one. The Standard Deviation of each of the two groups is 25.8199 in this problem. In the previous problem it was 258.199. Thus, the less the scores spread out, the more certain we can be that there is a real difference.

This example is a further extension of the last problem. The data looks as follows:

	Zippy	**Turtle**
	996	1096
	997	1097
	998	1098
	999	1099
	1000	1100
	1001	1101
	1002	1102
	1003	1103
	1004	1104
	----------	----------
Average	1000	1100
SD	2.5820	2.5820

This situation is depicted in Fig. 10-11. Notice that the Standard Deviations are now very small indeed (2.5819), while the difference between the Averages is 100 (change Program 10-1 to print them out), and hence we would have a greater expectation of finding a real difference. Figure 10-11 shows no overlap whatever between the groups. Running Program 10-1 shows that the t is 77.4597! That's a very high t, much higher than you'll normally get in real situations, certainly much higher than 2.12 or 2.92. The Standard Deviations are much smaller than the difference here. The difference is really significant.

The next example is a very different but still fictitious problem. The Ajax Bolt Company, mentioned in Chapter 9, makes quarter inch bolts. So does the Burred Bolt Company. Samples are taken of their products, with the following results:

	Burred	**Ajax**
	.240	.253
	.236	.258
	.251	.262
	.240	.265
	.250	.270
	----------	----------
Average	.2434	.2616
SD	.0059861	.0058175

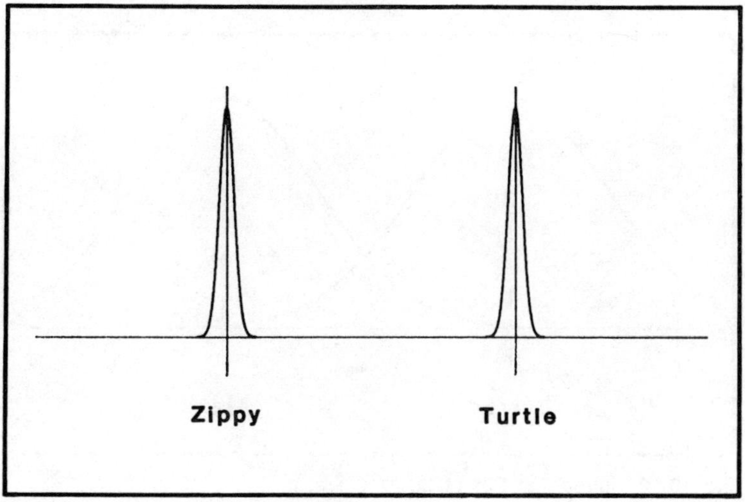

Fig. 10-11. Zippy-Turtle Printers with small SD.

Is there a difference between the two companies' products, other than what one would expect from errors in taking samples, and chance factors? The situation is depicted in Fig. 10-12.

Putting the data into Program 10-1, we discover that the t is 4.3607, with 8 Degrees of Freedom. With 8 Degrees of Freedom we need 2.31 for significance at the 95% level, and 3.36 at the 99% level. Our t of 4.3607 is greater than either value, and thus we may be 99% sure that Burred Bolts are indeed smaller than Ajax Bolts. The difference is indeed significant. This agrees with our reading of Fig. 10-11. There is some overlap between the groups, but not a great deal.

We'd like to find out whether Varoom Motorcycles are faster than the products of the Dragster Motorcycle Company. Consequently, we take samples of the products of each company, with the following results:

	Varoom	**Dragster**
	110	115
	120	110
	105	121
	100	126
		97
	----------	----------
Average	108.75	113.8
SD	7.3951	9.9880

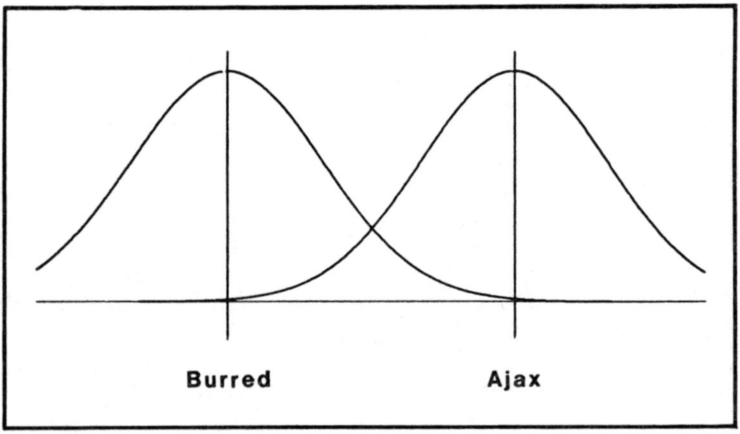

Fig. 10-12. Burred-Ajax Bolt t-test.

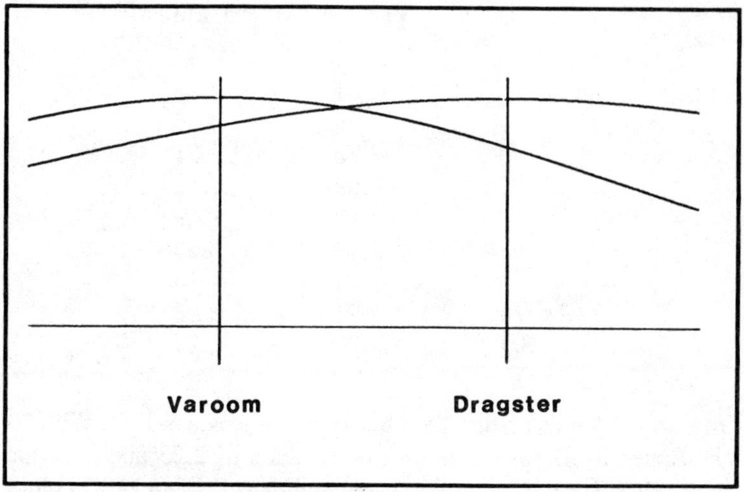

Fig. 10-13. Varoom-Dragster Motorcycle t-test.

Again, the question is that of whether there is a significant difference between the groups. Figure 10-13 seems to show a lot of overlap between the two samples. However, it's hard to judge by eye, as you've learned. Running Program 10-1, we discover that the t calculated is only 0.7435. With 7 Degrees of Freedom we need values of 2.36 and 3.50. Therefore, in point of fact, the difference isn't very large after all! That's interesting. The difference appears large, but statistically it isn't so large.

Notice that this is the first example with unequal numbers in the groups. It's perfectly permissible to do problems with unequal N's, although some people think that it's more aesthetic to have equal sample sizes. Also, certain assumptions involved in the t-test are better served if the sample sizes are equal. Still, problems with unequal N's are worked very, very frequently by research workers. Also, keep in mind the fact that, in real research, we should use much larger samples. We use small N's to make good illustrations, but this procedure would not make for good research practice.

SOME REAL DATA

Now, let's try some data that isn't fictitious.

(6) In a certain study men and women were measured for the length of their index fingers. The results (in centimeters) were as follows:

	Female	Male
	2.81	2.88
	2.69	3.00
	2.75	3.31
	2.69	2.81
	2.63	2.69
	2.69	2.63
	----------	----------
Average	2.71	2.88
SD	.06	.22

Program 10-1 tells us that the t between the sexes is 1.71. With 10 DF, Appendix B says that we need values of 2.23 and 3.17 for significance. Clearly, our value is not significant. Figure 10-14 plots the data. Notice that there is a great deal of overlap between the groups, and this must be the reason for the small t. The overlap is largely due to the higher Standard Deviations for the males. As we shall see later in the chapter, it is not a particularly good idea to use the t-test when the Standard Deviations of the two groups differ a great deal, but it's the sort of thing that happens frequently with real data.

In the next example, males and females are tested for strength in their right hand, using a hand dynamometer that reads in kilograms. The results are as follows:

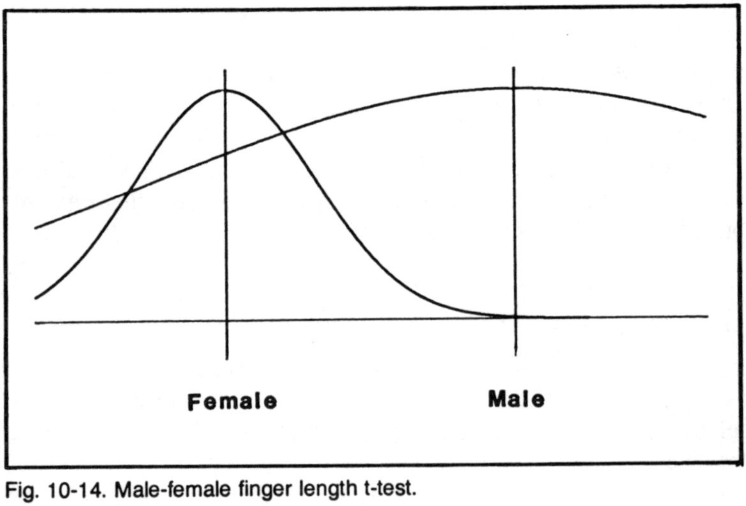

Fig. 10-14. Male-female finger length t-test.

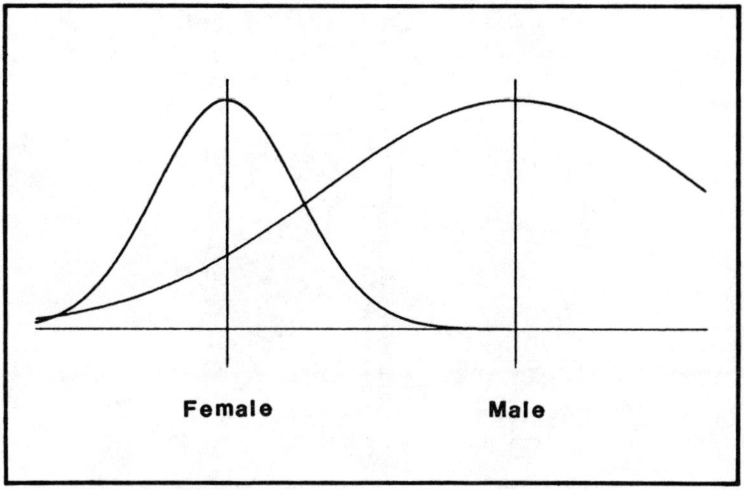

Fig. 10-15. Male-female strength t-test.

	Female	Male
	27.0	65.0
	18.0	78.5
	38.0	52.0
	29.0	42.5
	23.5	39.0
	26.5	31.0
	----------	----------
Average	27.0	51.33
SD	6.03	16.17

As in the last example, you'd expect males to be stronger than females. As Fig. 10-15 shows, the SD for the males is much greater than that for females (not a good situation for the t-test). Run Program 10-1. You'll find that the t between the two groups is 3.15. With 10 DF, one needs values of 2.23 and 3.17 or greater. We can be 95 percent sure that males are stronger than females. The difference is said to be significant at the 95 percent level.

In my classes in Beginning Psychology, I have long had the suspicion that the brighter students sit toward the front of the class. Folklore among college professors suggests that less serious students sit toward the back. Let's see if that's the case. Here's the data scores on the first examination:

	Back	Front
	34	53
	53	52
	51	56
	40	43
	47	51
	50	44
	42	50
	34	52
	52	51
	46	33
	-------	-------
Average	44.9	48.5
SD	6.74	6.38

These data are plotted as Fig. 10-16. The figure shows that there is a great deal of overlap between the groups, and thus probably no real difference. Program 10-1 tells us that the t is only 1.16. With 18 DF, Appendix B says that we need minimum values of 2.10 and 2.88. Clearly, there is not a significant difference between people who sit in the front and those who sit in the back. However, we might find it with a larger sample. The people in the front do have a somewhat larger Average.

Finally, I have in the past done some research on a test of femininity. It would be interesting to know whether women do, in

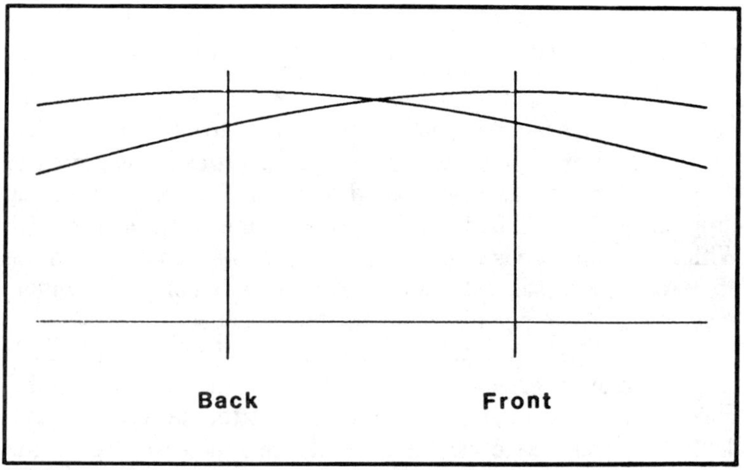

Fig. 10-16. Back-front t-test.

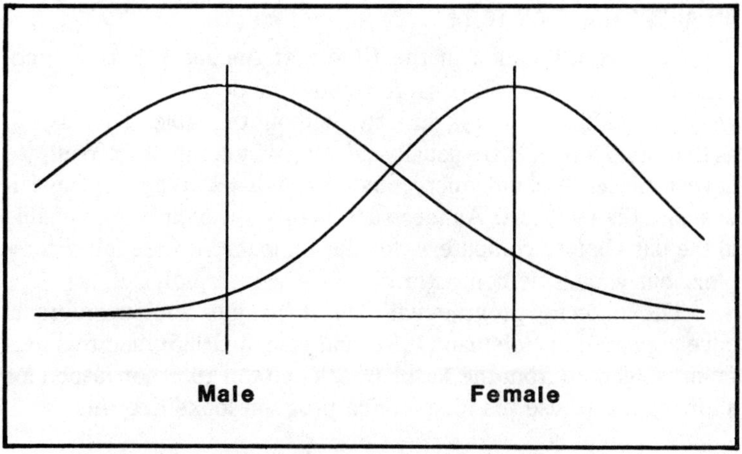

Fig. 10-17. Male-female scores on femininity test.

fact, score higher on this test than men. The test would be pretty useless if that were not the case. Data on 16 subjects follows:

	Males	Females
	6	10
	15	9
	2	14
	6	14
	2	13
	2	6
	6	16
	2	13
	---------	---------
Average	5.12	11.88
SD	4.17	3.06

These data are plotted as Fig. 10-17. Notice that there is a good deal of overlap between the two groups, but not really all that much. It's difficult to judge the situation by eye. Run the data in Program 10-1, and you'll find that the t is 3.46. With 14 DF, Appendix B tells us that we need values of 2.14 and 2.98 for significance. Thus, there is indeed a real difference between the two groups, one significant at the 99 percent level. Our test seems to work. Females get higher scores on the femininity test than males.

AN APPROXIMATION FOR t

Looking up values in the t-table of Appendix B is a time-consuming, awkward process. It would be far better to have a way to get the significance levels directly, without the table. Tables such as that of Appendix B are usually calculated by computers. Well, we have a perfectly good microcomputer at hand. Why not have it evaluate the t's for us? An idea such as this was simply unthinkable in the days before computers, for the computations are fairly complex, but your little computer can do it very rapidly.

The following program will do just that job. Similar programs have appeared in Veldman (1967) and Jaspen (1965), and this program is adapted from the latter (thanks go to Professor Jaspen for allowing me to use his ideas). The program looks like this:

```
3ØØØ REM PROGRAM 1Ø.2 T-VALUE SIGNIFICANCE
3Ø1Ø INPUT "T= ";KØ           :REM GET T
3Ø2Ø INPUT "DF=";K2           :REM GET DF
3Ø3Ø KØ=KØ*KØ:K1=1            :REM ADJUST VALUES
3Ø4Ø GOSUB 4ØØØ               :REM GOTO DISTRIB
3Ø5Ø PP=1-H                   :REM ADJUST H
3Ø6Ø PRINT PP                 :REM PRINT ANSWER
3Ø7Ø GOTO 3ØØØ                :REM KEEP IT UP
4ØØØ REM APPROXIMATION ROUTINE :REM BEGIN APPROX
4Ø1Ø IF KØ>=1 THEN 4Ø4Ø       :REM BIG ENOUGH?
4Ø2Ø PRINT "NOT SIGNIFICANT":REM SAY IF NOT
4Ø3Ø H=1:RETURN
4Ø4Ø K5=Ø.2222222             :REM SET ONE CONST
4Ø5Ø K6=K5/K1:K7=K5/K2        :REM SET MORE
4Ø6Ø K8=KØ^Ø.3333333          :REM SET MORE
4Ø7Ø K9=(1-K7)*K8-1+K6        :REM TOP
4Ø8Ø Z=K9/SQR(K7*K8*K8+K6)    :REM GET Z
4Ø9Ø IF Z>3 THEN 411Ø         :REM ADJUST IF <3
41ØØ Z=Z*(1.+Ø.Ø8*Z^4/K2^3)   :REM ADJUST
411Ø GOSUB 5ØØØ               :REM NORMAL APPROX
412Ø RETURN
5ØØØ REM NORMAL CURVE APPROXIMATION
5Ø1Ø H=1+Ø.Ø49867347Ø*Z       :REM CALCULATE
5Ø2Ø H=H+.Ø21141ØØ61*Z*Z      :REM  NORMAL CURVE
5Ø3Ø H=H+.ØØ32776263*Z^3      :REM  APPROXIMATION
5Ø4Ø H=H+.ØØØØ38ØØ36*Z^4      :REM  FROM PROGRAM
5Ø5Ø H=H+.ØØØØ4889Ø6*Z^5      :REM  6.2.
5Ø6Ø H=H+.ØØØØØ5383Ø*Z^6
5Ø7Ø H=H^-16/2              :REM FINISH APPROXIMAT.
5Ø8Ø RETURN
```

I won't try to explain the way the mathematics of the program works, for it's very, very complex. Consult the references above if you're curious. Notice, however, that lines 5000 through 5070 are exact duplicates of lines 30 through 90 of Program 6-2, the Normal Curve program. Program 10-2 takes the t-value, adjusts it so that it will work properly, and then inserts it into the Normal Curve function.

There is a new BASIC programming concept in Program 10-2. Notice that in line 3040, we have:

```
3040   GOSUB 4000       :REM GOTO DISTRIB
```

This is our first encounter with subroutines. We're going to be using the routine starting in line 4000 over and over in later chapters. It would be very nice indeed if we didn't have to reprogram it every time that it was needed. Subroutines are provided in your little computer for precisely that purpose. They allow a small program that is used time after time in a bigger program to be included only once. Whenever you want to use the smaller program, you simply say, "GOSUB XXXX," where XXXX is the line number of the subroutine, and the computer will go there. When the computer has finished the calculations in the subroutine, it encounters the statement, RETURN. RETURN means to return to the main program, right after the GOSUB statement.

In our program, line 3040 tells the computer to GOSUB 4000. The computer then jumps to line 4000, calculates the proper approximation, and then, in line 4120 encounters RETURN. At that point, the computer jumps back to line 3050, continues its work and finally prints out the answer in line 3060. The flowchart in Fig. 10-18 shows the action of the subroutines. If we had to evaluate the approximation many, many times in the course of our calculations, the little subroutine could be called again and again and again.

Another subroutine is called in line 4110. Here, the Normal Curve approximation of Program 6-2 is needed, and it's been made into a subroutine in starting at line 5000. When this subroutine completes its work, it returns to line 4120, which is another RETURN. You can put subroutines within subroutines. This may be done again and again, but there is a limit to how "deep" they may be nested. Your computer manual will give you exact information on this.

BASIC has many built-in-subroutines, things like the square root function, SQR. It would be extremely annoying to have to include in the program the mathematics for the square root approximation every time you wanted to extract one, and so BASIC provides it for you, in the form of a subroutine that comes with the

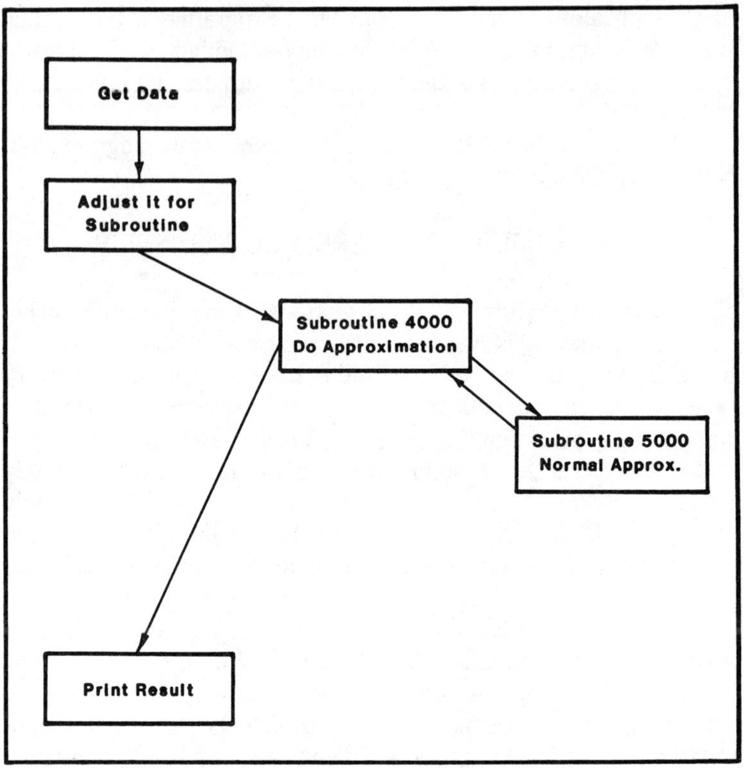

Fig. 10-18. Flowchart for t-value significance program.

machine, and we can do the same thing with GOSUB...RETURN statements in our own BASIC programs.

Notice that this is the first new BASIC concept to be found in this book in a long time. We have gotten an amazing amount of usage out of a very small number of BASIC terms.

Let's try the program. Run it, and it will ask for a t-value. Let's try 1.96, the Normal Curve value for 95 percent. Recall that the t-table approaches the Normal Curve values for large numbers of Degrees of Freedom, so when the program asks for the Degrees of Freedom, type in 1000, or some large number. After you push ENTER the computer thinks for quite a while, and then tells you that the answer is .9525. That is, we can be 95.25 percent sure that there is a real difference. We know that this is the 95 percent level, and that's what the computer tells us. Of course it's a little off, but not seriously so. If we try other t-values from Appendix B, we get the following results:

t-value	DF	Computer's Answer
2.31	8	.9520
2.06	25	.9524
2.00	60	.9526
1.98	125	.9528

In each case, Program 10-2 should give the value of 95 percent. It doesn't quite, but it's rather close. Kelley, Beggs, and McNeil (1969) give a much more accurate routine to calculate this value, but it's very long, and might frighten you if presented here. Consult it if you need more accuracy, although be fore warned that it's in FORTRAN and would have to be converted.

The idea of subroutines may be extended, by the way, to many other areas. Thus, we could make all of Program 10-2 a subroutine of Program 10-1, so that it would tell you the significance level of your t-test as it is calculated. That would be a depth of three subroutines. I thought about including this in the book, but it makes for a very long program, and it seems wise to keep our programs as short as possible. You have all the knowledge to do it yourself! An example of such a combination of a probability routine with a statistical program will be found in Chapter 13.

Program 10-2 yields the following values for the t's in the examples in this chapter:

Problem	t	DF	Computer's Answer
M-F IQ	3.00	8	.9836
Zippy #1	.78	16	NS
Zippy #2	7.75	16	.9999
Zippy #3	77.46	16	1.0000
Ajax Bolt	4.36	8	.9977
Varoom	.74	7	NS
Finger	1.71	10	.8848
Strength	3.15	10	.9900
Femininity	3.46	14	.9962

Using Program 10-2 is obviously much easier than looking in Appendix B, and, besides, the answers are much more accurate and useful.

ASSUMPTIONS IN THE t-TEST

Certain assumptions underlie the t-test:

1. It is assumed that people are randomly assigned to the two groups. This assumption is generally not a problem in most studies, but at the same time, you must always be alert to situations in which the assignment of subjects is not random.

2. It is assumed that the underlying distributions for the two groups (in the population) follow the Normal Curve. This is difficult to see with our tiny examples, but advanced statistics texts provide means of checking this assumption with samples of a size used in research.

3. It is assumed that the Standard Deviations of the two groups are the same. This assumption is really needed mainly so that Formula 10-2 will work properly. Again, advanced statistics textbooks provide ways to check this assumption.

In general, research has shown that violation of the assumptions does not appear to be a serious problem if the two groups in the t-test are the same size.

PROBLEMS

1. Two groups of students take a test like the Graduate Record Examination (GRE) (fictitious data, of course). Group 1 is a control group; it does nothing special. Group 2 takes a stimulant before going into the examination. Is there a real difference between the groups? (Yes. t = 2.7793, significant at the 95 percent level, not the 99 percent level.) Don't forget the negative ones (−1) to end the groups!

Control	500	400	300	300
Stimulant	750	740	900	400

2. Mentally retarded people are given an IQ test, and the test is also given to college freshmen. Are the retarded children really less bright than the college students? (Yes, if one wants to be 95 percent sure. t=2.4097, significant at the 95 percent level, not 99 percent level.)

Retarded	70	80	40	80	20	90
Freshmen	130	120	50	107	130	

3. A store has some oranges, and some fruit which looks a lot like oranges. The fruit is weighed, with the results (in pounds). Is it

reasonable that both fruits are really oranges? (That's a tough one. The t is 1.2481, not significant at either the 95 percent or 99 percent level. Consequently, we might assume that there's no difference between the groups, and thus that the fruits are the same. Professional statisticians, however, might object to this procedure, for they generally don't like to say definitely that there's no difference on the basis of the t-test.)

Oranges	1.0	1.3	1.5	0.9	0.96
Unknown	1.2	1.6	1.3	1.2	

4. Highflier Airplanes, Inc. and Left Brothers are competing for a new jet plane contract. The company who plans to buy the planes runs tests, and gets the speeds below for the two types of planes. Is one plane really faster than the other? (t=3.06, significant with 7 DF at the 95 percent level, not significant at the 99 percent level. Thus, if we're willing to be 95 percent sure, Left Brothers planes are indeed faster.)

Highflier	600	620	610	601	
Left Brothers	701	725	650	615	700

5. A group of students takes a Graduate School Admissions test under normal conditions. Another group takes the same test, but prepares beforehand with one of those $10 study guides that you see in the bookstores. Does previous study really help them? (t=2.571, significant at the 95 percent level. The guide seems to make things worse!)

No Guide	452.1	554.8	477.3	897.2	798.4
Study Guide	157.3	249.1	538.4	323.1	

IF r = +1.00 THE
PROGRAM MAY BLOW UP!

11
More About Significance Tests

The simple t-test described in the last chapter may well be the most widely used of all the significance tests, perhaps because it's fairly easy to understand and to use. However, people are never satisfied with the status quo; they always want something just a little bit better. In this case, it would be nice to have a statistical test that would be a bit more likely to find a significant difference, if there is one. The more likely a statistical test is to find significance, the more power it is said to have.

There are statistical tests available that are more powerful than the simple t-test. Indeed, there is a whole area of statistics called *experimental design*, in which many very complex and usually more powerful statistical techniques are discussed. Indeed, there are so many experimental designs that it is unlikely that any one person knows all of them. If you're curious about this, consult Federer's (1955) enormous compendium of experimental designs. He lists seemingly hundreds of them! Obviously, I can cover only a few of these.

One of the most commonly used experimental designs is the paired t-test, also called correlated or matched. We'll talk about it next. It's very widely used, and is rather easy to compute.

THE PAIRED t-TEST

The following example is very, very much like the first example of Chapter 10. We'd like to see the effect of various drugs on IQ test performance, so a first group is given sleeping pills, and a second group a stimulant (the data are fictitious, of course):

	Group 1 Sleeping Pills	Group 2 Stimulant
	102	104
	101	101
	100	102
	99	103
	98	100
Average	100	102
SD	1.4142	1.4142

Do a t-test of that data, using Program 10-1. You'll find that the t is +2.0000. Using Program 10-2, you'll find that with 8 Degrees of Freedom, 2.0000 is not significant at even the 95 percent level. Thus, you would have to conclude that drugs have little or no effect upon IQ, beyond what you would exect from chance factors. The situation is depicted in Fig. 11-1.

There is nothing whatever wrong with doing the experiment in that fashion. However, if we're clever enough, we can find ways to reduce the error in the experiment. And the less the error, the more powerful the experiment will be and the more likely we are to find a significant difference.

How can we reduce error? The main difficulty with the experiment as it exists is that we can't be certain that the two groups are equal in all respects except for the one variable in which we're

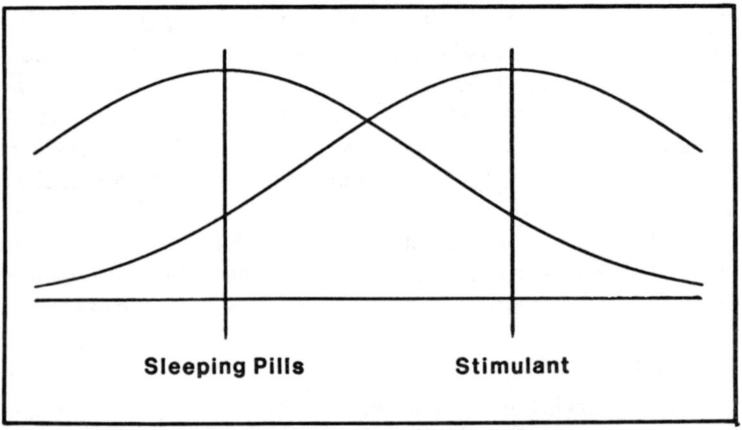

Fig. 11-1. Sleeping pills-stimulant t-test.

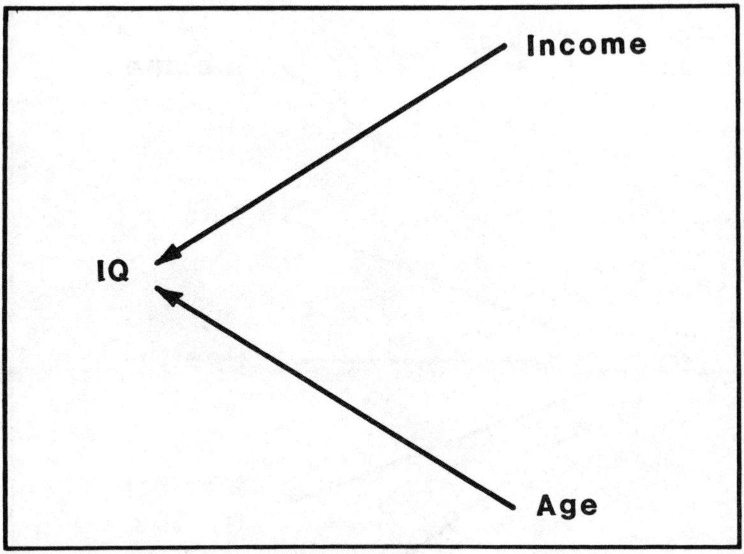

Fig. 11-2. Matching on two variables.

interested—the drugs. For example, it's quite possible that those people taking sleeping pills all come from a higher economic bracket than those taking stimulants. We haven't controlled for that, and it could confuse things. Or perhaps those in Group 2 are a great deal older than those in Group 1. It's difficult to say precisely what effect a difference in age would have on IQ, but obviously we should try to control such things as much as possible. If we don't match the groups carefully, we may very well find that uncontrolled variables such as income and age can affect the experiment in ways that we haven't considered (see Fig. 11-2).

In short, it makes good sense to match the subjects in both groups as closely as possible. We might match them on income, age, health, marital status, schooling, and so on. Say that our first subject in the first group makes $25,000 a year, is 42 years old, is in good health, is married, and has completed twelve grades of school. This is shown in Fig. 11-3.

Now, to do a proper job of matching, the first subject in the second group also should make $25,000 per year, should be 42, in good health, married, and a high school graduate. Every subject in the second group should be matched as closely as possible with a subject in the first group on all important variables. Whenever we match on a particular variable, we've reduced the potential error in the experiment.

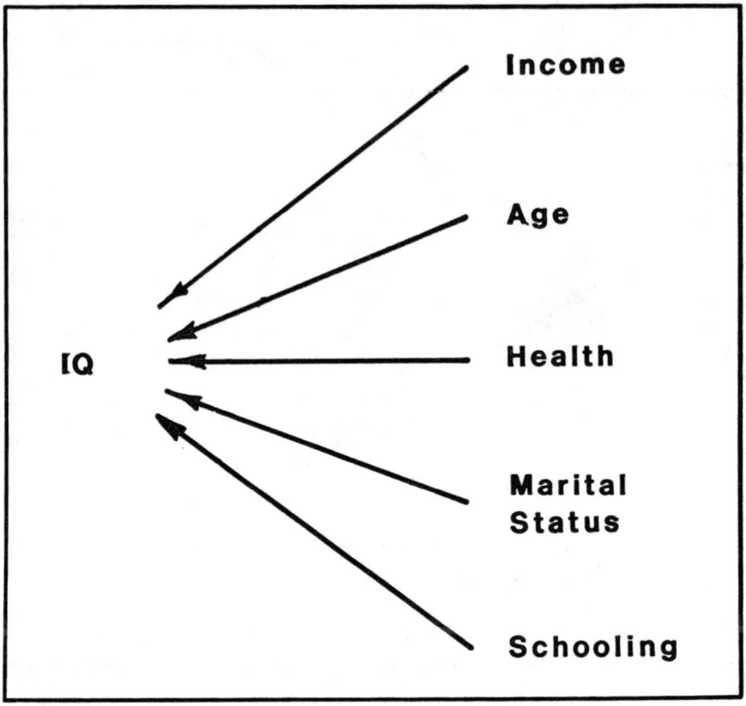

Fig. 11-3. Matching on five variables.

A perfect experiment would occur if the subjects were matched on every possible variation. In practice, this is impossible, for there are millions of variables on which to match people. Trying to match groups on even our five variables is a nasty job. Think of just how many friends you'd have to interview to find one who makes $25,000 per year, is 42, in good health, married, and who has completed exactly 12 years of schooling. The chances are that you know almost no one who meets all those criteria. You might have to interview thousands of people to find one in that category. Usually it's possible to match people on two or three very important variables, but to match them on more than that number becomes a very difficult task.

Thus, the ideal of perfect matching may well be unattainable. And yet the goal of science is to control all important variables. Thus we are in a quandary.

A SOLUTION

There is a clever solution to this problem. We can have two groups that match perfectly if we use the same person twice! That is, we use the same person once in each group. This is shown below:

Person	Group 1 Sleeping Pills	Group 2 Stimulant
A	102	104
B	101	101
C	100	102
D	99	103
E	98	100
	------	------
Average	100	102
SD	1.4142	1.4142

This seems almost too good to be true. By using the same person twice, we achieve almost perfect matching. This seems much too simple. There must be some disadvantages to such a procedure!

And of course there are. The principal problem lies in the fact that we must make very, very sure that what a person in Group 1 does does not affect his performance when he's a member of Group 2. Thus, we must be very careful to use different IQ tests for the two groups. Clearly, someone taking the same IQ test twice is probably going to be better at taking it the second time, and, indeed, may remember some of the items. Furthermore, he may be more comfortable in the second testing, after getting used to the experimental situation the first time. To get around this problem, we can have some subjects take sleeping pills first and stimulants second, while others take stimulants first and sleeping pills second.

With care, we can solve most of these problems, and retain the obvious advantage of the perfect matching obtained by using the same person. There's another advantage to this method, and perhaps this is the real reason that the paired t-test is so popular. We need only half as many people! That's a tremendous advantage in most situations. It doesn't sound like a lot of work to run 100 subjects in an experiment, but in practice it is a great deal of work indeed.

THE PAIRED t FORMULAS

To work out the paired t-test, we first calculate the difference between each person's pair of scores, as shown below:

Person	Group 1	Group 2	D
A	102	104	−2

Person	Group 1	Group 2	D
B	101	101	0
C	100	102	−2
D	99	103	−4
E	98	100	−2
		------	------
		ΣD	−10
		ΣD^2	28

In short, we simply subtract 104 from 102, yielding −2, 101 from 101, yielding zero, and so on. **It is essential that the subtractions always be made in the same direction.** It can also be done the other way, subtracting Group 1 from Group 2, but be certain that you are consistent about the direction in which you do the subtractions.

Once the differences, or D's, have been obtained, the rest is easy. You simply get the Average and Standard Deviation(s) of the D's, using the formula for s, as shown in Formulas 9-1 or 9-2. Then you get the Standard Error of the Mean, as shown in Formula 9-1. Then the Average of the D's is divided by the Standard Error, yielding the t.

I'd better explain that. First we get the s for the D's, as shown:

$$s_D = \sqrt{\frac{\Sigma D^2 - \frac{(\Sigma D)^2}{N}}{N-1}} \qquad \textbf{11-1}$$

$$= \sqrt{\frac{28 - \frac{(-10)^2}{5}}{4}} = 1.4142$$

Now, we must get the Standard Error for the D's, as follows:

$$s_{\bar{D}} = \frac{s_D}{\sqrt{N}} \qquad \textbf{11-2}$$

$$= \frac{1.4142}{\sqrt{N}} = 0.6325$$

Now, we know that the average D is (−10/5) or −2. To finish up, we divide the average D by the Standard Error:

$$t = \frac{\bar{D}}{S_{\bar{D}}} \qquad \textbf{11-3}$$

$$\frac{-2}{.6345} = 3.1623$$

Thus, the t in this case is 3.1623. Is it significant or not? To learn about this we must look it up in the table in Appendix B, or use Program 10-2. But first we must know the Degrees of Freedom. Notice here that the numbers that we're using are the differences between the pairs of scores. Consequently, the Degrees of Freedom in this case will be the number of pairs (or differences), minus one. There are 5 pairs, and so we have 4 Degrees of Freedom. With 4 DF, we find from Appendix B that we need a value of 2.78 for significance at the 95 percent level, and one of 4.60 for significance at the 99 percent level. Our value of 3.1623 is greater than 2.78, and hence the groups are really different if we're willing to accept 95 percent accuracy. However, 3.1623 is not greater than 4.60, and so we can't be 99 percent sure. Program 10-2 is much more convenient. It says that our t of 3.1623, with 4 Degrees of Freedom, is significant at the 96.70 percent level.

Notice that the difference between the groups, which was not significant if the people weren't matched, is now significant if we consider the groups to be paired. That's precisely what we had hoped for!

THE PAIRED t PROGRAM

Our microcomputers make it unnecessary to work out the problems by hand. The following program, which is a great deal like Program 3-6 (the old Standard Deviation program) will do the job:

```
10 REM PROGRAM 11.1. PAIRED T-TEST
20 READ X1,X2                   :REM GET DATA
30 IF X1=-1 THEN 90             :REM TEST FOR END
40 D=X1-X2                      :REM GET DIFFERENCE
50 S=S+D                        :REM SUM DIFFERENCES
60 Q=Q+D*D                      :REM SUM SQRD DIFFS
70 N=N+1                        :REM GET COUNT
80 GOTO 20                      :REM CONTINUE
```

```
90 V=Q-S*S/N                    :REM TOP OF FORMULA
100 SS=SQR(V/(N-1))             :REM GET S
110 SE=SS/SQR(N)                :REM GET SE
120 A=S/N                       :REM GET AVG DIFF
130 T=A/SE                      :REM GET T
140 DF=N-1                      :REM GET DF
150 PRINT "PAIRED T=",T,"WITH DF=",DF
160 DATA 102,104                :REM DATA TO USE
170 DATA 101,101
180 DATA 100,102
190 DATA 99,103
200 DATA 98,100
210 DATA -1,-1
```

The program is very similar to past programs; a flowchart is given in Fig. 11-4. Line 20 gets the data as two data points, X1 and X2. Line 30 checks for the end of the data. Line 40 computes D, as used in the Formulas 11-1 to 11-3 Lines 50 through 80 cumulate various needed sums, just as we've done many, many times before. Line 90 gets the top of the Standard Deviation equation, or the sum of little x squared. Line 100 gets s, and Line 110 the Standard Error of the Mean for D. Then the Average D is computed in 120, the t-value in 130, and the DF in 140.

Run the programs. You'll find that we're now in a position to do many paired t-tests, so that we can get a feeling for the way in which the test works. Again, notice that, depending on the direction of the difference, a negative t may be reported by the program. I like this feature, but you can remove it with an ABS statement, just as in the last chapter.

PAIRED t EXAMPLES

Jet Marketing Company makes a speedup kit for microcomputers that claims to make the units run more rapidly. The same program is run on a number of computers with and without the kit, yielding results as follows (the data in this and the following examples are, of course, fictitious):

Computer	Speedup	Normal
A	100	110
B	110	120
C	115	130
D	130	140
E	140	150
Average	119	130
	----	----
SD	14.2829	14.1421

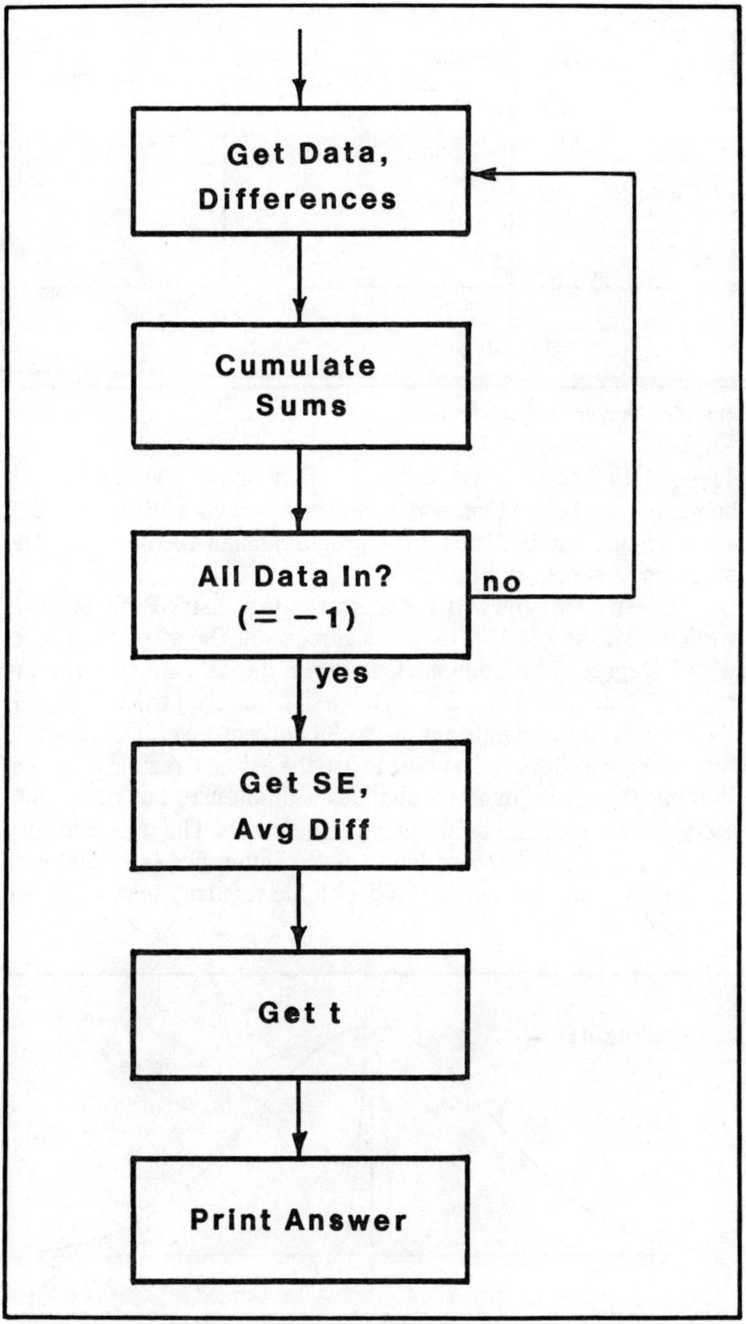

Fig. 11-4. Flowchart for paired t-test program.

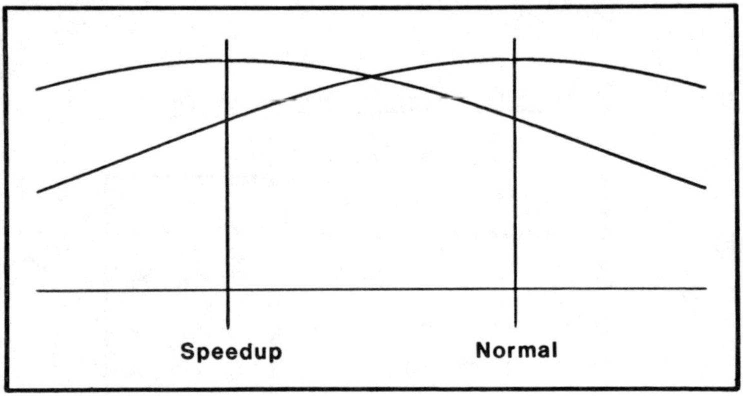

Fig. 11-5. Printer Speedup Kit t-test.

The question to be answered is whether or not the speedup kit helps. It would seem that it does, for the Average with the kit is 119 seconds, and without the kit the program takes 130 seconds. The situation is shown in Fig. 11-5.

If the data are run with the ordinary t-test, using Program 10-1, we find that the t is 1.0945, not significant at the 99 percent level with 8 Degrees of Freedom. However, if the data are run with the paired t-test, using Program 11-1, we find that the t has risen to 11! With 4 DF, that's significant at the 99 percent level. Obviously in this case it was to our advantage to use the paired t-test procedures. Without them, we found no significant difference, but taking into account for the matching, we found significance! The reason for this is simply that the Standard Error of the Difference is 1.0000 with the paired t-test, while it's 10.05 with the regular t-test, as shown in Fig. 11-6.

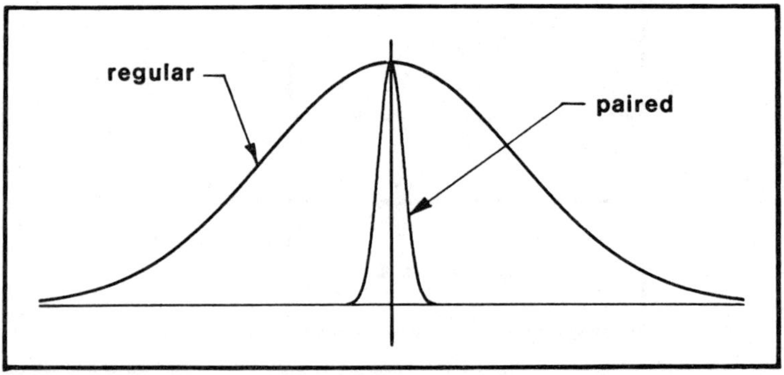

Fig. 11-6. Change in the Standard Error with paired t.

Unethical Pharmaceutical Company has developed some smart pills—pills that are said to increase one's intelligence. Scientists are of course skeptical of Unethical's claims, and conduct an experiment to determine the effect of the pills. The data are as follows:

Person	Control	Smart Pills
A	100	160
B	110	180
C	120	170
D	130	190
	----	----
Average	115	175
SD	11.1803	11.1803

The data, assuming that the data fit the Normal Curve, is shown in Fig. 11-7. Of course, in practice we should use bigger samples, but this remains a good example. If we run the data as a plain old t-test with Program 10-1, ignoring the pairing, we find that the t is only 6.573. With 6 Degrees of Freedom, that's significant at the 99 percent level. Hence, we conclude that there really is a difference between the groups.

Now run the data using Program 11-1. This tells us that if we consider the pairing, the t is now 14.6969. With 3 DF, the t is even more significant at the 99 percent level. Thus, the difference is significant, in either case.

Hank Slammer, the baseball player, feels that the spitball should be legalized, since it does indeed assist pitchers. He claims

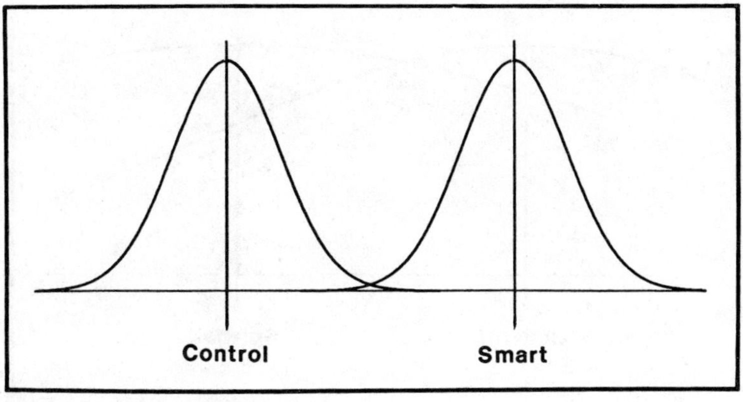

Fig. 11-7. Smart pills t-test.

that it makes the ball go faster. The umpires are skeptical of this claim, and run an experiment with many pitchers, yielding data as follows (speeds are in mph):

Pitcher	Control	Spitball
A	60	70
B	70	65
C	80	90
D	86	92
E	90	106
F	71	90
	----	----
Average	76.167	85.5
SD	10.2375	13.9254

The situation is depicted in Fig. 11-8.

Is Hank right? If we run the data with a normal t-test, using Program 10-1, we discover that he probably isn't, for the t is only 1.2075, not significant with 10 DF at even the 95 percent level. However, if we use the paired procedure, using Program 11-1, the paired t is 2.7119, significant at the 95 percent level with 5 DF. Again the matching assists us in finding a difference.

You may notice that we have fewer DF when using the paired t-test than when we use the ordinary t. Fewer DF should make it more difficult to find a difference, for the numbers needed from the t-table are higher—and that's the case, generally. But with experi-

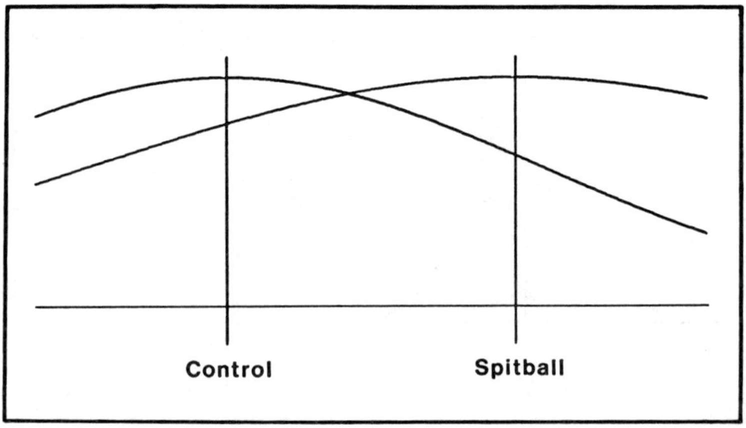

Fig. 11-8. Spitball t-test example.

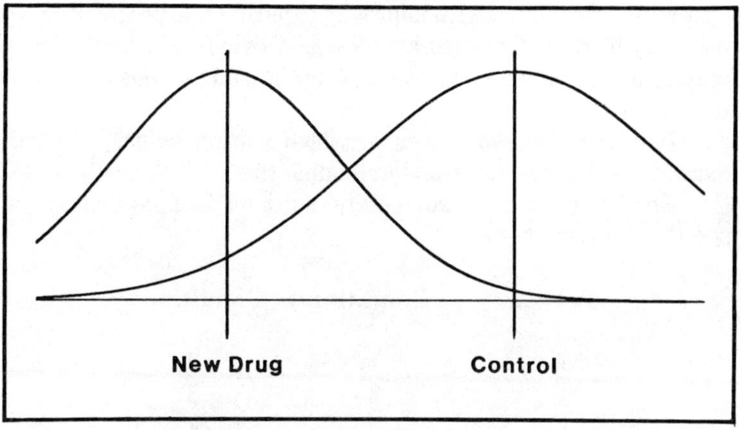

Fig. 11-9. Irregular heartbeat medicine t-test.

ence, you learn that you generally gain more than enough in the t-value to make up for the loss in DFs. Otherwise the paired-t procedure wouldn't be so popular!

"Aren't there any situations in which pairing doesn't help?" you query. It's hard to believe that any procedure would always make things easier. That is, of course, true. In the following situation, a drug company claims that a new drug will eliminate irregular heartbeats, and we'd like to see whether the drug works (really fictitious data!):

Subject	**New Drug**	**Control**
A	1	8
B	4	6
C	6	6
D	3	12
	----	----
Average	3.5	8
SD	1.8028	2.4494

In this example, the numbers are irregular heartbeats measured per second. The situation looks very much as shown in Fig. 11-9.

Is the medicine effective? If we do a simple t-test, using Program 10-1, we find that the t is 2.5627, significant at the 95 percent level with 6 DF. Thus, the drug helps with an unpaired t.

However, the paired t-value tells us a different story. Running Program 11-1, we find that the paired t is only 2.1412, not significant with 3 DF. Thus, with improved experimental procedures, we find that the drug doesn't seem to help. Obviously in a situation as important as this one, we want to be as sure as possible about things!

But what are the places in which pairing helps? The next example will assist us in understanding this.

This example uses completely fictitious data, so that we can see the situation better:

Person	Condition 1	Condition 2
A	1	2
B	2	3
C	3	4
D	4	5
E	5	6
	----	----
Average	3	4
SD	1.4142	1.4142

Running a normal t-test, with Program 10-1, we find that there is no significant difference, for the t is only 1. The t must always exceed 1.96, the Normal Curve value, as a minimum for signficance, and so the difference isn't significant. This is easily seen from the depiction of Fig. 11-10. It certainly doesn't look as though the difference should be significant. Perhaps using the paired t procedure will help. Try Program 11-1.

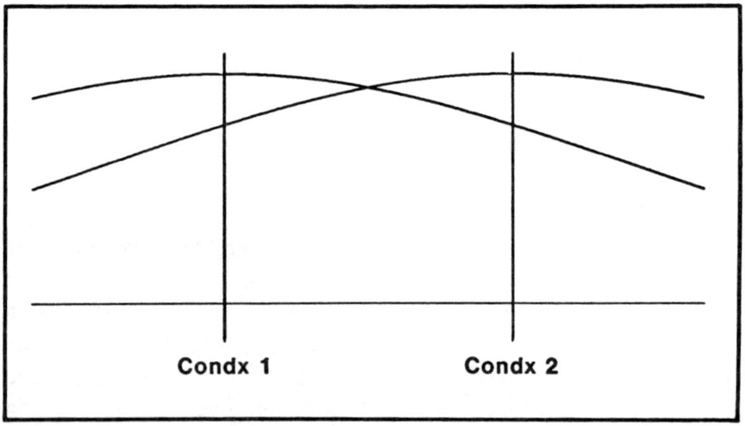

Fig. 11-10. Example of paired t that doesn't work.

Oops! Your microcomputer objects—Division by Zero Error. There must be something wrong with the program!

No, there isn't. Notice that the Correlation between the two groups is +1.00 (if you don't believe this, try Program 7-2). That is, the pairs of scores are perfectly correlated. The paired t-test, as we shall see in Chapter 15, actually uses the Correlation in the formula—and if the Correlation is perfect, the procedure won't work. It may very well blow up. Whether it will or not depends upon the precise nature of the data.

As a matter of fact, the Correlation is closely related to the effectiveness of the paired t procedure. In our examples in this chapter, we've had the following situations:

Example	Correlation	Result
First	+0.60	paired t helps
Example (1)	+0.99	paired t helps
Example (2)	+0.80	paired t helps
Example (3)	+0.84	paired t helps
Example (4)	−0.45	paired t hinders
Example (5)	+1.00	blow up!

Thus, if the Correlation is positive, and less than one, the paired t procedure is of use, but if the Correlation is negative, you're better off without the paired t. If the Correlation is a perfect positive one, it may not work at all.

In practice, when one does an experiment, the Correlation between the variables will almost always be a moderately positive one. Negative Correlations just don't seem to happen very often in the real world. And, of course, perfect positive Correlations don't happen either. In short, almost all of the time, we find that the paired-t procedure improves the accuracy of our experiments.

SIGNIFICANCE OF THE CORRELATION

In Chapter 7 we learned ways of computing Correlations, although we hadn't yet encountered any significance tests. Many times in statistical work it is useful to know whether a Correlation is significantly larger than zero. Take, for example, a Correlation of +0.28, taken on a sample of 50 people. A Correlation of +0.28 isn't very high. Perhaps it could have arisen purely by chance. How could we test for this?

For once, the procedure is a simple one. We simply use the following formula:

$$t = r \sqrt{\frac{N-2}{1-r^2}} \qquad 11\text{-}4$$

Putting in the Correlation of +0.28, and letting N=50 (for the 50 people), we have the following result:

$$t = 0.28 \sqrt{\frac{50-2}{1-(.28)^2}} = 2.02$$

Thus, the Correlation may be converted into a t of 2.02. We now look this value up in Appendix B, with Npairs −2 or 48 Degrees of Freedom. Appendix B tells us that with 48 DF we need values of 2.01 or 2.68. The t exceeds 2.01, so that we can be 95 percent sure that the Correlation is significantly bigger than zero. But it doesn't exceed 2.68, so we can't be 99 percent sure. You see, the procedure isn't very difficult. To make it even easier, Appendix C contains a table of values of the Correlation Coefficient needed for significance at the 95 percent and 99 percent levels. You simply look up the Correlation in question, along with the number of data pairs, and if your Correlation exceeds the value in the table, it's significant.

A CORRELATION SIGNIFICANCE PROGRAM

The table in Appendix C is handy, but it would be even handier to have your microcomputer evaluate the significance for you. After all, a computer generated the table, it seems senseless not to use the same technique with a computer at hand. The following program will do the job:

```
3000 REM PROGRAM 11.2. R SIGNIFICANCE
3010 INPUT "R=";R              :REM GET R
3020 INPUT "NPAIRS=";NP        :REM GET N OF PAIRS
3030 K2=NP-2                   :REM ADJUST TO DF
3040 K0=R*SQR(K2/(1-R*R))      :REM FORMULA 11.4
3050 K0=K0*K0:K1=1             :REM ADJUST VALUES
3060 GOSUB 4000                :REM GOTO DISTRIB
3070 PP=1-H                    :REM ADJUST H
3080 PRINT PP                  :REM PRINT ANSWER
3090 GOTO 3000                 :REM KEEP IT UP
4000 REM APPROXIMATION ROUTINE :REM BEGIN APPROX
4010 IF K0>1 THEN 4040         :REM BIG ENOUGH?
4020 PRINT "NOT SIGNIFICANT":REM SAY IF NOT
4030 RETURN
4040 K5=0.2222222              :REM SET ONE CONST
```

```
4050 K6=K5/K1:K7=K5/K2         :REM SET MORE
4060 K8=K0^0.3333333           :REM SET MORE
4070 K9=(1-K7)*K8-1+K6         :REM TOP
4080 Z=K9/SQR(K7*K8*K8+K6)     :REM GET Z
4090 IF Z>3 THEN 4110          :REM ADJUST IF <3
4100 Z=Z*(1.+0.08*Z^4/K2^3)    :REM ADJUST
4110 GOSUB 5000                :REM NORMAL APPROX
4120 RETURN
5000 REM NORMAL CURVE APPROXIMATION
5010 H=1+0.0498673470*Z        :REM CALCULATE
5020 H=H+.0211410061*Z*Z       :REM  NORMAL CURVE
5030 H=H+.0032776263*Z^3       :REM  APPRXIMATION
5040 H=H+.0000380036*Z^4       :REM  FROM PROGRAM
5050 H=H+.0000488906*Z^5       :REM  6.2
5060 H=H+.0000053830*Z^6
5070 H=H^-16/2                 :REM FINISH APPROX
5080 RETURN
```

Look carefully at this program. It bears a remarkable resemblance to Program 10-2, and works pretty much the same way. Line 3010 asks for the r, Line 3020 asks for the number of pairs. Formula 11-4 is calculated in lines 3030 and 3040, and the remainder is precisely like Program 10-2. We're going to use the routine beginning in line 4000 again and again in the remainder of the book. The concept of subroutines makes it very easy to change programs to meet new circumstances. See how simple their use makes programming?

If you give the program the r value of +0.28, and tell it that there are 50 pairs, it will tell you that the likelihood that the Correlation is greater than zero is 95.36 percent. That's extremely handy. Published tables, like that in Appendix C, never seem to have the precise values for Degrees of Freedom that one wants. This program will allow you to evaluate any value!

Let's try some other values. Put Program 11-2 in your computer, and do these, all taken from Appendix C:

r	**Number of People**	**Significance Level**
.273	52	.9524
.195	102	.9531
.254	102	.9903
.349	32	.9522
.997	3	.9513

In each case we should get either 95 percent or 99 percent, and the results are very, very close.

Be aware of one thing. Although we have said that the Correlations are greater than chance, or significant, that doesn't necessarily mean that they're of any use. For example, try a Correlation of 0.05, on a sample of 3000 people in Program 11-2. You'll find that this value is significant at the .9938 level. Yet recall from Chapter 7 that a Correlation of 0.05 is really so small as to be of no practical use. With large samples, many things will turn out to be significant that really are useless for prediction.

We've come to the end of this short chapter. Once again, no new BASIC concepts have been added to our repertoire, even to handle these complex new tests.

PROBLEMS

1. A company claims to make a substance that will make frogs jump better. Several frogs are tried in jumping contests, with and without the substance. Does it help? Work the problem with both paired and unpaired t procedures. Figures are distances in feet. (Unpaired t=1.3969, not significant. Paired t=3.5590, significance level at 95 percent level.)

Control	20	15	19	17	26
Drug	25	22	20	19	30

2. Rogers company makes a Citizens' Band Supercharger, said to increase the range of CB units. Several of these are tried with and without the Supercharger, with distances measured in miles. Does it make a difference? (Unpaired t=19.39, significant at 99 percent level. Paired t=14.52, significant at 99 percent level.)

Control	20	18	17	12
Supercharger	80	85	90	95

3. Group 1 consists of normal people, while those in Group 2 have been given a drug which temporarily induces mental retardation. Does the drug make a difference?

Group 1	100	70	120	116
Group 2	80	67	43	77

(Normal t, 2.4614, significant at 95 percent level. Paired t is 2.1873, not significant.)

4. Cracked Block Antifreeze Company makes cheap antifreeze. They have a new formula which is said to be far better than what they made before. Does it work? The figures are freezing points in Fahrenheit degrees. (Unpaired t=4.368, significance level 99 percent. Paired t=5.562, significance level 99 percent.) (Notice that if one figure in Group 1 were −1 degree, the program wouldn't work! Why?)

Regular	0	2	4	16
New	−12	−18	−16	−14

5. In a study of Correlation of 0.30 is achieved on a sample of 100 people. Is it significantly greater than zero? (Yes—significance level is 99.74 percent.)

6. In another study a Correlation of .99 is obtained on a sample of 3 people. Is this significant? (No, the significance level is only 90.7 percent.)

JOINT PROBABILITY??

12
Chi Square: Some New Programming Concepts

The sorts of statistical tests that we have used up to this point in the book have involved fairly precise data. In my field (psychology), we measure IQs to a precision of perhaps three digits, say 121. That isn't as accurate as the sort of measurements physicists obtain, which might be something like 121.2348, but it is more accurate than the type of measurement we're going to discuss in this chapter.

There is a kind of data, used in all sciences, which is low in accuracy by its very nature. This type of data is *frequency* data. Measurement of this sort involves only simple counts of things, and often very small counts, at that. Analysis of such data requires a new technique of analysis, and hence this chapter.

AN EXAMPLE OF FREQUENCY DATA

Let's jump right to the sort of problem that is usually analyzed with this form of data. Recall that a few years ago, the automobile companies were using advertisements like this: an advertising agency had a number of people drive two different cars. Each subject picked one of them as the best car. Then the manufacturer whose car had the highest number of votes usually ran an ad, with a table such as the following (all data is fictitious, of course):

Kaiser	Packard
400	0

(I'm using old car names, so I won't be sued!). From the data above, we discover that 400 people preferred the Kaiser, and no people at all liked the Packard. Would you have any difficulty in picking out the car that was liked best?

Of course not. Everyone preferred the Kaiser. There's no question that the car was liked best. Similarly, if a number of people were polled, and results like this were obtained:

Kaiser	**Packard**
0	400

there would be absolutely no difficulty in ascertaining people's preferences. They clearly preferred the Packard. Studies such as these might strike you as so simple as to be trivial.

The difficulty comes when things are not so clear-cut. Examine these data:

Kaiser	**Packard**
150	250

Now, which car is preferred? That decision is more difficult. It's clear that more people liked the Packard (250), but 150 people still preferred the Kaiser. That's a lot of people. If you were in charge of designing cars for Packard, there might be some things in the Kaiser car that you might want to include in your cars.

Even worse would be a situation like this:

Kaiser	**Packard**
190	210

Now, which car is liked best? That's very difficult. There's no question that the Packard was preferred, but there isn't much difference between the preferences. Clearly we need a method more sophisticated than simple inspection to make the decision.

"Why don't you use the t-test from Chapter 10?" you ask. "We used that to tell differences between groups." Ah, but for the t-test, we must have scores, or data like scores. Here we have only two numbers—190 and 210. Obviously, a new technique is needed.

THE CHI SQUARE TEST

The new technique is called Chi Square. Notice that that's not very good grammar, for it should be Chi Squared, but the test is universally referred to in that fashion. The formula for it is very simple:

$$\chi^2 = \Sigma \frac{(O - E)^2}{E} \qquad \textbf{12-1}$$

The terminology is very clear. O, the letter O (not zero) is an Observed frequency. E is an Expected frequency. Therefore, you simply take the Observed frequency for each group, subtract the Expected frequency from it, square the result, and divide by the Expected frequency. This is done for each group, and the big sigma sign means to add up the result for all groups.

The Observed frequency for the data just above is 190 for the Kaiser, and 210 for the Packard. That's all clear, but what are the Expected frequencies? That will take some thought.

To make the situation somewhat like the t-test, assume that we'd expect as a general rule that there wouldn't be any preference between the two groups. After all, even knowing nothing about any given two cars, we'd still expect that people would like them about the same. If there are 400 people to make the choices, and they are to be divided equally between the two groups, then each Expected Frequency would be 200. Then our problem would look like this:

	Kaiser	**Packard**
Observed	190	210
Expected	200	200

Putting the data into Formula 12-1, we'd have:

$$\chi^2 = \Sigma \frac{(0 - E)^2}{E}$$

$$= \frac{(190 - 200)^2}{200} + \frac{(210 - 200)^2}{200}$$

$$= 1.00$$

Thus, Chi Square is 1.00. That's easy enough. Now, what is done with the result?

That's simple enough. We look it up in Appendix D, in the back of the book. To do this, we need the number of Degrees of Freedom.

For problems such as this one, where the Expected frequencies are assumed to be equal, the Degrees of Freedom are calculated as the number of groups minus one. We have two groups, and therefore DF = 2 − 1 = 1. From Appendix D, we discover that the values needed for 95 percent and 99 percent certainty are 3.84 and 6.64. If our value of 1.00 exceeds either of these values, we can be 95 percent or 99 percent sure that the Observed frequencies differ.

However, as we see, 1.00 is not larger than 3.84 or 6.64. Consequently, we have no real reason for thinking that preferences for the two groups are really different.

A SIMPLE CHI SQUARE PROGRAM

As you can see, calculating Chi Square is really simple, but, lazy as I am, I'd prefer to have my microcomputer do it for me. The following program will do the job.

```
10 REM PROGRAM 12.1. SIMPLE CHI SQUARE
20 READ O,E                          :REM GET DATA
30 IF O=-1 THEN 80                   :REM TEST FOR END
40 X2=(O-E)^2/E                      :REM GET CHI SQ
50 S=S+X2                            :REM CUMULATE IT
60 N=N+1                             :REM COUNT ENTRIES
70 GOTO 20                           :REM CONTINUE
80 DF=N-1                            :REM GET DF
90 PRINT "CHI SQ=",S," WITH DF=",DF :REM ANSWER
100 DATA 190,200                     :REM DATA TO USE
110 DATA 210,200
120 DATA -1,0
```

The program works as shown in Fig. 12-1. The Observed and Expected frequencies are read from the DATA statements in line 20. Line 30 checks to see whether the data is all in or not, looking for a −1 as usual. Formula 12-1 is calculated in line 40, and the Chi Square values summed in line 50. Line 60 keeps count of the number of pairs of values entered. Line 70 jumps back to get more data. The Degrees of Freedom for this type of problem are calculated in line 80, and the result printed in line 90. It's a very short program.

Run it. It gives the correct answer, which is one. Simply as a convenience, it also gives the Degrees of Freedom for this particular problem. Many other sorts of Chi Square problems may be worked with this simple program, but be warned that in other problems the DF may well be different.

Program 12-1 is an acceptable way of doing things, but to use it

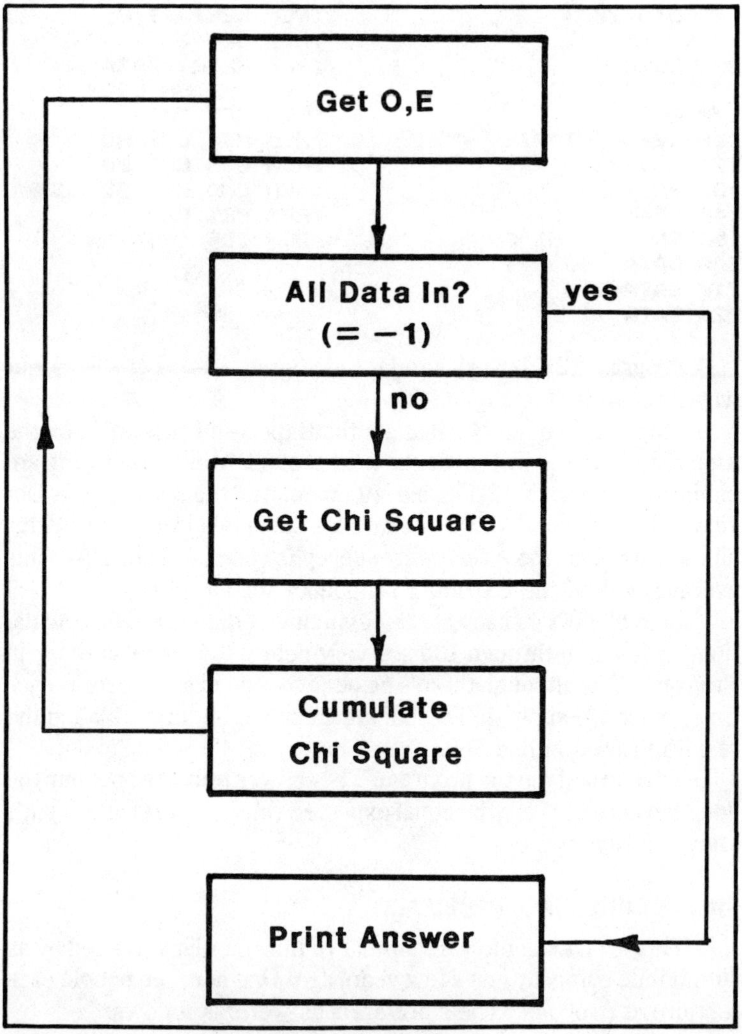

Fig. 12-1. Flowchart for simple Chi Square program.

we must calculate the expected values first. This is a nuisance. Program 12-2 does the whole job for us at one time:

```
10 REM PROGRAM 12.2. AUTOMATIC CHI SQ
20 READ O                           :REM GET DATA
30 IF O=-1 THEN 70                  :REM TEST FOR END
40 T=T+O                            :REM CUMULATE SUM
50 N=N+1                            :REM COUNT ENTRIES
```

```
60 GOTO 20                          :REM KEEP IT UP
70 E=T/N                            :REM GET E
80 RESTORE                          :REM RESET DATA
90 FOR J=1 TO N                     :REM CHI SQ LOOP
100 READ O                          :REM GET DATA
110 X2=(O-E)^2/E                    :REM GET CHI SQ
120 S=S+X2                          :REM SUM CHI SQ
130 NEXT J                          :REM END CHI SQ LOOP
140 DF=N-1                          :REM GET DF
150 PRINT "CHI SQ=",S," WITH DF=",DF :REM ANS!
200 DATA 190
210 DATA 210
220 DATA -1
```

Program 12-2 is very similar to Program 12-1. Let's work our way through it (see Fig. 12-2).

Lines 10 through 70 calculate the Expected Frequency, on the assumption that it will be equal in all groups. Most statements are similar to Program 12-1. Line 40 cumulates the total of the Observed Frequencies, line 50 cumulates the number of groups. After all the data is in, the Average frequency is obtained in line 70. This average will be the Expected Frequency for all groups.

Line 80 puts us back at the beginning of the data. The calculations in lines 90 through 130 are very nearly the same as those in Program 12-1, although there's no need to read in a value for E each time, since we know it. The DF are calculated in line 140, and the results printed in line 150.

Type in and run the program. The answer remains one, but you don't have to calculate the equal expected values beforehand. That's a great advantage.

SIMPLE CHI SQUARE PROBLEMS

This is an example very similar to the one that was used by an automobile company just a few years ago. One hundred people each test-drove two cars. Their preferences were as follows:

Cord	Reo
60	40

On the assumption that the Expected Frequency would be equally divided between the two groups, the expected values would each be 50:

	Cord	Reo
Observed	60	40
Expected	50	50

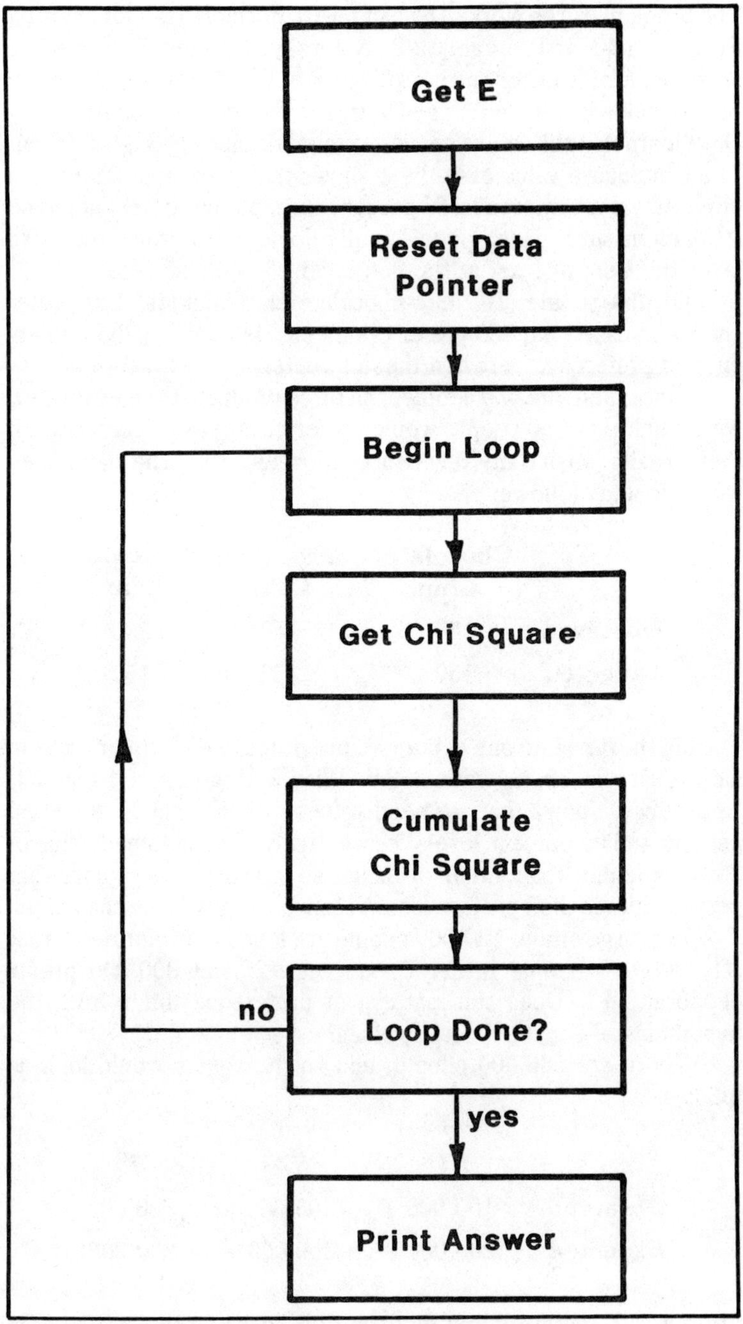

Fig. 12-2. Flowchart for automatic Chi Square program.

The problem can be worked by hand, with Formula 12-1, or by using Program 12-1 or Program 12-2. Naturally, I prefer the computer program, for it's faster and more accurate. We discover that the Chi Square value is 4.00, with one Degree of Freedom. From Appendix D we learn that the values needed for significance are 3.84 and 6.64. The Chi Square value of 4, then, shows that there appears to be a difference if we want to be 95 percent sure, but not if we want to be 99 percent sure. Thus, people really do seem to prefer the Cord Over the Reo, at least at the 95 percent level.

Of 450 people, 100 like chocolate chip cookies, 150 prefer chocolate cake, and 200 prefer cream pie. Is there really a difference in preference between these sweets?

Since there are 450 people, and three products to be evaluated, we'd think that 150 people would prefer each type of confection if there really wasn't any difference in preference. The data, then would look as follows:

	Chocolate Chip	Chocolate Cake	Cream Pie
Observed	100	150	200
Expected	150	150	150

Putting the data into one of our two programs, we discover that the Chi Square is 33.33, with 2 DF. With 2 Degrees of Freedom, Appendix D shows that we need values of 5.99 or 9.21 at the 95 percent and 99 percent levels, respectively. The obtained value of 33.33 is higher than either of these, so we can be very sure that people differ in their preferences. Cream pie really is the champion.

In a large study, 100,000 people are found to prefer deodorant #1, 200,000 people prefer deodorant #2, and 300,000 prefer deodorant #3. Does this pattern of preference differ from the hypothesis of equal expected values?

There are 600,000 people, and so the values would look as follows:

	#1	#2	#3
Observed	100,000	200,000	300,000
Expected	200,000	200,000	200,000

Program 12-2 will tell you that the Chi Square in this situation is

10,000. With two DF, we again need 5.99 and 9.21, so that we can be very, very certain that there is a difference in preference. This problem is a difficult one to work out by hand: most inexpensive calculators don't like numbers as large as those you'll encounter with this problem.

Here's a problem that will show you the virtues of doing problems on your microcomputer, for the numbers are quite nasty. Of 17,692 people voting in an election, 7,653 prefer the Democrats, 4,971 prefer the Republicans, 2,346 prefer the Socialists, and 2,722 prefer the Populists. Is there a difference in preference beyond the assumption of equal linking? The data looks like this:

	Democrats	**Republicans**	**Socialists**	**Populists**
O	7653	4971	2346	2722
E	4423	4423	4423	4423

Program 12-2 tells us that the Chi Square in this case is 4,056.1913. With 3 DF, we need values of 7.81 and 11.34, and thus we can be much more than 99 percent sure that there is a difference in preference.

It's probably a good idea to give an example in which there is no difference in preference at all. Ninety cats are tested for preference for three cat foods, and each cat food is preferred by 30 cats, as shown:

	Ham	**Beef**	**Fish**
Observed	30	30	30
Expected	30	30	30

Running these data through Programs 12-1 or 12-2 we find that the Chi Square value is zero. That's what you'd expect, and the value is certainly less than the 5.99 or 9.21 from Appendix D, and so we must conclude that the cats like all these foods equally.

SMALL EXPECTED FREQUENCIES

In using Formula 12-1, you must make very certain that the Expected Frequencies don't become very small. Look at Formula 12-1. If E becomes 0, then the Chi Squared value will become infinitely large. Generally, one doesn't get tiny Expected Frequencies with the simple Chi Squared problems that we've been discussing, but the problem of small Expected Frequencies will be an issue in what follows. In general, calculation of a Chi Squared value in which an Expccted value less than 5 or 10 occurs is considered improper.

TWO-WAY CHI SQUARE

Assume that you have 200 people; 92 are old and 108 are young. Of the same 200 people, 96 like big-bad swing music and 100 like rock and roll. The situation looks as shown in Fig. 12-3. The question to be answered is that of whether there is some sort of relationship between the sort of music one likes, and one's age. Casual inspection certainly seems to show that, for 76 old people like swing, and only 16 old people like rock. Conversely, only 20 young people like swing, while 88 young people like rock.

"Let's do a Correlation between age and music preference," says the unwary reader. "Simply use Formula 7-2, and the job is done." Aha, just try it! Formula 7.2 requires scores, and all that we have in this problem is frequency data. Formula 7-2 won't work. What can we do?

To repeat, the data in Fig. 12-3 is frequency data. Thus, the values in the table must be Observed Frequencies. In short, Fig.

	OLD	YOUNG
SWING	76	20
ROCK	16	88

Fig. 12-3. A two-way Chi Square example.

	OLD	YOUNG
SWING	O=76	O=20
ROCK	O=16	O=88

Fig. 12-4. Observed frequencies in two-way Chi.

12-3 may be redrawn as Fig. 12-4. Thus, we have Observed Frequencies in our two-dimensional table. If we could only arrive at some Expected Frequencies for each cell, we could calculate a Chi Square value for each cell, add them up, and deal with them just as we did in the previous problem. How to arrive at the Expected Frequencies is the subject of the next section.

EXPECTED FREQUENCIES IN TWO-WAY TABLES

To calculate the Expected Frequencies we need a bit of probability theory. Although the theory of probability underlies all of statistics, we haven't talked about it to any extent in this book, for it is a very tricky business, and an extended discussion is not necessary in a simplified text such as this one. But we can no longer avoid it.

First of all, what is the probability of being old? Well, of the 200 people in our group, 92 are old. Thus the probability of being old is 92/200 or 0.46.

Next, what's the probability of liking swing? Again, of the 200 people in the group, 96 like swing. The probability of liking swing, then, 96/200 or 0.48. Up to this point the probabilities are:

$$p(\text{old}) = 0.46 \qquad p(\text{swing}) = 0.48$$

Now, somehow from these values we want to arrive at the Expected Frequency for each cell. We'd like the Expected Frequencies to be set up so that they would reflect the appearance of the table if there were no relationship between age and preference for music. It is here that a bit of probability theory comes to our rescue.

Look at the upper left hand cell of Fig. 12-2. This is the cell for old people who like swing music. In formal terms, this is the cell for old people and swing music.

There is a law of probability that states that, if the variables are independent, or not related in any way, we can get the probability of both together, or the joint probability, as it is called, by simply multiplying the two probabilities together.

$$p(\text{joint}) = p(\text{old}) * p(\text{swing}) = 0.46 * 0.48 = 0.2208$$

Thus, the probability of being old and liking swing at the same time is 0.2208. That's under the assumption, which we need for our Expected Frequencies, that there's no relationship between age and preference for music.

However, the value we just calculated is a probability. Figure 12-2 is in terms of frequencies. How can we get from the probability to the frequency? We obtain it simply by multiplying the result by the total number of people. In short, the Expected Frequency for the upper right-hand cell is as follows:

$$0.2208 \times 200 = 44.16$$

I realize that this seems (and is) very complicated. In practice, however, to find the Expected Frequency for any cell, you simply multiply the total number of people in one category by the total number of people in the other category, and divide by the total number of people.

$$\begin{aligned} \text{Expected Freq.} &= \frac{\#\text{old} \times \#\text{swing}}{\text{total } \#} \\ &= \frac{92 \times 96}{200} \\ &= 44.16 \end{aligned} \qquad \textbf{12-2}$$

If you're totally lost by this procedure, don't worry, for very shortly we're going to develop a computer program to do the whole job automatically.

For the upper right-hand cell, then, the young swing lovers, we multiply 108 by 96 and divide by 200, yielding an Expected Frequency of 51.84. The remainder of the Expected Frequencies is calculated in the same fashion, resulting in Fig. 12-5. In that table, we have, then, the Observed and Expected Frequencies. All we need to do is use Formula 12-1 to complete the job:

$$\chi^2 = \Sigma \frac{(0 - E)^2}{E}$$

$$\frac{(76-44.16)^2}{44.16} \quad \frac{(20-51.84)^2}{51.84} +$$

$$\frac{(16-47.84)^2}{47.84} + \frac{(88-56.16)^2}{56.16}$$

$$= 81.75$$

	OLD	YOUNG	
SWING	O=76 E=44.16	O=20 E=51.84	96
ROCK	O=16 E=47.84	O=88 E=56.16	104
	92	108	200

Fig. 12-5. Expected frequencies in two-way Chi.

Thus the Chi Square for that table (they're always called "Contingency Tables") is 81.75. The Degrees of Freedom for such a table are found by multiplying the number of rows minus one by the number of columns minus one. Thus we have DF = (2−1)*(2−1)=1. From Appendix D, we find that the values needed for significance are 3.84 and 6.64, and so we may say that there is some sort of relationship between age and music preference, since 81.75 exceeds both values.

The term most often used to express the relationship between two variables in a Contingency Table is *association*, and one says that they are *associated*. Obviously, we can't use the term correlated, since a real Correlation isn't possible in that situation. Consequently, the term association is often used.

Notice, also, that working out one of these tables requires a lot of effort. This would be especially true if the table were, say, four columns wide, and five rows high! Fortunately, our microcomputer will come to the rescue.

A CONTINGENCY TABLE PROGRAM

The following program will calculate the value of Chi Square for a Contingency Table. It involves some new programming concepts, so look at it carefully!

```
10 REM PROGRAM 12.3. CONTINGENCY TABLE
20 DIM IN(6,6),R(6),C(6)      :REM SAVE SPACE
30 READ NR,NC                 :REM GET SIZE
40 FOR J=1 TO NR              :REM ROW LOOP
50 FOR K=1 TO NC              :REM COLUMN LOOP
60 READ IN(J,K)               :REM GET DATA
70 T=T+IN(J,K)                :REM GET GRAND TOTAL
80 NEXT K                     :REM END COLUMN LOOP
90 NEXT J                     :REM END ROW LOOP
100 FOR J=1 TO NR             :REM ROW LOOP
110 R(J)=0                    :REM ZERO ROW SUMMER
120 FOR K=1 TO NC             :REM COLUMN LOOP
130 R(J)=R(J)+IN(J,K)         :REM GET ROW SUM
140 NEXT K                    :REM END COLUMN LOOP
150 NEXT J                    :REM END ROW LOOP
160 FOR K=1 TO NC             :REM COLUMN LOOP
170 C(K)=0                    :REM ZERO COL SUMMER
180 FOR J=1 TO NR             :REM ROW LOOP
190 C(K)=C(K)+IN(J,K)         :REM GET COL SUM
200 NEXT J                    :REM END ROW LOOP
210 NEXT K                    :REM END COL LOOP
220 FOR J=1 TO NR             :REM ROW LOOP
230 FOR K=1 TO NC             :REM COL LOOP
240 E=R(J)*C(K)/T             :REM EXPECTED VALUE
```

```
250 X2=(IN(J,K)-E)^2/E          :REM CHI SQUARE
260 S=S+X2                      :REM GET SUM
270 NEXT K                      :REM END COL LOOP
280 NEXT J                      :REM END ROW LOOP
290 DF=(NR-1)*(NC-1)            :REM GET DF
300 PRINT "CHI SQ=",S," WITH DF=",DF
400 DATA 2,2                    :REM DATA TO USE
410 DATA 76,20
420 DATA 16,88
```

You may think that this is a frightening problem. There are some new ideas, loops within loops, and so on, but when you analyze it, you'll discover that it isn't so bad. It's illustrated in Flowchart 12-6.

Line 20 immediately introduces a new concept. We've encountered DIM statements before, in Chapter 5. Recall that they tell the computer to save some space, and it's not difficult to see that R(6) means save 6 spaces for row sums. Similarly, C(6) means save 6 spaces for column sums. Recall from the previous example that we need the row and columns sums to get the Expected Frequencies.

But what about IN(6,6)? This statement means save space for a table with 6 rows and 6 columns. A table of this type is shown as Fig. 12-7, although it shows only a 4 by 4 table. In Program 12-3, we'll adopt the convention that the first number in the IN variable is the row, and the second number is the column, sort of like IN(R,C). For each R and each C combination, there is a unique location in the table, and one must be very, very careful not to get the location of R and C mixed up. Notice that the variable, IN, uses up 36 memory locations. Actually, in most microcomputers, each decimal point number is stored in four or more bytes, so that the statement has really taken 144 or more storage locations. One must be very careful not to make the dimensions of IN too large. Were we to permit a Dimension (DIM) statement of IN(100,100), the table would take up 10,000 locations, or, really, 40,000 bytes in most microcomputers. That is very likely more room than you have in your computer. At any rate, line 20 sets up the table.

Line 30 reads in the number of rows and the number of columns. The program must know these values, and puts them into variables NR and NC.

Lines 40 through 90 read in the data, putting it into the table associated with variable IN. We want to read each row first (that makes sense), and the outer loop, starting in line 40 and ending up in line 90 handles the rows with variable J. Then, within each row, we must scan across the column, and that is handled by the loop starting

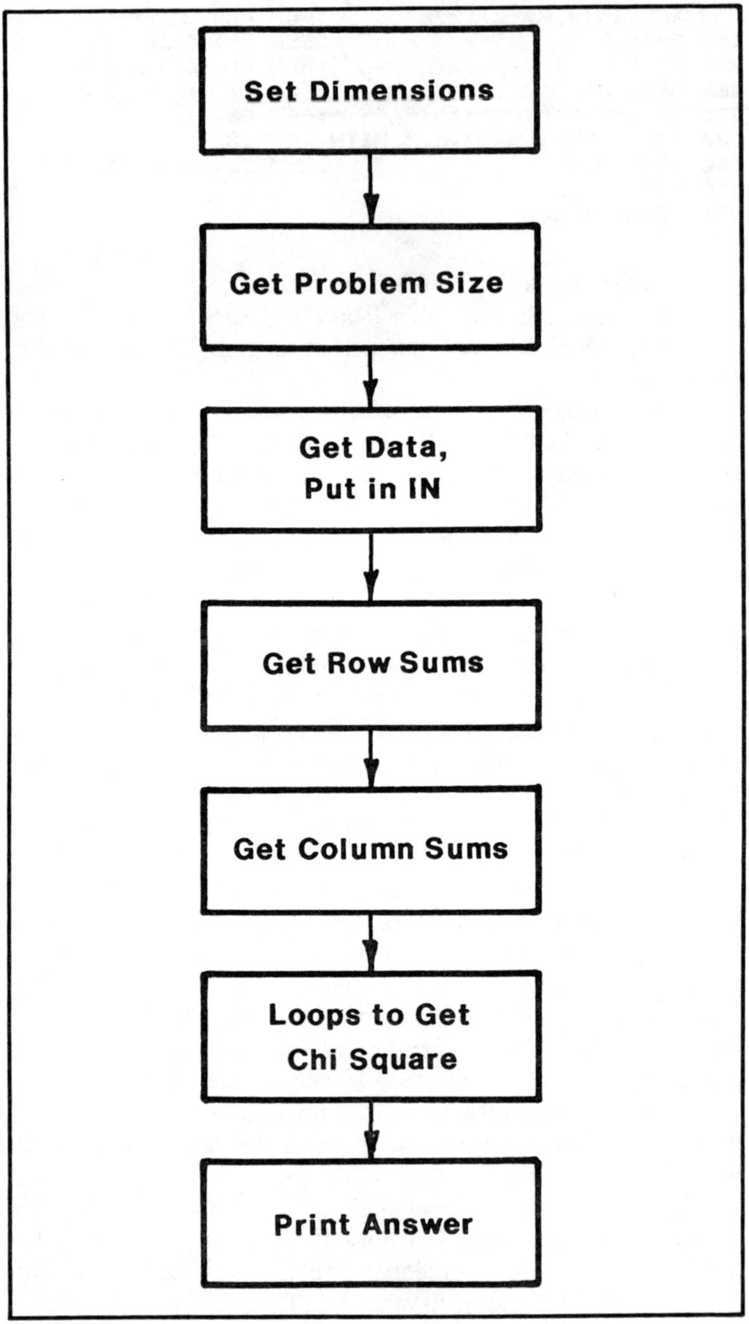

Fig. 12-6. Flowchart for Chi Square contingency table program.

		COLUMNS			
		1	2	3	4
ROWS	1	(1,1)	(1,2)	(1,3)	(1,4)
	2	(2,1)	(2,2)	(2,3)	(2,4)
	3	(3,1)	(3,2)	(3,3)	(3,4)
	4	(4,1)	(4,2)	(4,3)	(4,4)

Fig. 12-7. An example of dimensioning.

in line 50 and ending in line 80, with variable K. Line 60 reads in the data, putting it into the table IN(J,K)—remember, in these loops, J is rows, and K columns. In line 70, for good measure, we get the total of all the numbers in T. This could be gotten in a number of places, but it's handy to have it here.

Thus, when line 90 has ended, all the data has been read in, and stored in the table, IN.

To work out the problem, we must have the row and column totals. So, in lines 100 through 150, we get the sum for each row. Consequently, it makes sense that the first loop, starting in line 100, scans across rows, as shown in Fig. 12-8. Then, in line 110, a place to do the summing is cleared. Clearing probably isn't necessary in your computer (it is in my Sharp!), but it highlights the location into which the summing will be done. We want the sum for each row to wind up in R(J), so why not use it for the summing variable? Then, in lines 120 to 140, the summing for each row is done by adding up the values across the columns, with the row value constant. That's a mouthful. You may have to read through it a number of times to figure out what was said. When lines 110 through

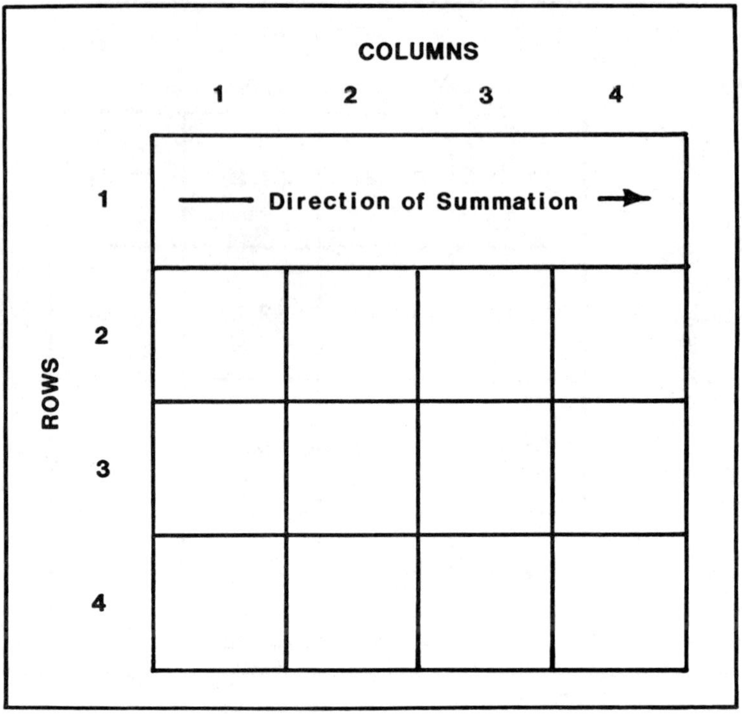

Fig. 12-8. Summation across rows.

150 have completed their work, we have the row sums in R(1), R(2), . . ., and so on up to the last row needed.

The operation of these summing loops is terribly important. Study them if you don't understand them at first, and, if necessary, put in some print statements so that you can watch their operation. If you're an alert reader, you may become aware that the row sums could have been obtained while the data was being read in, in lines 40-90. But I prefer to do one thing at a time in a book of this sort.

The column sums are gotten in lines 160-210, in precisely the same fashion (see Fig. 12-9). They're put in variable C, but if you can understand the row sum loops, these should be easy also.

Calculation of the Chi Square value is found in lines 220 through 280. Lines 220 and 230 scan through the table. Line 240 calculates the Expected Frequency for each cell by multiplying the proper row and column sums together, and dividing by the total. The Chi Square is calculated in line 250 by Formula 12-1, and summed in line 260. Lines 270 and 280 simply end the loops. Line 290 calculates the DF, and line 300 prints the results.

That looks frightening. Run the program. You'll find that it gives you the Chi Square value that took ten minutes of hand computation in the twinkling of an eye. Better yet, it will do much larger tables that are almost impossible to do by hand. If you'd like to view the Expected Frequencies as they are calculated, add line 245 to the program.

```
245 PRINT E    :REM PRINT EXP FREQ
```

TWO-WAY CONTINGENCY TABLE EXAMPLES

It is instructive to have some examples of pure cases, as usual with fictitious data. Look at Fig. 12-10. Notice that the frequencies in the cells are all equal (we've retained the old labels). Because of this, one would expect the Chi Square value to be very low. Run the data in Program 12-3. You'll find that the value is 0.00. That's not surprising, for it's evident to the eye that there isn't any association between the variables. Appendix D shows that with 1 DF, we need values of 3.84 (95 percent) and 6.64 (99 percent) for significance.

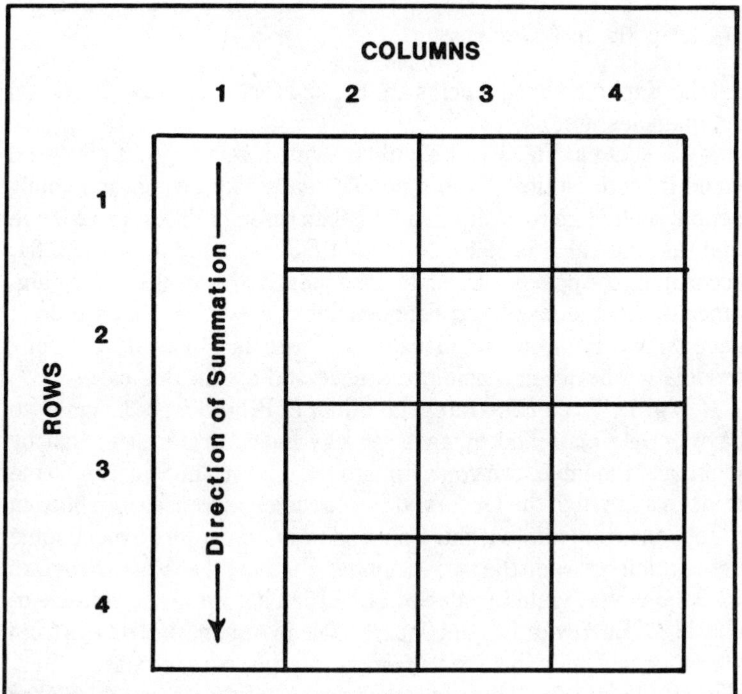

Fig. 12-9. Summation down columns.

	OLD	YOUNG
SWING	80	80
ROCK	80	80

Fig. 12-10. No Chi Square at all!

All the Expected Frequencies are 80, and that's what the Observed Frequencies are.

The data in Fig. 12-11 should provide a very high Chi Squared value indeed. Notice that old people really like swing, and young people really like rock. Running the data through Program 12-3, we find that the value is 1940.79. With 1 DF, we need 3.84 and 6.64, according to Appendix D, and consequently the value is very significant. All the Expected Frequencies are 505, so that we don't have to worry about small values. There is obviously a strong association between music preference and age in this case.

Fig. 12-12 demonstrates the utility of Program 12-3. Again we have people classified by ages, and by music preference. Just by looking at the table, can you estimate the size of Chi-Square? To me it looks as though the Observed Frequencies increase from bottom to top, and from left to right. Consequently, you might expect some association between the two variables. Running the data in Program 12-3, however, yields a value of 11.8815. That value seems reasonably high. However, the Chi Square table in Appendix D differs from our other tables, in that the values grow larger with increasing DFs. With 9 DF, we find that we need values of 16.92 and 21.67 for significance. Thus, Fig. 12-12, which looks as though an association

	OLD	YOUNG
SWING	1000	10
ROCK	10	1000

Fig. 12-11. A very high Chi Square.

	AGE			
	10–19	20–29	30–39	40–49
CLASSICAL	250	300	350	400
COUNTRY	200	250	300	350
SWING	150	200	250	300
ROCK	100	150	200	250

Fig. 12-12.A difficult two-way Chi Square.

	COFFEE	TEA	COLA	ORANGE
FEMALE	500	600	700	800
MALE	100	200	300	400

Fig. 12-13. A 2 × 4 Chi Square example.

is present, in point of fact has none. Notice that the computer saves an enormous amount of time here. Without it, you'd have to calculate sixteen Expected Frequencies, and sixteen components to the Chi Square. That would be a formidable task. All Expected Values are above 400.

Figure 12-13 shows another problem that would be difficult to compute. Here we have beverage preference versus sex, again with fictitious data. Again it looks as though there should be a significant Chi Square, for the data appears to proceed in an orderly fashion from left to right. Use Program 12-3, or work it out by hand if you have the stamina, and you'll discover that the Chi Square value is 60.2931. Appendix D shows that with 3 DF, we need 7.81 (95 percent) or 11.34 (99 percent) for significance. Consequently, we may be 99 percent sure that there is an association between sex and beverage preference.

Figure 12-14 is an example constructed specifically to yield a

	COFFEE	TEA	COLA	ORANGE
FEMALE	501	503	499	504
MALE	490	510	501	503

Fig. 12-14. A tiny Chi Square.

	M	F
SKIRTS	90	10
PANTS	10	90

Fig. 12-15. A 2 × 2 table ordered one way.

	F	M
SKIRTS	10	90
PANTS	90	10

Fig. 12-16. The same table ordered another way.

very low Chi Square. The categories are the same as last time. Notice that the values in the table are all very similar to one another. Running the data in Program 12-3, we discover the Chi Square value is only .1732. Again, with 3 DF we need 7.81 (95 percent) or 11.34 (99 percent). Clearly, there is no association between the two dimensions in this case, as one would expect.

Finally, look at Fig. 12-15 and 12-16. Notice that the tables are the same, except for the way in which they're ordered. It may not be obvious that both tables should yield the same value of Chi-Square. Run both in Program 12-3. You'll discover that the value calculated in both cases is 128. With 1 DF, one needs values of 3.84 (95 percent) and 6.64 (99 percent). Clearly, both tables are significant, and there is an association between dress preference and sex. A little thought will show that even though the tables are arranged differently, the information in them should be the same, and hence the answer.

A CHI SQUARE SIGNIFICANCE TEST PROGRAM

As in previous chapters, it would be handy to have a program that would give us the significance level of various Chi Square values directly. Program 12-4 is just such a program:

```
3000 REM PROGRAM 12.4. CHI SQUARE SIGNIFICANCE
3010 INPUT "CHI SQ=";X2          :REM GET CHI SQ
3020 INPUT "DF=";K1              :REM GET DF
3030 K0=X2/K1                    :REM TRANSFORM
3040 K2=100000                   :REM VERY LG #
3050 GOSUB 4000                  :REM GOTO DISTRIB
3060 PP=1-H                      :REM ADJUST H
3070 PRINT PP                    :REM PRINT ANSWER
3080 GOTO 3000                   :REM KEEP IT UP
4000 REM APPROXIMATION ROUTINE :REM BEGIN APPROX
4010 IF K0>1 THEN 4040           :REM BIG ENOUGH?
4020 PRINT "NOT SIGNIFICANT":REM SAY IF NOT
4030 H=1:RETURN
4040 K5=0.2222222                :REM SET ONE CONST
4050 K6=K5/K1:K7=K5/K2           :REM SET MORE
4060 K8=K0^0.3333333             :REM SET MORE
4070 K9=(1-K7)*K8-1+K6           :REM TOP
4080 Z=K9/SQR(K7*K8*K8+K6)       :REM GET Z
4090 IF Z>3 THEN 4110            :REM ADJUST IF <3
4100 Z=Z*(1.+0.08*Z^4/K2^3)      :REM ADJUST
4110 GOSUB 5000                  :REM NORMAL APPROX
4120 RETURN
5000 REM NORMAL CURVE APPROXIMATION
5010 H=1+0.0498673470*Z          :REM CALCULATE
5020 H=H+.0211410061*Z*Z         :REM  NORMAL CURVE
```

```
5030 H=H+.0032776263*Z^3      :REM  APPROXIMATION
5040 H=H+.0000380036*Z^4      :REM  FROM PROGRAM
5050 H=H+.0000488906*Z^5      :REM  6.2
5060 H=H+.0000053830*Z^6
5070 H=H^-16/2                :REM FINISH APPROX.
5080 RETURN
```

This program is very similar to Programs 10-2 and 11-2. You simply give it a Chi Square value and the Degrees of Freedom, and it provides a close estimate of the significance level. Given the following values from Appendix D, Program 12-4 gives the following values:

Chi Square	DF	Significance level
3.84	1	.9527
18.31	10	.9503
26.30	16	.9502
38.88	26	.9500

As you can see, the values are rather close to 95 percent in each case. Of course, since an approximation is used, the values are not as accurate as those in Appendix D, but they're close enough to be usable. As a demonstration of the use of the Program with the data in our chapter, take the data of Fig. 12-5. The Chi Square there was 1940.79, with 1 DF. Program 12-4 says that we can be 100.00 percent certain there that the Chi Square is significant. That's a bit extreme, but clearly the significance level is very high.

In this chapter, we finally have a new programming concept—that of doubly subscripted variables. It's the first one in a long, long time, but it was absolutely necessary for the programs in this chapter.

PROBLEMS

1. Thirty-four cats like fish. Twenty-four cats like beef. Twenty-one dogs like fish. Thirty-six dogs prefer beef. Evaluate the association between animal type and food preference in this group of 115 animals (two-way problem). (Chi Square = 5.4644, significant at 95 percent level).

2. Twelve people like marmalade. 100 people prefer jelly. Do these frequencies differ from the hypothesis of equal Expected

Frequencies (one-way problem). (Chi Square = 69.1429, significant at 99 percent level).

3. Twenty-five intellectuals prefer the indoors; thirty-five intellectuals prefer the outdoors. Fifteen athletes prefer the indoors. Forty-seven athletes prefer the outdoors. Evaluate the relationship between outdoor preference and membership in intellectual-athletic groups. (Two-way problem). (Chi Square = 4.2244, significant at 95 percent level).

4. Forty-seven students prefer essay exams. Forty-seven professors prefer essay exams. Forty-five students prefer multiple choice exams. Twenty-three professors prefer multiple choice exams. Evaluate the association between examination preference and student or faculty status (two-way problem). (Chi Square = 4.2076, significant at 95 percent level).

5. Of fifty-four people, twenty-six prefer television, and twenty-eight prefer the radio. Do these frequencies differ from the idea of equal Expected Frequencies. (One-way problem) (Chi Square = .0741, not significant).

4,950 t TESTS?

13
The Analysis of Variance

The t-test covered in Chapters 10 and 11 is a very widely used statistic. However, it is limited to only two groups, and there are times when this limitation becomes very troublesome indeed. An example will help to explain this.

Say that you want to find out whether there are differences in the learning of geometry in all of the high school classes in that subject in St. Louis. A geometry test is given to all of the students in the 100 classes. The experimental design would look as follows:

Class 1	**Class 2**		**Class 100**
Student A	Student A	. . .	Student A
Student B	Student B	. . .	Student B
. . .	. . .	. . .	. . .
Student n	Student n	. . .	Student n
-------------	-------------		-------------
Average 1	Average 2	. . .	Average 100

Note that the same students are not in the various schools, so that Student A is not the same person from school to school.

This would seem to be a straightforward arrangement. The t-tests just like those in Chapter 10 could be conducted between all possible pairs of groups to discover whether there are differences between the various classes in knowledge of geometry. Unfortunately, there are serious difficulties with this design.

The first of the problems lies in the fact that simple t-tests would really not provide a solution to our problem. We're not really interested in knowing whether the members of Class 1 know more than the members of Class 2, or whether Class 77 knows more than Class 93. The overall question is that of whether there is a difference in learning of geometry over all of the schools, considered at one time. A single t-test will not give that information; something more general is needed.

Second, there is a very real practical problem involved. With 100 classes to be compared, you'd have to do an enormous number of t-tests. Comparing every group with every other group, with no duplication, would require 4,950 t-tests. Calculation of a t-test by hand requires a considerable amount of time (as you know if you've tried it!), especially if the samples are of substantial size. Assume that each t-test takes two hours. Working eight hours per day, five days a week, 52 weeks per year, you'd have to work for about five years to do them all! Even your microcomputer would have to work for a long, long time. If each t-test took it 30 seconds, it would take 41.25 hours to complete the job. That's a long time. (A big computer could do it very quickly, of course.) Thus, what is needed is a general indicator of whether geometric proficiency differs throughout the 100 classes.

A third, even more serious problem exists. Having to calculate 4,950 t-tests is bad enough, but I would expect 5 percent of these to be significant by pure chance. Recall from Chapter 10 that our t-values were set so that we could be 95 percent sure about our results. That leaves 5 percent margin for error. Thus, 5 percent, or 247.5 t-tests would turn out to be significant by pure chance! (Five percent of 4,950 t-tests is 247.5.) Thus, nearly 248 of the t's would be significant purely because of chance factors. Of course this situation would be totally unacceptable. For even if more than 247.5 significant t's were found, how could you tell which t's were significant by chance, and which were significant because of a real difference?

The more t's that are calculated, the more likely you are to find at least one difference that is significant by pure chance, as shown in the following table for the 95 percent level of confidence:

Number of t's	Probability of at least one significant t
1	.05

Number of t's	Probability of at least one significant t
2	.10
3	.14
4	.19
5	.23
10	.40
14	.51
100	.994
1000	.99999999999+

In short, if we calculate enough t's, at least one will almost always turn out to be significant, just by chance.

Consequently, a method is needed that will determine whether the classes differ overall, one which will take into account the fact that a certain number of t's would be significant by chance alone. That method is called the *Analysis of Variance.* The Analysis of Variance is one of the most widely used techniques of modern statistics, and there are seemingly endless varieties of vastly complicated versions of it. In this chapter, only the simplest form of the technique will be presented.

A SIMPLE EXPLANATION

In Chapter 10 we explained the development of the t-test formulas with simple diagrams, and we'll try the same technique with the Analysis of Variance. However, the ANOVA (the universally-used abbreviation for Analysis of Variance) equations are rather complicated, and we won't be able to develop them directly here. However, you'll be able to get a feeling for the way in which the process works—ANOVA is a direct extension of the ideas behind the t-test. Indeed, as we shall see later, there is a direct mathematical relationship between the two, especially when there are only two groups to be considered.

In the fictitious examples below, we'll have only three groups at first. Look at Fig. 13-1. Is it likely that there is an overall difference between the three groups in that situation? As in the first example in Chapter 10, it's difficult to say, for the hypothetical normal curves are jammed together at the left-hand side of the drawing. However, there's a good deal of overlap, and it seems at least doubtful that there really is a difference.

In Fig. 13-2, it seems somewhat more likely that there is an overall difference. The groups overlap much less. Still, there's

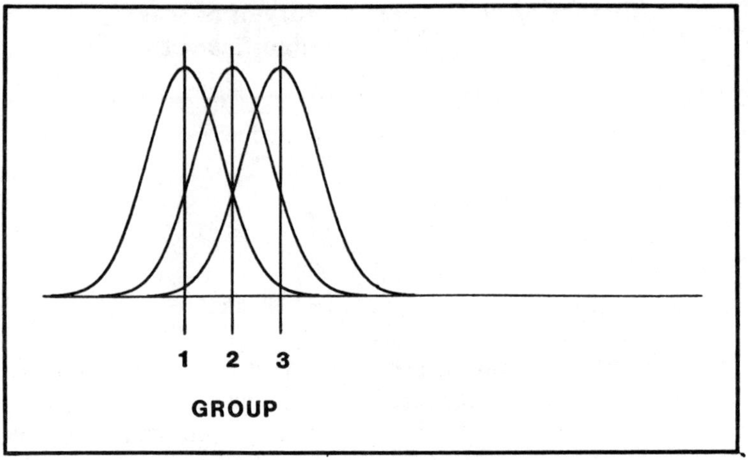

Fig. 13-1. ANOVA with three Means close together.

quite a bit. But in Fig. 13-3, there's almost no overlap. In this last example, it seems very probable that a real difference exists.

Thus, just as with the t-test, we find that the further apart the Averages are, the more likely we are to find a real difference. Unfortunately for our understanding, we must deal with more than two Averages, and many more differences between them, so that things are not so easy to understand. Just as in the case of the t-test, the top of the formula will be set up so that it increases as the composite difference between the Averages increases. The top of the formula will be called the *Between Groups* Mean Square, and is closely related to the top of the t-test formula.

A similar progression of increasing Means is shown in Fig.

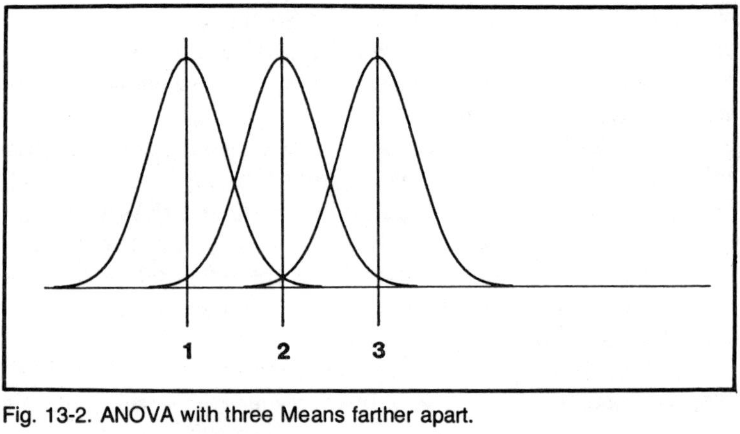

Fig. 13-2. ANOVA with three Means farther apart.

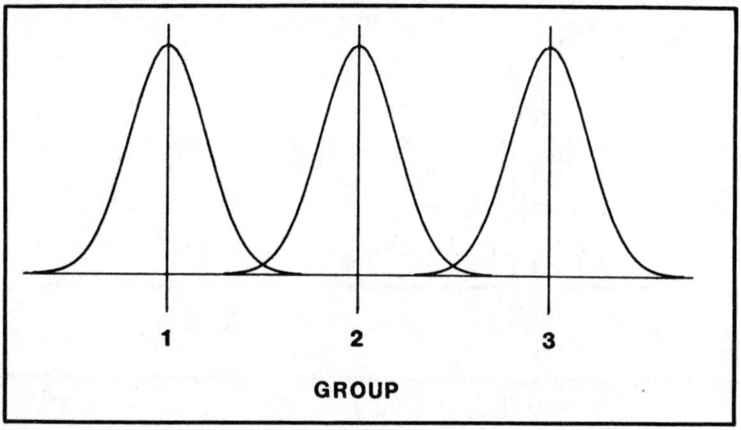

Fig. 13-3. ANOVA with threeMeans far apart.

13-4. Here the Standard Deviation of the various groups is smaller. But as the Averages get farther apart, we can see once more that there is much more likely to be an overall difference.

Recall from Chapter 10 that the Standard Deviations of the various groups also entered into the matter. Essentially, the more the individual groups spread out, the less likely it was that there was a real difference. In Fig. 13-5 we see this progression. Notice that the Standard Deviations are fairly small in Fig. 13-5A. Here there might indeed be a real difference. But in Fig. 13-5B, things spread out a lot more, and in Fig. 13-5C the overlap is quite large. Clearly, in Fig. 13-5C, there probably isn't a significant overall difference.

Thus, the more the individual groups spread out, the less likely we are to assume an overall difference. With the t-test, the term involving spread between the variables was on the bottom of the formula, and such will be the case with the Analysis of Variance. The generalized measure of spread will be called the *Within Groups* Mean Square. We'll get to it in a moment.

These examples are rather easy to understand, and it might be thought that you could simply look at the data to make a decision about whether groups really differ. However, a look at Fig. 13-6 should convince you that things are not always so easy! In this figure, there are three groups compressed at the left, and one rather far away at the right! Will the one on the right make up for the compression of the ones on the left, and create a real difference?

Even more confusing is Fig. 13-7. Here we have the same three groups at the left, and one on the right, but the Standard

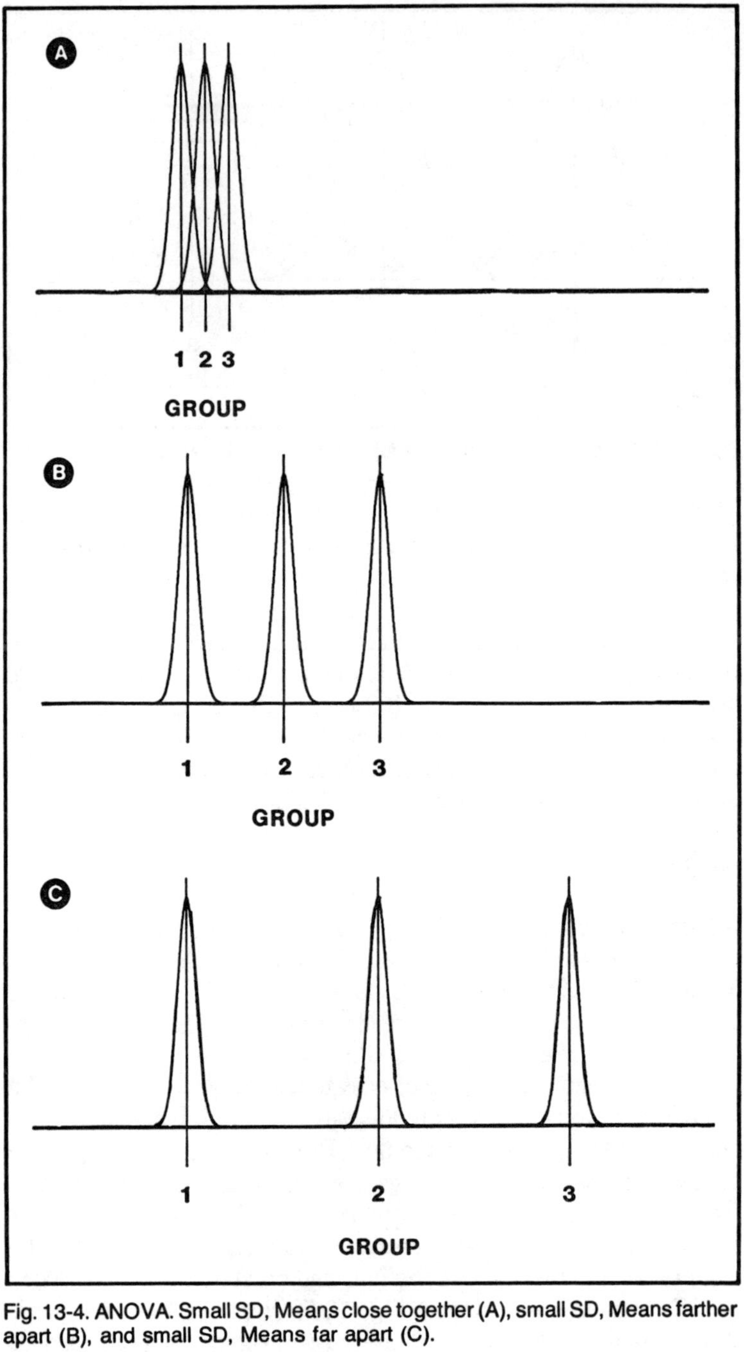

Fig. 13-4. ANOVA. Small SD, Means close together (A), small SD, Means farther apart (B), and small SD, Means far apart (C).

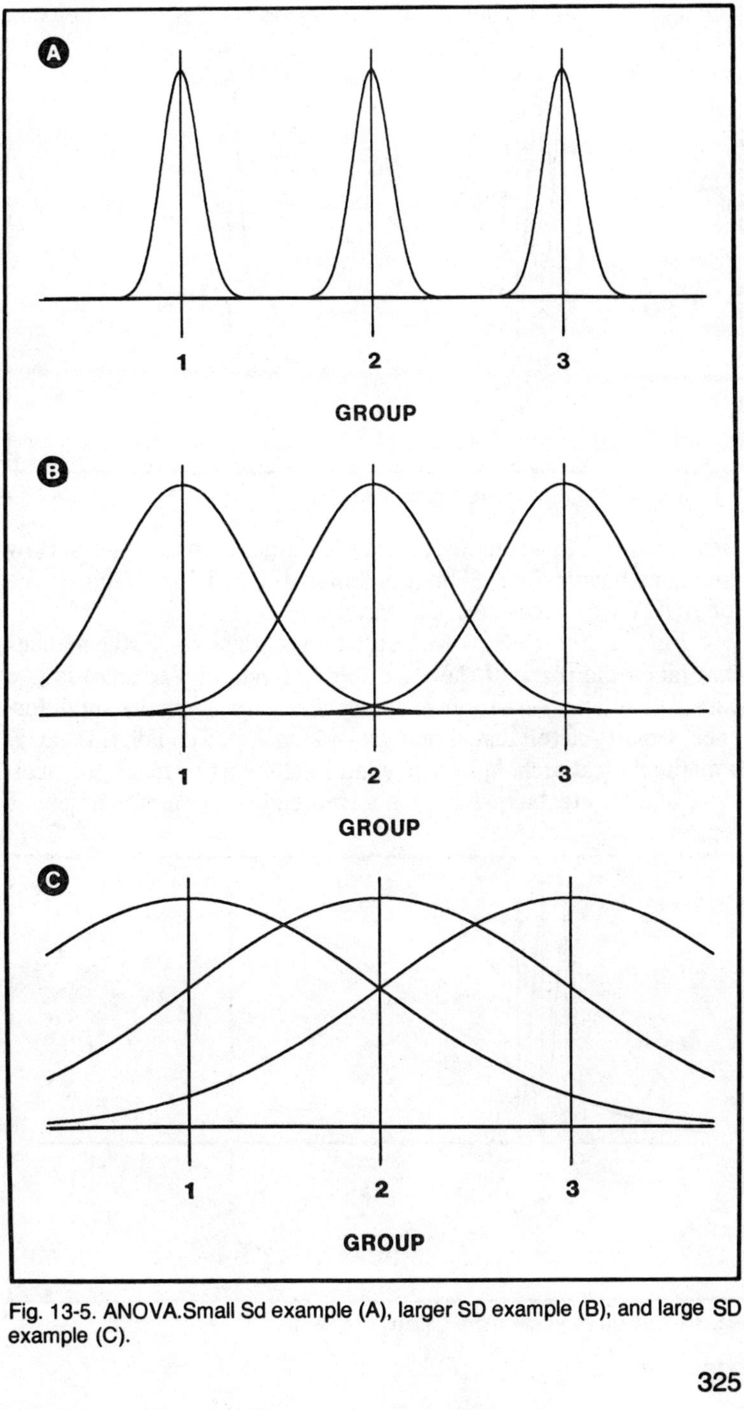

Fig. 13-5. ANOVA.Small Sd example (A), larger SD example (B), and large SD example (C).

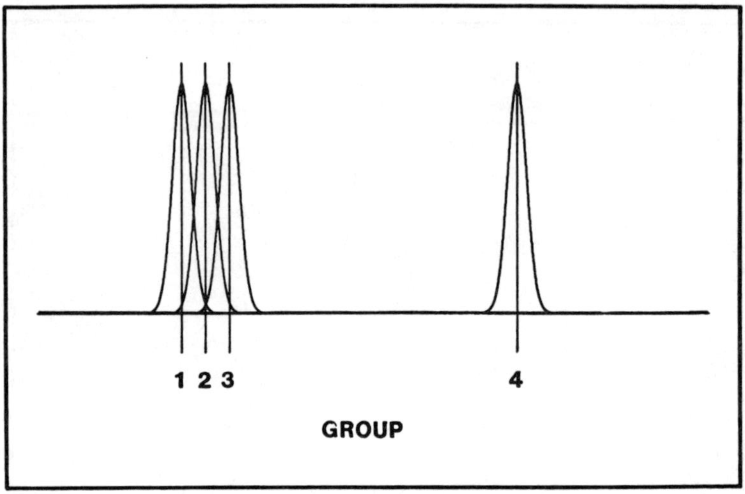

Fig. 13-6. An ANOVA situation difficult to guess.

Deviation of the one on the right is a lot larger than the others! How could one handle this? (The problem in Fig. 13-7 is a difficult one for ANOVA to handle, by the way).

Finally, Fig. 13-8 shows a situation in which the F (the number that takes the place of the t for the Analysis of Variance) is just barely significant—there is just barely an overall significant difference. Could you tell that simply by looking at it? I couldn't. Clearly, a mathematical technique is needed to allow us to make the decision, and this technique will be developed in the next section.

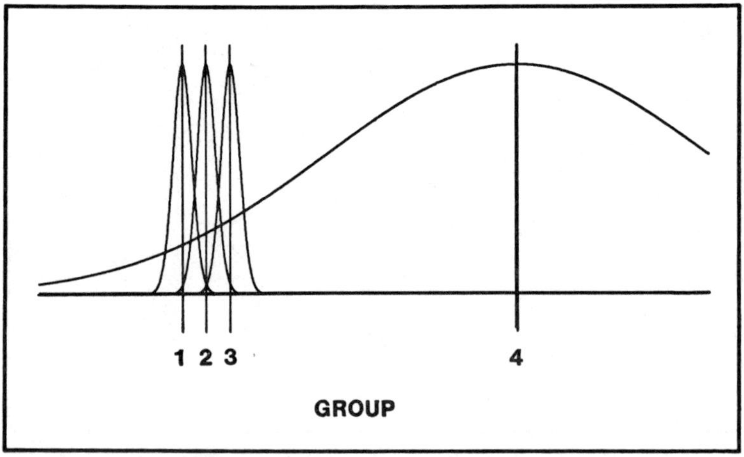

Fig. 13-7. A situation still more difficult to guess.

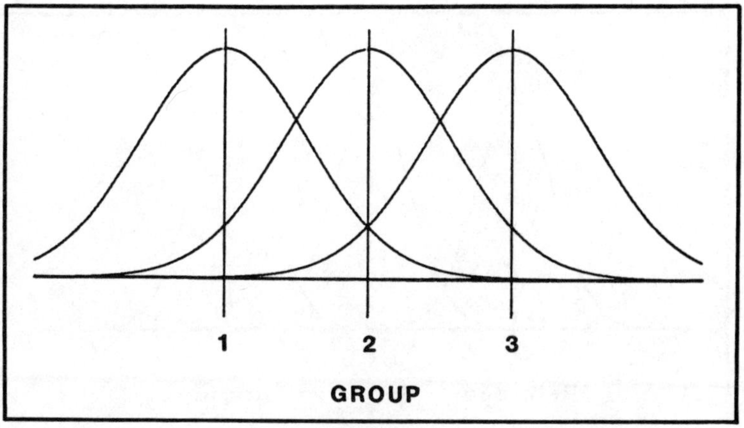

Fig. 13-8. A just barely significant ANOVA.

AN EXAMPLE

Assume that a perfect IQ test is given to three groups of people: mentally retarded persons, normals, and bright people. Scores are obtained as follows:

	Retarded	Normal	Bright
	60	90	120
	70	100	130
	80	110	140
	----------	----------	----------
Average	70	100	130
SD	8.16	8.16	8.16

Data conforming to these Averages and Standard Deviations are plotted in Fig. 13-9. From this figure, it appears that there is not a good deal of overlap. It seems reasonable that there is a real difference somewhere in the system of numbers. Now to the formulas.

THE SUMS NEEDED

Analysis of Variance problems are without question the most complex problems that we have yet encountered. One needs many, many pieces of information to work them out. Let's add a few totals to the data, above:

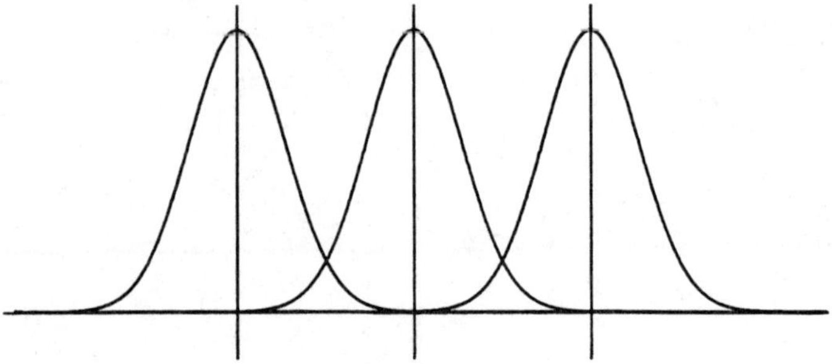

Fig. 13-9. Retarded-Normal-Bright ANOVA.

	Retarded	Normal	Bright	GRAND
	60	90	120	
	70	100	130	
	80	110	140	
	----------	----------	----------	
ΣX	210	300	390	900
ΣX^2	14,900	30,200	50,900	96,000
N	3	3	3	9

Notice that a great many sums are required: The sum for each group, the number for each group, the overall sum (900, called GRANDX), the overall sum of the squared numbers (96,000, called GRANDX2), and the total number of people (9, called GRANDN). Obviously, computation of an ANOVA will require a good deal of work.

THE FORMULAS

The first value needed is called the Total Sum of Squares, which we shall abbreviate SST. The formula looks like this:

$$SST = GRANDX2 - \frac{(GRANDX)^2}{GRANDN} \qquad \textbf{13-1}$$

Putting in the numbers from our little example, above, we find SST as follows:

$$SST = 96,000 - \frac{(900)^2}{9}$$

$$= 96,000 - 90,000$$
$$= 6,000$$

Thus, the Total Sum of Squares (SST) is 6,000. Formula 13-1 should look familiar. Although changed somewhat, it's really the sum of little **x** squared, the top of the Standard Deviation formula, Formula 3-5. Indeed, Formula 13-1 is the formula that underlies most of the science of statistics.

Some value of Degrees of Freedom is associated with all of the Sums of Squares in the Analysis of Variance. For the Total Sum of Squares, the formula is easy:

$$DFT = GRANDN - 1 \qquad \textbf{13-2}$$

DFT is simply the number of people in the experiment, minus one. Thus, DFT = 9 − 1 or 8 in our little example.

THE BETWEEN GROUPS SUM OF SQUARES

This is the difficult one. Actually, it's not really so hard, but the formula is quite frightening:

$$SSB = \Sigma \frac{(\Sigma X_j)^2}{N} - \frac{(GRANDX)^2}{GRANDN} \qquad \textbf{13-3}$$

That's surely our most complicated formula to date, but it's really rather simple. The second half of the formula is precisely the same as the second half of the SST formula (that is, Formula 13-1); so we know how to work that out. We know already that the value is 90,000.

The first half of Formula 13-3 is the frightening part, but what it says is rather simple. To get the desired value, simply take the sum for each group, square that sum, and divide by the number in the group. Then add those values up. That is,

$$SSB = \frac{(210)^2}{3} + \frac{(300)^2}{3} + \frac{(390)^2}{3} - 90,000$$
$$= 14,700 + 30,000 + 50,700 - 90,000$$
$$= 95,400 - 90,000 = 5,400$$

Study these computations very carefully. Once you figure them out, they're not so bad. It's Formula 13-3 that's scary!

SSB is the Between Groups Sum of Squares. Associated with it is a value for Degrees of Freedom.

$$DFB = NGROUPS - 1 \qquad \textbf{13-4}$$

That is, the Between Groups DF is found by subtracting one from the number of groups. We have 3 groups, so DFB will be equal to 3 – 1 or 2. Thus, DFB is 2.

THE WITHIN GROUPS SUM OF SQUARES

Finally, a simple one! The Within Groups values are calculated by subtracting the Between Groups values from the Total values.

$$SSW = SST - SSB \qquad \textbf{13-5}$$

$$DFW = DFT - DFB \qquad \textbf{13-6}$$

In our case, SSW will be SST – SSB or 6,000 – 5,400 or 600. And DFW will be DFT – DFB or 8 – 2 = 6.

Now we have all the basic values. What may we do with them?

FINISHING UP THE CALCULATIONS

There are even more calculations to come! We now need two Mean Squares: the Mean Square Between Groups, and the Mean Square Within Groups. The formulas are as follows:

$$MSB = \frac{SSB}{DFB} \qquad \textbf{13-7}$$

$$MSW = \frac{SSW}{DFW} \qquad \textbf{13-8}$$

In our case MSB will be 5,400/2 or 2,700, while MSW will be 600/6 or 100. Hang on! We're almost done!

Finally, we must compute the F. That's easy. The formula is as follows:

$$F = \frac{MSB}{MSW} \qquad \textbf{13-9}$$

In our case F is 2,700/100 or 27.

And that's it. Nine formulas! Surely, this is our most complex project to date. Of course the computations are not difficult, but it is difficult indeed to keep track of all these things. To help with this, statisticians always gather the results into a Summary Table, as follows.

Source	SS	DF	MS	F
Between	5,400	2	2,700	27
Within	600	6	100	
Total	6,000	8		

The figure that we want is the F, at the right of the table. The F will tell us whether there is a difference greater than chance between the various means. But how do we interpret it?

It's simple. We simply look it up in Appendix E in the back of the book (later on in the chapter, we'll have a program to do the job for you). Turn to Appendix E for a moment. Notice there that the columns are marked Degrees of Freedom for Between, while the rows are marked Degrees of Freedom for Within. In the little summary table above, there were 2 DF for the Between Mean Square, while there were 6 DF for the Within Mean Square. Thus, we look down column 2 and across row 6. At the intersection we find the value 5.14. Be sure to check this in the table. It's easy to do this backwards.

Now, if our F of 27 exceeds 5.14, we may be 95 percent sure that there is a real difference somewhere between the means of our three groups. Well, 27 certainly exceeds 5.14, and thus we may be at least 95 percent sure that there is a difference. Now we may properly proceed to look among the means with t-tests to ferret out the real differences.

What a long, tedious process we went through to arrive at the F! I'd hate to work many Analyses of Variance by hand. Fortunately, long, tedious processes are the specialty of your microcomputer. We shall now turn the job over to it.

A LIMITED ANOVA PROGRAM

The following program will do the job. It is not really a very good, or very efficient program, but it is very similar to Program 10-1, the t-test program. Program 13-1 was made very similar to Program 10-1 so that you could easily generalize from one to the other. Let's look at it:

```
10 REM PROGRAM 13.1. LIMITED ANOVA PROGRAM
15 REM COMPUTATIONS FOR GROUP 1 FOLLOW
20 READ X                         :REM 1ST GROUP DATA
30 IF X=-1 THEN 70                :REM TEST FOR END
40 S1=S1+X:Q1=Q1+X*X              :REM GET SUMS
50 N1=N1+1                        :REM GET COUNT
60 GOTO 20                        :REM CONTINUE
70 A1=S1/N1                       :REM GET AVERAGE
115 REM COMPUTATIONS FOR GROUP 2 FOLLOW
120 READ X                        :REM 2ND GROUP DATA
130 IF X=-1 THEN 170              :REM TEST FOR END
140 S2=S2+X:Q2=Q2+X*X             :REM GET SUMS
150 N2=N2+1                       :REM GET COUNT
160 GOTO 120                      :REM CONTINUE
170 A2=S2/N2                      :REM GET AVERAGE
215 REM COMPUTATIONS FOR GROUP 3 FOLLOW
220 READ X                        :REM 3RD GROUP DATA
230 IF X=-1 THEN 270              :REM TEST FOR END
240 S3=S3+X:Q3=Q3+X*X             :REM GET SUMS
250 N3=N3+1                       :REM GET COUNT
260 GOTO 220                      :REM CONTINUE
270 A3=S3/N2                      :REM GET AVERAGE
300 REM COMPUTATIONS FOR SS TOTAL
310 GS=S1+S2+S3:GQ=Q1+Q2+Q3  :REM GET GRAND SUMS
320 GN=N1+N2+N3                   :REM GET GRAND N
330 ST=GQ-GS*GS/GN                :REM GET SS TOTAL
340 DT=GN-1                       :REM GET DF TOTAL
400 REM COMPUTATIONS FOR SS BETWEEN
410 LH=S1*S1/N1+S2*S2/N2+S3*S3/N3 :REM LH SIDE
420 SB=LH-GS*GS/GN                :REM GET SS BETWEEN
430 DB=3-1                        :REM GET DF BETWEEN
500 REM COMPUTATIONS FOR SS WITHIN
510 SW=ST-SB                      :REM GET SS WITHIN
520 DW=DT-DB                      :REM GET DF WITHIN
600 REM GET MEAN SQUARES
610 MB=SB/DB                      :REM GET MS BETWEEN
620 MW=SW/DW                      :REM GET MS WITHIN
630 F=MB/MW                       :REM GET F-RATIO
700 PRINT "F=",F                  :REM PRINT ANSWER
1000 DATA 60,70,80,-1              :REM DATA TO USE
1010 DATA 90,100,110,-1            :REM (GROUP 2)
1020 DATA 120.130.140.-1           :REM (GROUP 3)
```

Consult the flowchart in Fig. 13-10 for assistance with this long program. We'll need all the help we can get! Also, it would be a good idea to review Program 10-1, the t-test program, to refresh your memory. We'll go through the program line by line.

Lines 10-70 cumulate the data for the first group. Line 20 reads in the data, while line 30 checks to see whether all of the data for Group 1 has been read. A value of minus one signals the end of the data; be sure to include it in your data, or the program will become very confused. Lines 40 and 50 cumulate the sums (in S1), sums of

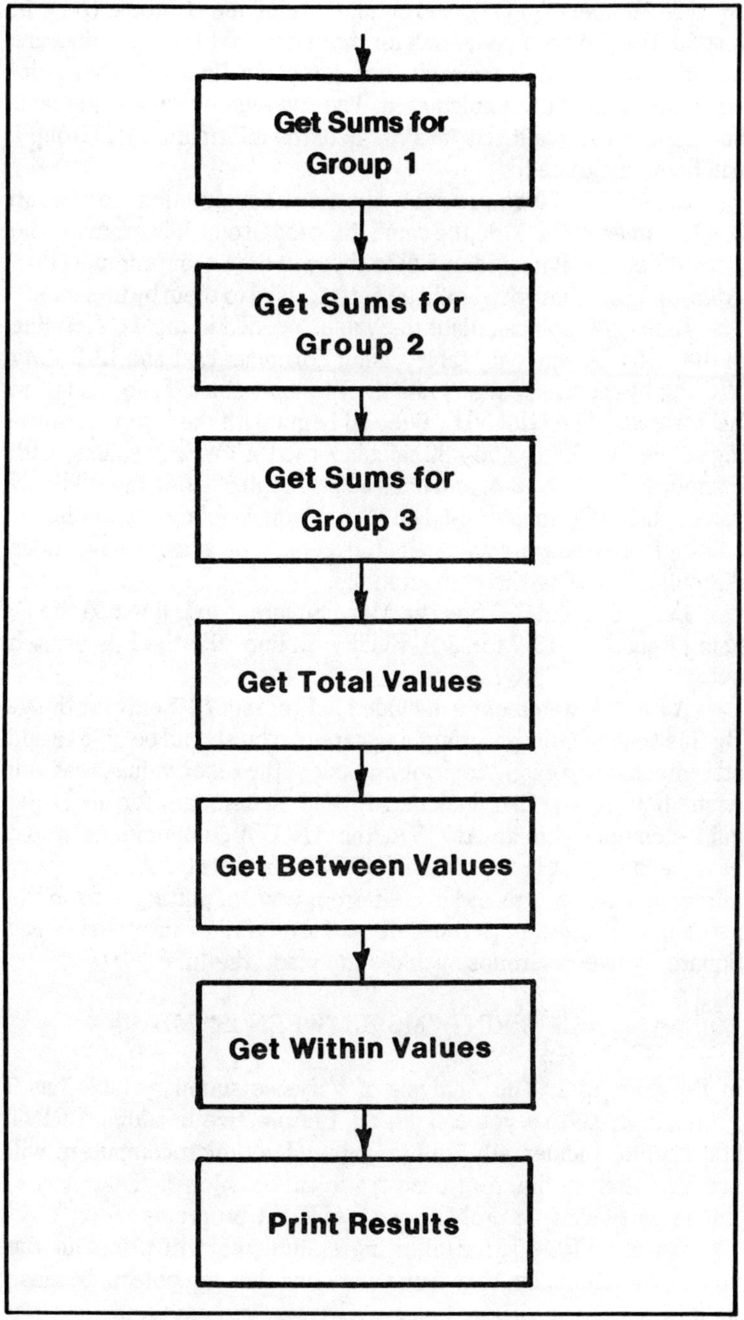

Fig. 13-10. Flowchart for ANOVA program.

squared numbers (in Q1), and count the number of people (N1). In line 60, the program goes back for more data. When all the data are in for Group 1, the program then jumps to line 70, where the Average for Group 1 is calculated. The Average isn't really needed, but it's nice to have it. By line 70, all of the information for Group 1 has been cumulated.

Lines 115-170 do precisely the same computations for Group 2, while lines 215-270 do the same thing for Group 3. Moreover, the procedures are almost identical to those of the t-test program (Program 10-1), so that you should be getting used to them by this time.

Lines 300-340 calculate the values needed in the TOTAL line of the ANOVA summary table, using Formulas 13-1 and 13-2. (My VIC-20, by the way, doesn't like the variable name ST, and so I used SU instead.) The GRAND values all begin with the letter G, starting in line 310. Lines 400-430 calculate the BETWEEN values, with Formulas 13-3 and 13-4, and lines 500 through 520 find the WITHIN values, using Formulas 13-5 and 13-6. If you've followed our discussions of previous programs in the book, you should have little difficulty following the computations.

Lines 610 and 620 find the Mean Squares, and, line 630 the F, using Equations 13-7 to 13-9. Finally, in line 700, the F is printed out.

A PRINT statement is included in Program 13-1 only for the F. By this time in your programming career, you should be able to add statements very easily to print out any of the other values that you wish. If Program 13-1 included PRINT statements for all of the values computed in an ANOVA, the ANOVA computations would have become lost in the mass of PRINT statements. Worse, every microcomputer seems to have different ways of putting data on the screen, and different screen sizes. However, to print the Mean Square Between Groups, you'd simply add the line:

```
615 PRINT "MS BETWEEN IS",MB
```

to Program 13-1. The Analysis of Variance summary table has 9 values in it, so that you can get a lot of practice in adding PRINT statements! Incidentally, different brands of microcomputers will produce slightly different answers in the smaller decimal places; this is particularly a problem with ANOVA programs.

On my TRS-80 the following statements will print out the necessary values, but they won't work on other computers, because their screens are different sizes.

```
640  PRINT "B",SB,DB,MB
650  PRINT "W",SW,DW,MW
660  PRINT "T",ST,DT
```

Program 13-1 has many deficiencies. It can handle only three groups. Thus, you could add more statements like those in lines 20-70 to handle other groups, but it would be awkward—and if there were 100 groups, it would be very awkward indeed! Furthermore, the program is not as efficient as it should be. It should be possible to write such a program in far fewer lines, to optimize it, as computer professionals would say. These deficiencies will be cleared up later in the chapter, but, for the moment, let's use the program to learn about the Analysis of Variance. Later, we'll modify Program 13-1 to learn more about efficient programming methods.

MORE EXAMPLES

The following set of data is interesting:

	Retarded	Normal	Bright
	90	90	90
	100	100	100
	110	110	110
	-------	-------	-------
Average	100	100	100
SD	8.16	8.16	8.16

Notice that the three groups are the same! That means that the Averages are the same, and, of course, there shouldn't be any differences at all between the groups. This is the limiting situation. If our procedures work correctly, the F should be zero! It's difficult to plot this one, for the curves would all be in the same place, but a simulation of the plot is found in Fig. 13-11:

If you put the data in the above example into Program 13-1, you'll obtain the following summary table:

Source	SS	DF	MS	F
Between	0	2	0	0
Within	600	6	100	
Total	600	8		

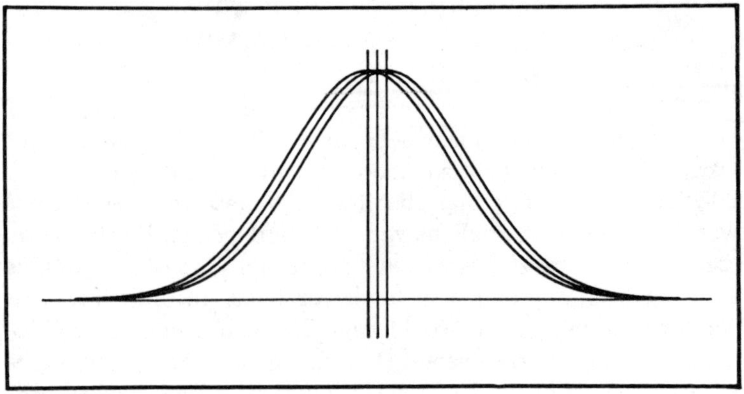

Fig. 13-11. An ANOVA with no Mean difference.

Precisely as we suspected! The F is zero. We again have 2 and 6 Degrees of Freedom, and so, from Appendix E, we discover that a value of 5.14 must be exceeded if there's a real difference somewhere between the means. Clearly, the F of 0 is less than 5.14. As you can see, the procedure works. With identical groups, ANOVA says that there is no difference. Notice also that the values for the Within Groups remain the same as in the last example. That's because the Standard Deviations have not changed. The spread in each group remains the same. Notice, also, that various microcomputers will produce slightly different answers; thus, you might get 599.9994 for the Sum of Squares Within Groups, rather than 600. This is normal, and is especially a problem in ANOVA procedures.

In the following example, there is only a very slight difference between the groups:

	Retarded	**Normal**	**Bright**
	89	90	91
	99	100	101
	109	110	111
	--------	-------	-------
Average	99	100	101
SD	8.16	8.16	8.16

The Averages are very close together, as shown in Fig. 13-12 (the plot program, alas, shows the problem by showing very wide SDs). Hence, we wouldn't expect a very high F, and such is the case, as shown in the summary table.

Source	SS	DF	MS	F
Between	6	2	3	.03
Within	600	6	100	
Total	606	8		

From Appendix E, we still need an F value for 5.14 for significance. 0.03 is clearly not significant, as we would expect. Thus, our preconception is borne out—there isn't a real difference between means so close together. Again, notice that the Within values remain the same. The spread hasn't changed.

In the following data, the Means remain the same as in the original data, but the Standard Deviations are increased:

	Retarded	Normal	Bright
	20	50	80
	70	100	130
	120	150	180
	-------	-------	-------
Average	70	100	130
SD	40.83	40.83	40.83

The data are plotted in Fig. 13-13. If these data are put into Program 13-1, we find the following summary table:

Source	SS	DF	MS	F
Between	5400	2	2700	1.08
Within	15000	6	2500	
Total	20400	8		

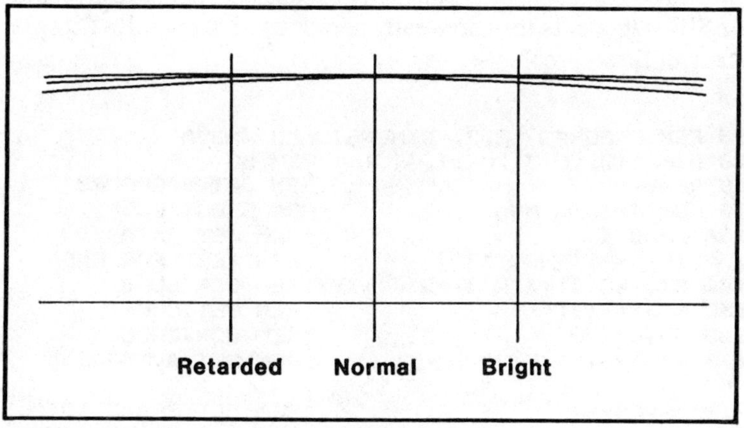

Fig. 13-12. ANOVA with a very slight difference.

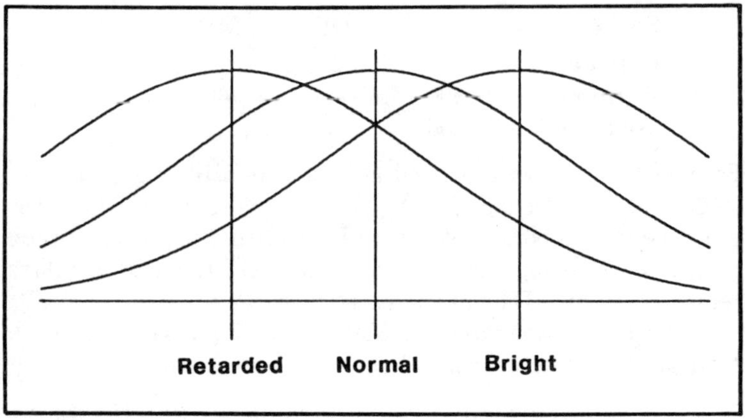

Fig. 13-13. ANOVA with wider SDs.

Again, Appendix E says that we need 5.14 for significance, and again the F is not significant. Notice that in this instance, the changes in values occur with the Within values. They've increased a lot. That makes sense, since the Standard Deviations have increased markedly. The Between values are identical with the ones in the first example. Thus, the Between values are associated with the Means, and the Within values are controlled by the Standard Deviations, as noted earlier. In our previous examples, the Within values were always the same.

A MORE EFFICIENT PROGRAM

As mentioned above, Program 13-1 is far from ideal. It handles only three groups, is very difficult to expand, and needs many more BASIC statements than are really necessary. Program 13-2 solves many of these problems:

```
10 REM PROGRAM 13.2. DIMENSIONED ANOVA
20 DIM S(5),Q(5),N(5),A(5)  :REM SET UP
30 READ NG                  :REM GET # GROUPS
40 FOR J=1 TO NG            :REM GROUP LOOP
120 READ X                  :REM GET DATA
130 IF X=-1 THEN 170        :REM TEST FOR END
140 S(J)=S(J)+X:Q(J)=Q(J)+X*X:REM GET SUMS
150 N(J)=N(J)+1             :REM GET COUNT
160 GOTO 120                :REM CONTINUE
170 A(J)=S(J)/N(J)          :REM GET AVERAGE
200 REM CUMULATE NEEDED SUMS
210 GS=GS+S(J)              :REM GET GRAND TOTAL
220 GQ=GQ+Q(J)              :REM GET GD SQR SUM
```

```
230 GN=GN+N(J)                :REM GET GRAND N
240 LH=LH+S(J)*S(J)/N(J)      :REM GET BETWEEN SUM
250 NEXT J                    :REM END GROUP LOOP
300 REM COMPUTATIONS FOR SS TOTAL
330 ST=GQ-GS*GS/GN            :REM GET SS TOTAL
340 DT=GN-1                   :REM GET DF TOTAL
400 REM COMPUTATIONS FOR SS BETWEEN
420 SB=LH-GS*GS/GN            :REM GET SS BETWEEN
430 DB=NG-1                   :REM GET DF BETWEEN
500 REM COMPUTATIONS FOR SS WITHIN
510 SW=ST-SB                  :REM GET SS WITHIN
520 DW=DT-DB                  :REM GET DF WITHIN
600 REM GET MEAN SQUARES
610 MB=SB/DB                  :REM GET MS BETWEEN
620 MW=SW/DW                  :REM GET MS WITHIN
630 F=MB/MW                   :REM GET F
700 PRINT "F=",F              :REM PRINT ANSWER
1000 DATA 3                    :REM # OF GROUPS
1010 DATA 60,70,80,-1          :REM DATA TO USE
1020 DATA 90,100,110,-1
1030 DATA 120,130,140,-1
```

Let's examine the program, using the flowchart in Fig. 13-14.

Program 13-2 uses dimensioned variables, and so in line 20, space is saved for these variables. As arranged in line 20, up to 5 groups may be used. The program is set up so that it must know the number of groups that it must expect before it can do anything else, and that value (NG) is obtained in line 30.

Line 40 begins a loop that will get the data for each of the groups. use of a FOR-NEXT loop for this purpose means that we need only one set of statements to do the job. There is no need to repeat the same statements over and over again, as with Program 13-1. Program 13-2 is set up so that the line numbers coincide as much as possible with those of Program 13-1. Notice that the statements in line numbers 120 through 170 generally correspond to those in the earlier program. Usually, simple variables are replaced with dimensioned variables, to allow use of the FOR-NEXT loop.

When all the data has been read in for one group, the grand sums are obtained in lines 210 through 230. Line 240 cumulates the value needed for the left-hand side of Formula 13-3, the Between Group Sum of Squares. In the present program, these are done in the initial FOR-NEXT loop, rather than later in the program (you could use another FOR-NEXT loop farther on in the program for this purpose).

Lines 330 through 700 are precisely the same as those in Program 13-1.

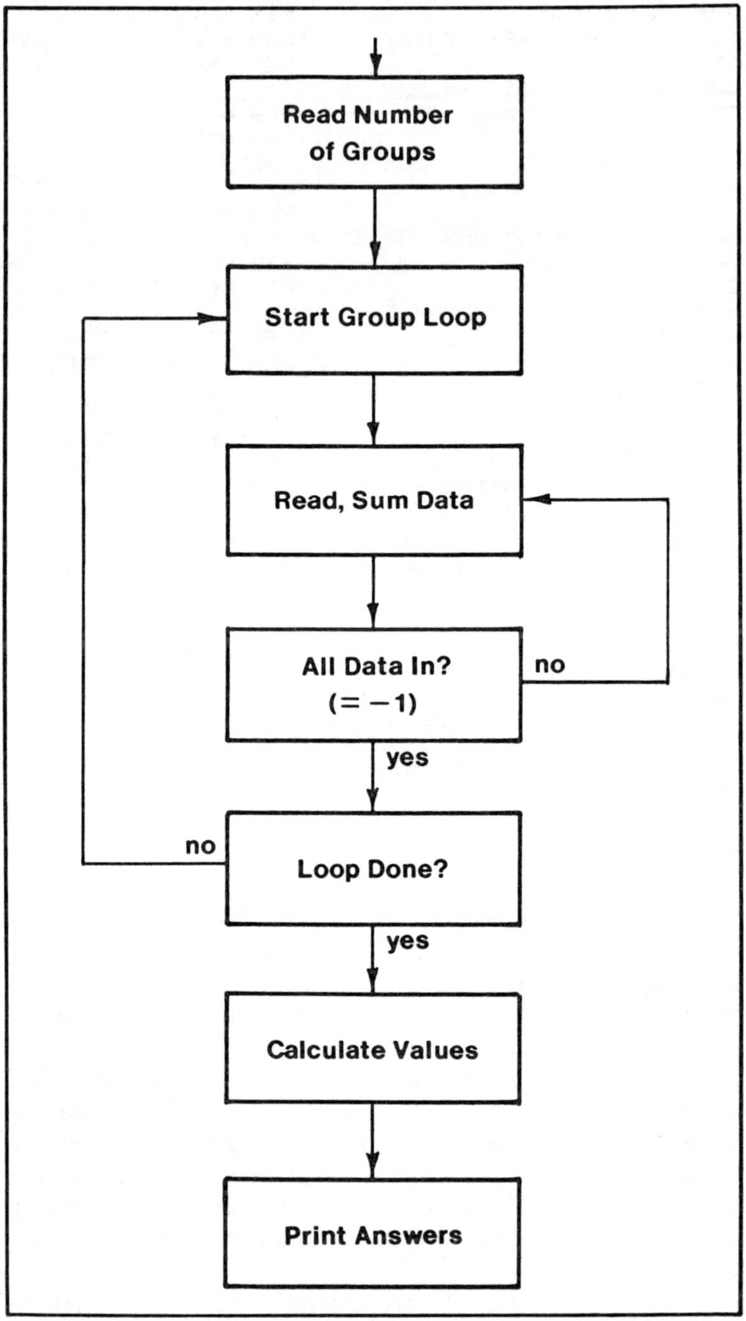

Fig. 13-14. Flowchart for the dimensional ANOVA program.

Finally, the first DATA statement must contain the number of groups, as shown in line 1000. The program can know the number of groups only through the use of this statement. The remaining DATA statements are the same as before.

This program allows the use of any number of groups, up to the number set up in the Dimension statement in line 20. Run the previous problems with Program 13-2. Notice that it gives the same answers, with only about half the number of statements found in Program 13-1. That's a considerable savings!

A STILL MORE EFFICIENT PROGRAM

Program 13-3 is about half again as long as Program 13-2! Furthermore, it counts the number of groups for you. I find this feature to be very convenient.

```
10 REM PROGRAM 13.3. EFFICIENT ANOVA
20 S=0:N=0                          :REM CLEAR SUMS
30 READ X                           :REM GET DATA
40 IF X=-1 THEN 90                  :REM GROUP DONE?
50 IF X=-2 THEN 120                 :REM ALL DONE?
60 S=S+X:N=N+1                      :REM GET GROUP SUMS
70 GS=GS+X:GQ=GQ+X*X:GN=GN+1:REM CUMULATE SUMS
80 GOTO 30                          :REM CONTINUE
90 NG=NG+1                          :REM COUNT # GROUPS
100 LH=LH+S*S/N                      :REM GET BETWEEN
110 GOTO 20                          :REM TO NEXT GROUP
120 ST=GQ-GS*GS/GN:DT=GN-1            :REM TOTAL VALUES
130 SB=LH-GS*GS/GN:DB=NG-1            :REM BETWEEN VALUES
140 SW=ST-SB:DW=DT-DB                :REM WITHIN VALUES
150 MB=SB/DB:MW=SW/DW                :REM MEAN SQUARES
160 F=MB/MW                          :REM GET F
170 PRINT "F=",F                     :REM PRINT ANSWER
1000 DATA 60,70,80,-1                 :REM DATA TO USE
1010 DATA 90,100,110,-1
1020 DATA 120,130,140,-1
1030 DATA -2                          :REM "ALL GROUPS IN"
```

Its operation is detailed in the flowchart in Fig. 13-15.

The program knows that all the data for a group have been read in when it encounters a −1. Similarly, the signal that all the groups have been read in is a −2. Each group must end with a −1, by the way, including the last one. The program isn't smart enough to figure out that a −2 also means the end of a group. Clearly, this scheme eliminates −1 and −2 as valid data points, but of course these signals may easily be changed to other values if necessary.

In this program variables S and N cumulate the sum and count

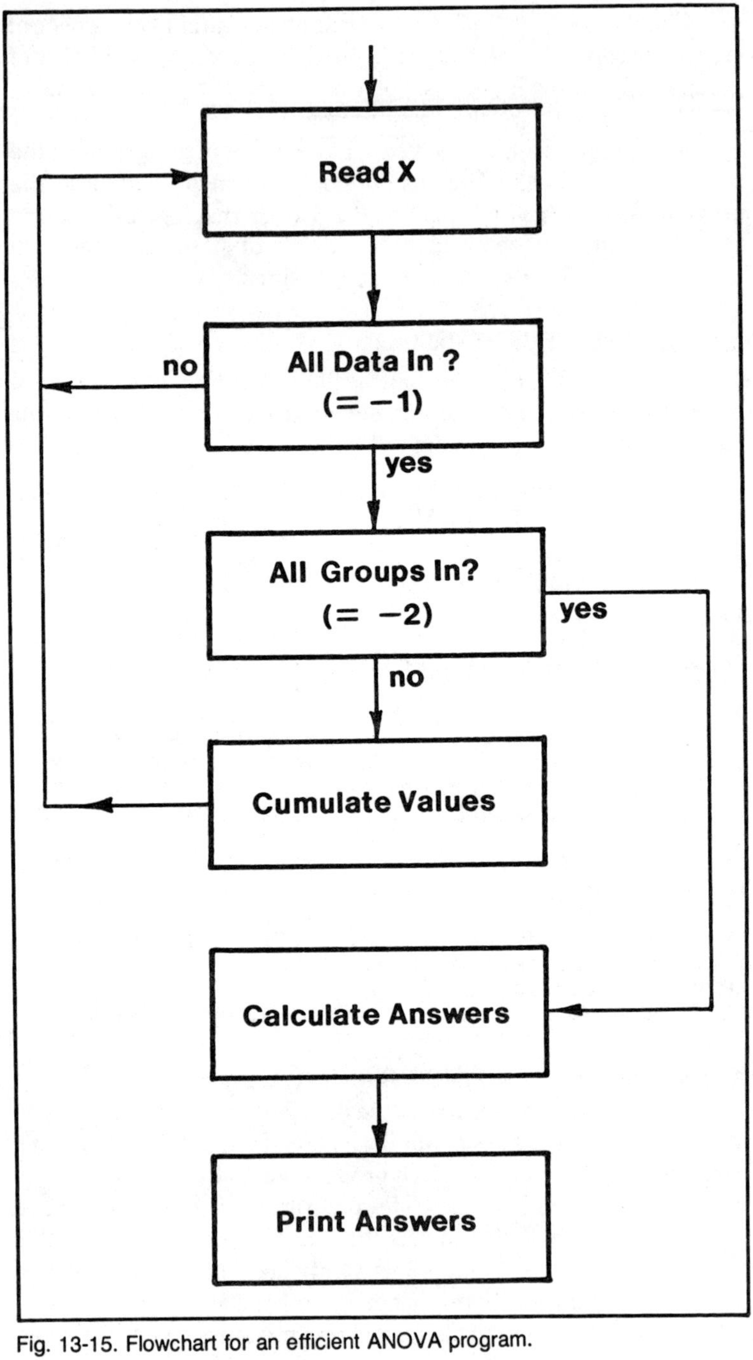

Fig. 13-15. Flowchart for an efficient ANOVA program.

for each group. Consequently, they must be set equal to zero before each group is read in; this is done in line 20. Line 30 reads the data. Line 40 checks for the end of the data in a given group, and line 50 checks for the end of all groups.

Line 60 cumulates the sums needed to get the Between Group sum of squares, just for one group. Line 70 directly computes the grand sums needed by the program.

When all the data in a group have been read in, the program transfers control to line 90, where the NG, the number of groups, is increased by one. Line 100 cumulates the sum needed for the left hand side of Formula 13-4, and line 110 returns to get another group.

You should be able to figure out the remaining statements in the program. The line numbers are different, but they're just like the ones in the previous programs. Run the program. You'll find that it gives the same answers as the previous ones.

As you can see, Program 13-3 is very compact. Of course, this last program doesn't save as many values as does Program 13-2 (for example, the sum and sum of squared numbers for each group), but generally you don't need those anyway.

Programmers generally like to make their programs as efficient and compact as possible. The progression toward this sort of optimization is clearly evident in Program 13-3.

MORE EXAMPLES

Now that we have developed some more general ANOVA programs, we can study some more examples. The data in all these examples is fictitious, of course, and keep in mind that we should use samples much larger than the ones mentioned here when doing real research. The small samples shown here are given so that you won't have to poke in so many numbers.

This time we have three sorts of bolts—those made by the Burred Bolt Company, the Ajax Bolt Company, and the Rolled Fastener Company. Samples are taken of various ¼-inch bolts, and their diameters measured, with the following results:

Burred	Ajax	Rolled
.275	.241	.250
.253	.246	.255
.263	.220	.245

	Burred	Ajax	Rolled
			.248
	-------	-------	-------
Average	.264	.236	.250
SD	.009	.011	.004

The data are plotted in Fig. 13-16 (as in Chapter 10, normal curves scaled to the SDs are plotted, rather than the actual data points.) Run the above data with either Program 13-2 or 13-3. You should obtain the following results:

Source	SS	DF	MS	F
Between	.00118	2	.00059	6.09
Within	.00068	7	.00009	
Total	.00185	9		

According to the Table in Appendix E, with 2 and 7 DF we must exceed a value of 4.74 to be 95 percent sure that a real difference exists somewhere between the three means. 6.09 is indeed significantly greater than 4.74, so that we can be 95 percent sure that one company's bolts are larger than at least one other company's. Now we may proceed to use t-tests to find the precise location of these differences, as shown in Chapter 10.

Let's stick with bolts for a while. Assume that we measure bolts of three sizes—one-quarter inch, one-half inch, and three-quarters of an inch. It seems reasonable that the Analysis of Variance ought to tell us that there is a real difference in this instance,

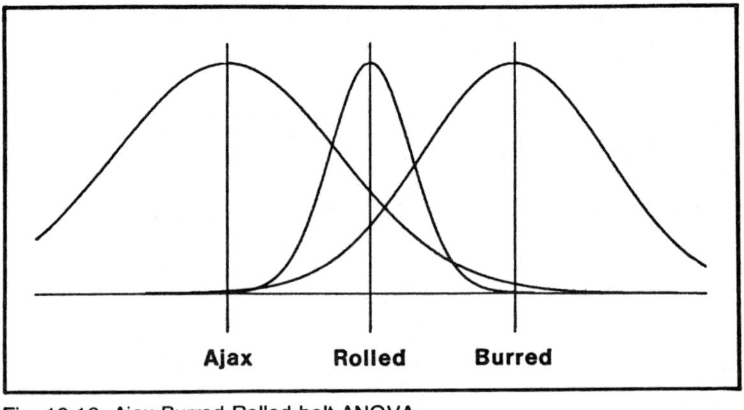

Fig. 13-16. Ajax-Burred-Rolled bolt ANOVA.

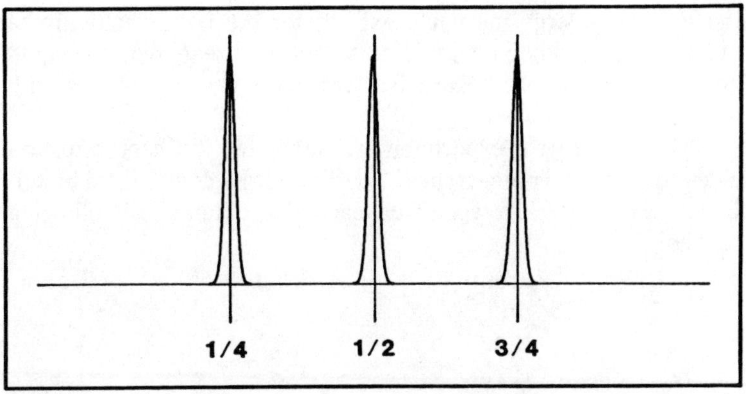

Fig. 13-17. Quarter, half, and three-quarter inch bolts.

since we know beforehand that there should be. The data are as follows:

	¼″	½″	¾″
	.275	.525	.775
	.253	.503	.753
	.263	.513	.763
	-------	-------	-------
Average	.264	.514	.764
SD	.009	.009	.009

If you're sharp, you'll notice that the data were constructed by simply adding .25 or .50 to the data of the first group. Plotted, the data look as shown in Fig. 13-17.

Put the data into Program 13-2 or Program 13-3. You should obtain these answers:

Source	**SS**	**DF**	**MS**	**F**
Between	.375	2	.1875	1545.56
Within	.0007	6	.0001	
Total	.3757	8		

With 2 Between and 6 Within Degrees of Freedom, Appendix E says that we need a value of 5.14. Clearly 1545.56 exceeds this value. Therefore, we may be 95 percent sure that there is a real difference somewhere between the various bolt sizes. Notice that

the F is very large, as you'd expect, for the bolts really are of different sizes to begin with. If this were real research, one could now proceed to find the precise location of the differences with t-tests.

Our motorcycle experimenter, mentioned in Chapter 10, decides to extend his researches. To this end, he adds a third brand, Beast, measuring the speed on each. He obtains the following results:

	Varoom	**Dragster**	**Beast**
	120	121	127
	122	123	129
	126	127	133
		124	
	-------	-------	-------
Average	122.7	123.8	129.7
SD	2.49	2.16	2.49

As may be seen in Fig. 13-18, the Beast is a lot faster than the others. Running the data in Program 13-3 yields the following table:

Source	**SS**	**DF**	**MS**	**F**
Between	87.5156	2	43.7578	5.46
Within	56.0781	7	8.0112	
Total	143.5940	6		

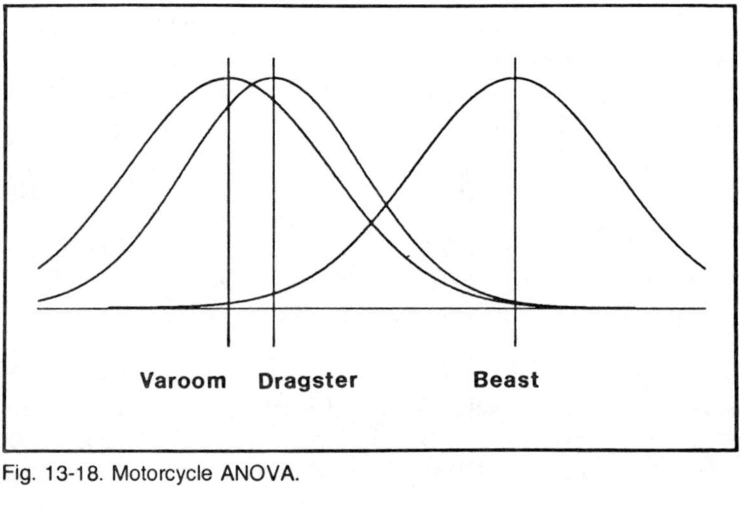

Fig. 13-18. Motorcycle ANOVA.

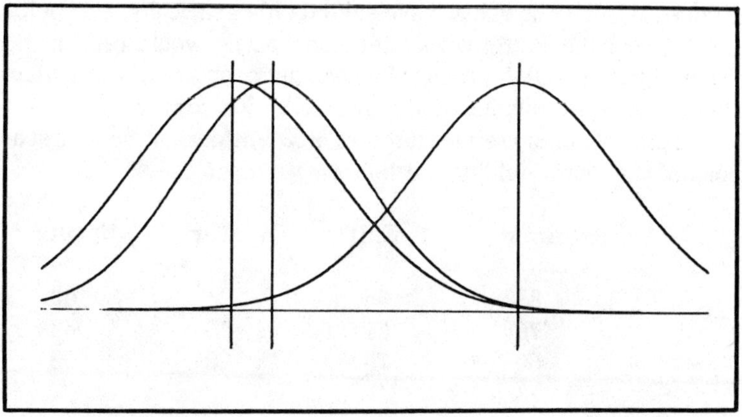

Fig. 13-19. ANOVA with numbers divided by 100.

With 2 and 7 DF, Appendix E shows that we need a value of 4.74 for 95 percent certainty. Our value of 5.46 exceeds 4.74, and so there is indeed a real difference. A set of t-tests will show clearly where it is.

The next example is the same problem as the last one, but the numbers have all been divided by 100. It's interesting to see what happens to the ANOVA in such a situation. The data will be as follows:

	Varoom	Dragster	Beast
	1.20	1.21	1.27
	1.22	1.23	1.29
	1.26	1.27	1.33
		1.24	
	-------	-------	-------
Average	1.23	1.24	1.30
SD	.02	.02	.02

Notice that the plot is the same (Fig. 13-19). The Summary Table, however, changes a bit:

Source	SS	DF	MS	F
Between	.00875	2	.00438	5.46
Within	.00561	7	.00080	
Total	.01436	9		

Notice that all of the values have shifted rather radically, except for the DF and the F. However, the conclusions would remain the same. Thus, ANOVA produces a constant conclusion, even when the numbers are divided (and multiplied) by a constant.

Temperatures are measured at 6:00 PM during the four seasons of the year, yielding the following temperatures:

	Summer	Fall	Winter	Spring
	83	65	31	60
	76	61	26	63
	60	70	20	59
	78			
	------	------	------	------
Average	74.25	65.33	25.67	60.67
SD	8.61	3.68	4.50	1.70

The data are plotted as Fig. 13-20. The ANOVA table appears as follows:

Source	SS	DF	MS	F
Between	4374.94	3	1458.31	32.2672
Within	406.75	9	45.19	
Total	4781.69	12		

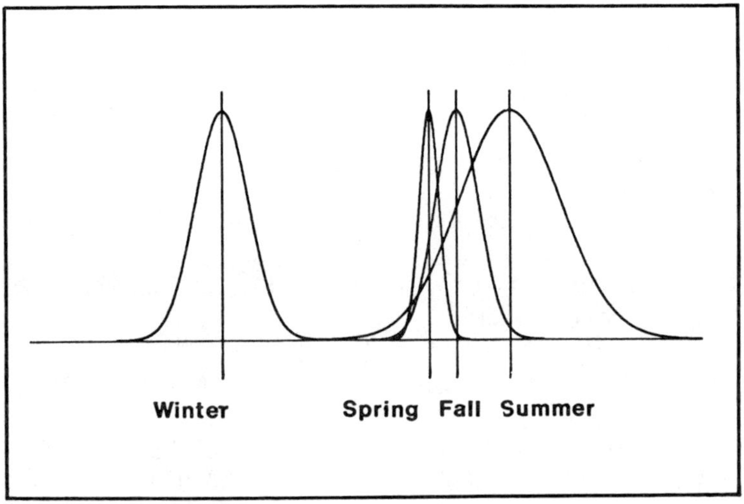

Fig. 13-20. Temperature ANOVA.

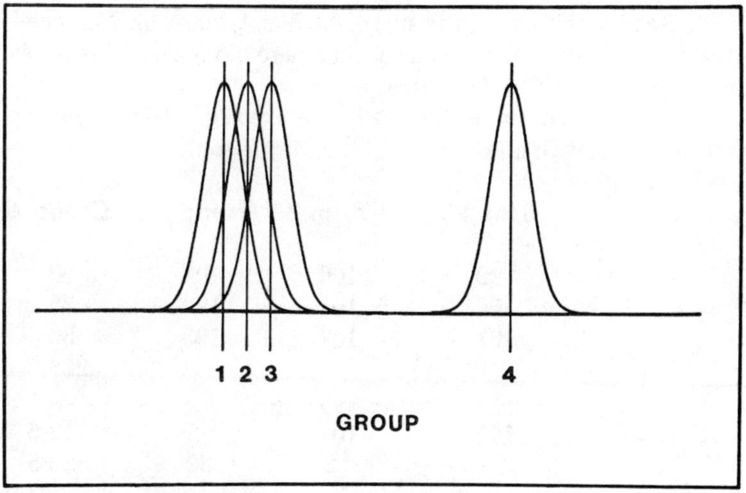

Fig. 13-21. A peculiar ANOVA.

Appendix E shows that with 3 and 9 DF, one needs an F of 3.86. Our F of 32.27 clearly exceeds that value, and so we may be at least 95 percent sure that the temperature does indeed have different averages at different seasons of the year.

Here are some data designed to look a lot like plot in Fig. 13-16. We'll compare them with the next set.

	Group 1	Group 2	Group 3	Group 4
	99	100	101	111
	100	101	102	112
	101	102	103	113
	------	------	------	------
Average	100	101	102	112
SD	.82	.82	.82	.82

The data are plotted as Fig. 13-21. If you run the data through one of the ANOVA programs, you'll find that the summary table looks like this:

Source	**SS**	**DF**	**MS**	**F**
Between	278.25	3	92.75	92.75
Within	8	8	1	
Total	286.25	11		

With 3 and 8 DF, we need a value of 4.07, and clearly 92.75 exceeds that value! Thus, ANOVA tells us that there is a real difference. We may now proceed to do t-tests.

Here's one example designed to be like Fig. 13-7. It's just like the last one, with a larger SD in the fourth group:

	Group 1	Group 2	Group 3	Group 4
	99	100	101	0
	100	101	102	75
	101	102	103	150
				225
	------	------	------	------
Average	100	101	102	112.5
SD	.82	.82	.82	83.85

This is plotted as Fig. 13-22 (obviously we overdid the SD on the last group a bit, so that the curve looks like a horizontal line). Notice the difference in the ANOVA summary table!

Source	SS	DF	MS	F
Between	372.234	3	124.078	.04
Within	28131	9	3125.67	
Total	28503.2	12		

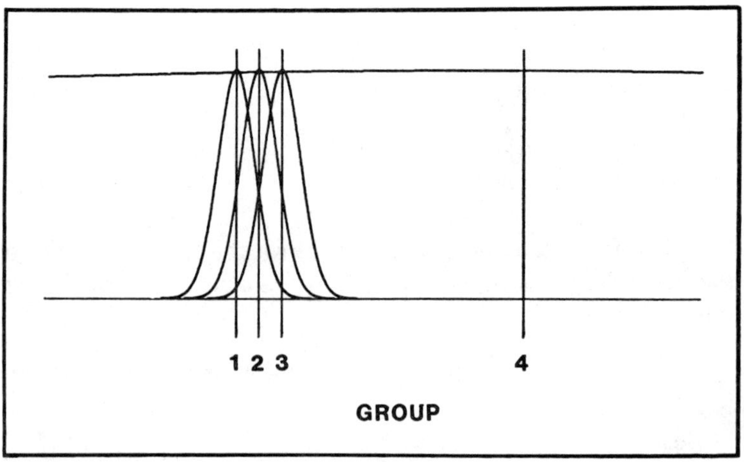

Fig. 13-22. A still more peculiar ANOVA.

Notice that the F is not significant (any F of less than 1 is not significant). Notice also that the principal difference here lies in the SS Within, which is much larger than the last example.

It is considered very poor practice in the Analysis of Variance to use groups whose Standard Deviations vary as much as those shown here, but it does make an interesting example!

Finally, let's give an example of an ANOVA using only two groups. Let's use the first example from Chapter 10:

	Group 1 Males	**Group 2 Females**
	102	105
	101	104
	100	103
	99	102
	98	101
	------	------
Average	100	103
SD	1.4142	1.4242

Putting these data in any of our ANOVA programs, we obtain the following summary table:

Source	**SS**	**DF**	**MS**	**F**
Between	22.5	1	22.5	9
Within	20	8	2.5	
Total	42.5	9		

With 1 and 8 DF, Appendix E shows that we need a value of 5.31, so that our F is significant at the 95 percent level.

Now, notice that the t we obtained in the t-test in Chapter 10 was 3. The F of 9 is 3 squared! This is true for any situation in which we have two groups. ANOVA will give precisely the same answer as the t-test, but the answer will be squared. This is because of the close mathematical relationship between the Analysis of Variance and the t-test. This also means that if you have a good ANOVA program, you really don't need the t-test program. However, the t-test procedure is still so widely used that most computer statistical packages still include it.

DETERMINING SIGNIFICANCE LEVELS

Rather than having to depend upon Appendix E, it would be nice to have a program which would determine the significance levels for us. Such a program is Program 13-4:

```
3000 REM PROGRAM 13.4 F-VALUE SIGNIFICANCE
3010 INPUT "DF BETWEEN ";K1  :REM GET DFB
3020 INPUT "DF WITHIN ";K2   :REM GET DFW
3030 INPUT "F= ";K0          :REM GET F
3040 GOSUB 4000              :REM GOTO DISTRIB
3050 PP=1-H                  :REM ADJUST H
3060 PRINT PP                :REM PRINT ANSWER
3070 GOTO 3000               :REM KEEP IT UP
4000 REM APPROXIMATION ROUTINE :REM BEGIN APPROX
4010 IF K0>=1 THEN 4040      :REM BIG ENOUGH?
4020 PRINT "NOT SIGNIFICANT":REM SAY IF NOT
4030 H=1:RETURN
4040 K5=0.2222222            :REM SET ONE CONST
4050 K6=K5/K1:K7=K5/K2       :REM SET MORE
4060 K8=K0^0.3333333         :REM SET MORE
4070 K9=(1-K7)*K8-1+K6       :REM TOP
4080 Z=K9/SQR(K7*K8*K8+K6)   :REM GET Z
4090 IF Z>3 THEN 4110        :REM ADJUST IF <3
4100 Z=Z*(1.+0.08*Z^4/K2^3)  :REM ADJUST
4110 GOSUB 5000              :REM NORMAL APPROX
4120 RETURN
5000 REM NORMAL CURVE APPROXIMATION
5010 H=1+0.0498673470*Z      :REM CALCULATE
5020 H=H+.0211410061*Z*Z     :REM  NORMAL CURVE
5030 H=H+.0032776263*Z^3     :REM  APPROXIMATION
5040 H=H+.0000380036*Z^4     :REM  FROM PROGRAM
5050 H=H+.0000488906*Z^5     :REM  6.2.
5060 H=H+.0000053830*Z^6
5070 H=H^-16/2            :REM FINISH APPROXIMAT.
5080 RETURN
```

This program looks a great deal like Program 10-2. Indeed, it's pretty much the same program. The approximation used will handle the t-test, Chi-Square, and the Analysis of Variance equally well. Again, for details see Jaspen (1965). The precise manner in which the program works is a bit too complex for description in a simple book such as this one.

Let's try some values from Appendix E in Program 13-4. For example, if we put look in DF Between Column 2, and in the DF Within Column 6, we find the entry 5.14 in Appendix E. Putting those values into Program 13-4, my TRS-80 says that the probability that there is a real difference is .950235. The value should be .95 exactly, and .950235 is fairly close. Other values yield the following results:

DF Between	DF Within	F	Prob
2	6	5.14	.9502
1	1	161	.9505
9	60	2.04	.9502
1	100	3.94	.9528
2	6	27	.9990

Notice that all of these values are rather close to the precise values. Program 13-4 is an approximation, and so it doesn't give exact values. However, the values that it does provide are good enough for most practical purposes.

The last entry in the little table above is the F from the first real example in this chapter, the one involving retarded, normal, and bright people. Remember that the Degrees of Freedom in that case were 2 and 6, while the F was 27. Program 13-4 tells us that we can be 99.90 percent sure that there is a real difference somewhere between those three means. It's a handy program indeed.

ADDING THE SIGNIFICANCE LEVEL PROGRAM TO THE ANOVA PROGRAM

I have hinted in previous chapters that it is perfectly possible to combine significance level programs, such as Program 13-4, with other programs. Such an arrangement would be very handy indeed. Let's try it in this case. Program 13-5 combines Program 13-3 with Program 13-4, and calculates both the F and the significance level of the F.

```
10 REM PROGRAM 13.5. ANOVA WITH SIGNIF LEVEL
20 S=0:N=0                          :REM CLEAR SUMS
30 READ X                           :REM GET DATA
40 IF X=-1 THEN 90                  :REM GROUP DONE?
50 IF X=-2 THEN 120                 :REM ALL DONE?
60 S=S+X:N=N+1                      :REM GET GROUP SUMS
70 GS=GS+X:GQ=GQ+X*X:GN=GN+1:REM CUMULATE SUMS
80 GOTO 30                          :REM CONTINUE
90 NG=NG+1                          :REM COUNT # GROUPS
100 LH=LH+S*S/N                      :REM GET BETWEEN
110 GOTO 20                          :REM TO NEXT GROUP
120 ST=GQ-GS*GS/GN:DT=GN-1           :REM TOTAL VALUES
130 SB=LH-GS*GS/GN:DB=NG-1           :REM BETWEEN VALUES
140 SW=ST-SB:DW=DT-DB                :REM WITHIN VALUES
```

```
15Ø MB=SB/DB:MW=SW/DW           :REM MEAN SQUARES
16Ø F=MB/MW                     :REM GET F
17Ø PRINT "F=",F                :REM PRINT ANSWER
18Ø K1=DB:K2=DW:KØ=F            :REM PREPARE VALUES
19Ø GOSUB 3ØØØ                  :REM GET SIGN LEVEL
2ØØ PRINT "SIGN=",PP            :REM PRINT IT
21Ø END
3ØØØ REM F VALUE SIGNIFICANCE ROUTINE
3Ø4Ø GOSUB 4ØØØ                 :REM GOTO DISTRIB
3Ø5Ø PP=1-H                     :REM ADJUST H
3Ø6Ø RETURN                     :REM GO BACK TO PGM
4ØØØ REM APPROXIMATION ROUTINE :REM BEGIN APPROX
4Ø1Ø IF KØ>=1 THEN 4Ø4Ø         :REM BIG ENOUGH?
4Ø2Ø H=1                        :REM CORRECT H
4Ø3Ø RETURN
4Ø4Ø K5=Ø.2222222               :REM SET ONE CONST
4Ø5Ø K6=K5/K1:K7=K5/K2          :REM SET MORE
4Ø6Ø K8=KØ^Ø.3333333            :REM SET MORE
4Ø7Ø K9=(1-K7)*K8-1+K6          :REM TOP
4Ø8Ø Z=K9/SQR(K7*K8*K8+K6)      :REM GET Z
4Ø9Ø IF Z>3 THEN 411Ø           :REM ADJUST IF <3
41ØØ Z=Z*(1.+Ø.Ø8*Z^4/K2^3)     :REM ADJUST
411Ø GOSUB 5ØØØ                 :REM NORMAL APPROX
412Ø RETURN
5ØØØ REM NORMAL CURVE APPROXIMATION
5Ø1Ø H=1+Ø.Ø49867347Ø*Z         :REM CALCULATE
5Ø2Ø H=H+.Ø2114100061*Z*Z       :REM  NORMAL CURVE
5Ø3Ø H=H+.ØØ32776263*Z^3        :REM  APPROXIMATION
5Ø4Ø H=H+.ØØØØ38ØØ36*Z^4        :REM  FROM PROGRAM
5Ø5Ø H=H+.ØØØØ4889Ø6*Z^5        :REM  6.2.
5Ø6Ø H=H+.ØØØØØ5383Ø*Z^6
5Ø7Ø H=H^-16/2                :REM FINISH APPROXIMAT.
5Ø8Ø RETURN
6ØØØ DATA 6Ø,7Ø,8Ø,-1           :REM DATA TO USE
6Ø2Ø DATA 9Ø,1ØØ,11Ø,-1
6Ø3Ø DATA 12Ø,13Ø,14Ø,-1
6Ø4Ø DATA -2                    :REM LAST GROUP IN
```

If you run this program, you'll find that it provides the F, and then tells you the significance level. There's no need for Appendix E at all, and no need to run another program to find the significance level. That's real convenience.

The program works as follows (see Fig. 13-23). Lines 10-170 are precisely the same as the corresponding lines in Program 13-7. In line 180, the two DFs and the F are equated to the values needed in the significance level program. Then in line 190, the program jumps to subroutine 3000. Recall subroutines? We first encountered them in Chapter 10. When Program 13-5 jumps to subroutine 3000, the significance level of the F will be calculated in variable PP, and then control will be returned to the main program, in this case to

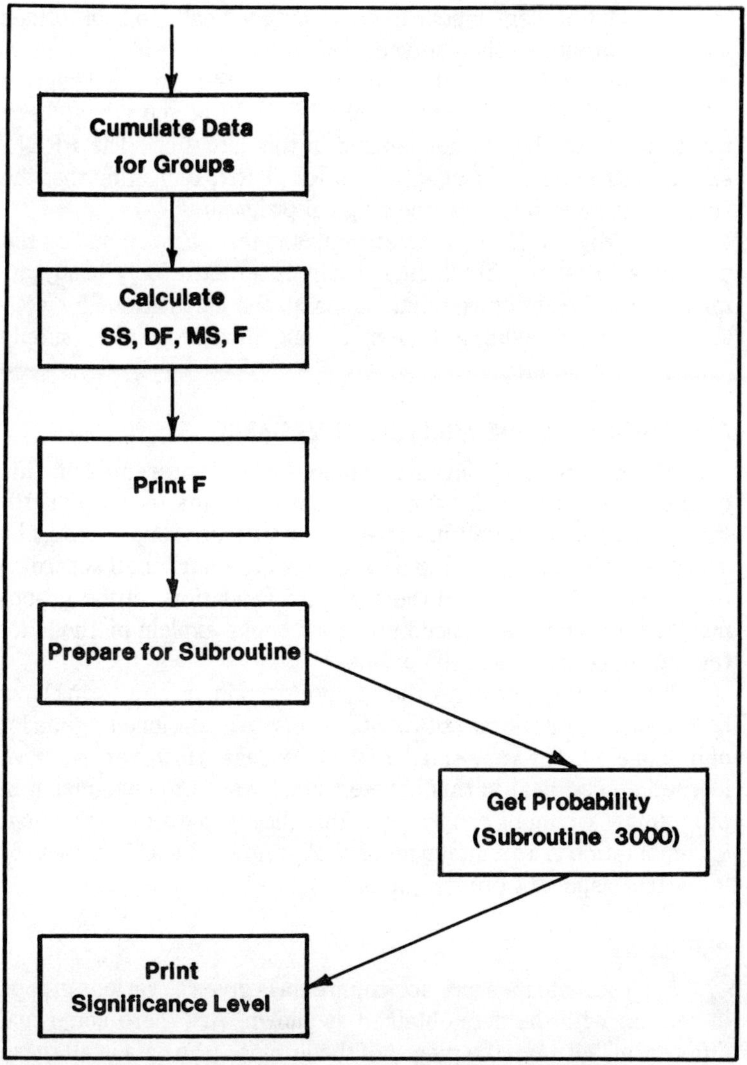

Fig. 13-23. Flowchart for ANOVA with significance level program.

line 200. In that line, the significance level will be printed out. Finally, the END statement is needed at line 210 to prevent the program from proceeding on to Subroutine 3000. Generally, we haven't needed END statements in our simple examples; they simply indicate the end of a program, as you'd imagine.

Subroutines are very important in advanced programming applications. If you look at many complex computer programs, you'll find that a lot of them appear to be nothing but calls to subroutines and the subroutines themselves.

Subroutine 3000 itself is almost exactly the same as Program 13-4. The INPUT statements, and a PRINT statement have been removed, since they're not needed in this situation. The PRINT statement has been replaced with a RETURN, indicating that the subroutine must return to the original program.

Try Program 13-5 on the examples in this chapter, and on the problems to follow. You'll find that it is an extremely handy arrangement. The microcomputer does all the calculations for you, and avoids your having to look up the significance level by simply calling a subroutine.

ASSUMPTIONS IN THE ANALYSIS OF VARIANCE

The assumptions for the simple ANOVA presented in this chapter are the same as for the t-test problems in Chapter 10: People must be randomly assigned to the various groups, it must be assumed that the underlying populations are distributed according to the Normal Curve, and the Standard Deviations of the groups must be the same. Advanced statistics books explain methods for testing these various assumptions.

We are at the end of another chapter. Again, no new terms have been added to our library of BASIC terms. We continue to get a lot of mileage out of a very small number of ideas. However, we have learned a good deal in this chapter about ways of economizing in programs—arranging programs so that they do a maximum amount of computation in a minimum number of program lines. That's a very important aspect of programming.

PROBLEMS

1. A scale to measure schizophrenia is given to various groups of people, with the data obtained as shown. Are there some real differences between the means of the groups? (the data in all these problems is, of course, fictitious)

Normals	56.1	13.1	13.4		
Phobics	67.3	5.7	35.6		
Obsessives	98.4	55.4	88.2	44.2	54.5
Schizophrenics	87.6	43.5	87.2	87.3	88.3

(F is 3.98, significant at 95 percent level. Try this one by hand!)

2. An intelligence test is given to students at the following schools, with data as shown:

State College	99	101	108	
Community College	108	99	99	
Harvard	182	169	180	120

(F=11.6441, significant at 99 percent level).

3. Intelligence tests are given to laborers, clerical workers, and engineers. Do the groups indeed differ?

Laborers	100	99	80	101	
Clerical	110	109	90	113	108
Engineers	120	119	100	122	125

(F=5.8970, significant at 95 percent level).

4. Here are some numbers designed to yield a rather large F. They are completely fictitious, but they're interesting.

Group 1	1	2	3	4
Group 2	50	60	70	80
Group 3	100	110	120	130

(F=113.8060, significant at 99 percent level).

5. A study is done to see whether cows, cats, or dogs are the brightest. An animal intelligence test is used, which has higher scores corresponding to higher intelligence.

Cows	101	103	105	
Cats	105	107	109	
Dogs	111	113	115	117

(F=21.2333, significant at 99 percent level).

18 FORMULAS ARE NOTHING - FOR A COMPUTER

14
Two-Way Analysis of Variance

In this chapter, we will take up a rather tricky statistical topic that shows the usefulness of the microcomputer admirably. The following procedures are very, very complex to calculate, yet your little computer can complete all of the calculations very quickly without the mistakes that a human calculator would be sure to make. Even better, the microcomputer will enable us to talk about the topic without being perpetually bogged down in computations. Only those of us who grew up in the days before computers can really appreciate this benefit.

In the present chapter, we will perform an Analysis of Variance procedure which is very similar to that discussed in the last chapter. However, our groups will be classified on two dimensions, rather than on just one. An example is essential to explain this. Look at Fig. 14-1. These fictitious data show, better than any other example I've found, the utility of the Two-Way ANOVA. In addition, all of the numbers work out evenly! In this diagram, we have IQ scores on eight people. Four are old and four are young. Further, four have had psychotherapy, and four haven't. People who haven't had a treatment are always called the *control group*. Notice that the figure is arranged so that the people are classified on two dimensions. We'd like to know whether age has any effect on IQ, and at the same time whether being in psychotherapy has any effect. With the experimental design shown in Fig. 14-1, we can test both ideas at one time, and do a better job of it in the bargain.

	Therapy	Control
Old	110 120	130 140
Young	130 140	110 120
Sums	500	500

Fig. 14-1. A two-way ANOVA.

A data analysis such as that shown in Fig. 14-1 is called, formally, a Two-Way Factorial Design, but we'll just call it a Two-Way Analysis of Variance in this book. Notice that the term Two-Way implies that there might be a Three-Way analysis, or a Four-Way analysis, and such things do exist. We cannot, of course, get into these complex topics here.

Look at Fig. 14-1 again. It's a bit difficult to interpret the data in a table of this type, especially if there are a great many pieces of data. Therefore, a table of the various Averages is almost always provided, as shown in Fig. 14-2. This figure provides the Average for each cell of Fig. 14-1, and also the Means for the rows and columns. All of these values will be used in the interpretation of the results.

INTERPRETATION WITH SIMPLE PROCEDURES

"Wait," you say. "Why couldn't we simply analyze the data with an Analysis of Variance like that found in the last chapter, or with a t-test? It would certainly work, and we wouldn't need any new procedures!"

And that is true, up to a point. Let's look at the data, ignoring the old-young dimensions:

	Therapy	Control
	110	130
	120	140
	130	110
	140	120
	-------	-------
Average	125	125
SD	11.18	11.18

Now, this set of data could be analyzed as a simple Analysis of Variance, using Program 13-3, or a simple t-test, using Program 10-1. Recall that an ANOVA with two groups is the same thing as a t-test. Run the programs. You'll find that Program 13-3 tells you that the F is zero, and Program 10-1 tells you that the t is zero. A zero value for either t or F indicates that there's no difference between the groups. That's rather obvious, since the Averages are

MEANS

	Therapy	Control	
Old	115	135	125
Young	135	115	125
	125	125	

Fig. 14-2. Means in a two-way ANOVA.

the same for both groups. Thus, we'd conclude that there isn't any difference between the IQs of those who have been in therapy and those who haven't.

If we look at the data from the other point of view, ignoring the therapy-no-therapy dimension, we obtain the following table:

	Old	Young
	110	130
	120	140
	130	110
	140	120
	------	------
Average	125	125
SD	11.18	11.18

Again, using Program 13-3 or Program 10-1, we find that the t and the F are zero. We'd conclude that there's no difference between the IQs of people who are old, and people who are young. The Averages are the same, and, indeed, the numbers are the same.

In short, by using simple ANOVA techniques, or the t-test, we would conclude that there is no difference between people with therapy and people with no therapy, or between old and young people. Thus, we might conclude that there is nothing to be found in Fig. 14-1 or Fig. 14-2. The techniques that we have found up to this time would force us to accept this conclusion.

THE CONCEPT OF INTERACTION

Is it really the case that there is nothing whatsoever in Fig. 14-2? Let's look at the Means a bit; indeed in Two-Way ANOVA it's customary to plot them, as shown in Fig. 14-3. In Fig. 14-3, we plot the means for the therapy and no-therapy group. Some people find diagrams of this sort confusing, but they're really simple. The therapy-no-therapy dimension is plotted on the horizontal (x) axis, while the values of the means are plotted on the vertical (y) axis. Then a line (dotted) is drawn between the points for the old group, and another one for the young group. Now watch what follows carefully:

What Fig. 14-3 (or Fig. 14-2, for that matter) says is this: old people do better, have higher IQs, without therapy. That's shown by the fact that with no therapy, old people's Average is 135, while with therapy it's only 115.

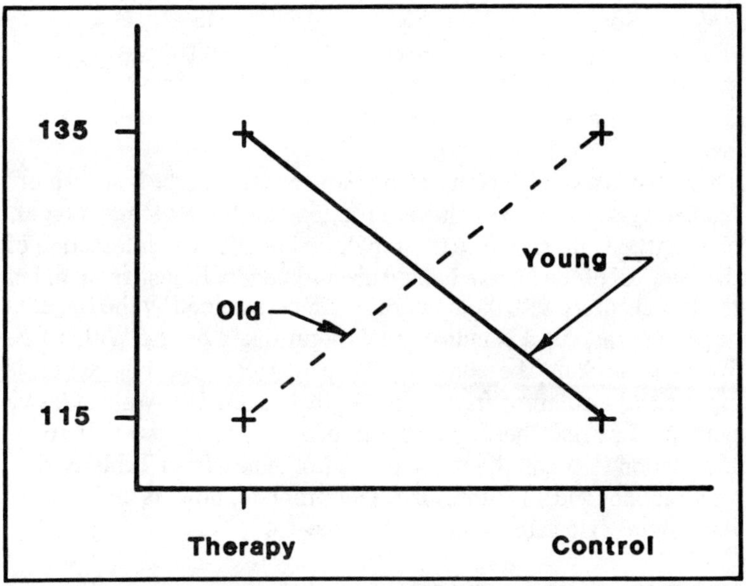

Fig. 14-3. A plot of the Means in two-way ANOVA.

At the same time, young people do better with therapy. With therapy, their IQ is 135 points, but without therapy, it's only 115 points.

In other words, there's more in Fig. 14-1 than can be learned with our previous techniques. The two dimensions of age and therapy work together to affect the people in different ways. Technically, it is said that the two dimensions interact, and the result is said to be a *significant Interaction*. The Two-Way ANOVA allows us to get at this effect; it would not be possible to do so with simpler techniques. Study Figs. 14-1 through 14-3 thoroughly until you understand the meaning of Interaction. You'll probably never see an example of Interaction that is this pure, ever, ever again!

THE SUMMARY TABLE

The Two-Way ANOVA produces a summary table, just as did the simple Analysis of Variance of the last chapter. For Fig. 14-1, it looks as follows:

Source	SS	DF	MS	F
Rows	0	1	0	0
Columns	0	1	0	0

Source	SS	DF	MS	F
Interaction	800	1	800	16
Within	200	4	50	
Total	1000	7		

This table is a good deal more complex than that of the last chapter. Instead of one F, we now have three, one for the Row effect (age), one for the Column effect (therapy), and one for the Interaction of the two. All of the F's are looked up in Appendix E, just as we did in the last chapter, with the numerator DF determined by the Effect in question, and the denominator DF determined by the Within DF. We can see from the summary table that only the Interaction is signficiantly different from zero. With 1 and 4 DF, we can be 95 percent sure that there is a real Interaction, since the F of 16 for Interaction exceeds the value of 7.71 obtained from Table E for 1 (Interaction) and 4 (Within) DF. Our problem, now, is understanding how to calculate all of these numbers!

TWO-WAY ANOVA FORMULAS

There are so many calculations needed to obtain the summary table that you might wonder whether anyone can really comprehend them. (Indeed, if you're the sort of person whose mathematics is rather shaky, you might want to skip over this section for the moment.) Still, we must present these formulas. As mentioned above, soon they will be incorporated into a computer program, and that will ultimately be the definition of the terms. We'll work out the data for Fig. 14-1 at the same time that we present the formulas. We'll also make the designations on the formulas conform to the computer program to come, to assist your understanding.

First, we're going to need the Correction Term, or the last part of Formula 13-1. This term is used so often in advanced ANOVA procedures that it is always defined first. We'll call it C to make it fit the computer program:

$$C = \frac{(\text{GRANDX})^2}{\text{GRANDN}} \qquad \textbf{14-1}$$

That formula says, simply, "take the sum of all the numbers in the table, square this sum, and divide by the number of numbers." The sum of all the numbers in our table (GRANDX) is 1,000, and thus C is 1,000*1, 000/8, or 125,000. Work through the numbers until you understand that.

Next, we must find the Total Sum of Squares, which we'll again call SST. This formula is precisely like Formula 13-1 of the last chapter, except, of course, for the fact that we've defined C, and that will simplify things a bit:

$$\text{SST} = \text{GRANDX2} - \text{C} \qquad \textbf{14-2}$$

That simply says, "take the sum of all the numbers, squared before summing (GRANDX2), and subtract C from it." The sum of all the squared numbers in the table (check this!) is 126,000, and thus SST is 126,000 – 125,000 or 1,000. Notice that this agrees with our summary table above.

The Degrees of Freedom, Total, is found as before:

$$\text{DFT} = \text{GRANDN} - 1 \qquad \textbf{14-3}$$

in our case, DFT will be 8 – 1, or 7.

ROW AND COLUMN EFFECTS

The Sums of Squares for the Row and Column effects are calculated precisely as they were in the last chapter for the Between values, although things look a bit different.

For the Row Sum of Squares, or SSR, we simply take the sum in each row, square it, and divide by the number of people in the row. Then we subtract the Correction Term. The formula looks a good deal more frightening than it needs to be:

$$\text{SSR} = \frac{1}{\text{NC} * \text{NN}} \Sigma\,(\text{ROWSUM})^2 - \text{C} \qquad \textbf{14-4}$$

ROWSUM is simply the sum for each row. NC*NN is the number of people in each row. NC is the number of columns, and NN is the number of people in each cell. If there are two people in each cell, and two columns, then there will be NC*NN people in each row. You'll have to think about that for a while before it makes sense! SSR, then, will be equal to the following:

$$\text{SSR} = \frac{1}{4}\,(500^2 + 500^2) - 125{,}000 = 0$$

If you look at Formula 13-3 from the last chapter, you'll discover that that's exactly what we did there.

The formula for Columns is almost identical:

$$SSC = \frac{1}{NR*NN} \cdot \Sigma(COLSUM)^2 - C \qquad \textbf{14-5}$$

$$= \frac{1}{4} (500^2 + 500^2) - 125{,}000 = 0$$

Thus, we square the sum for each column, divide by the number in the column (NR*NN), add them up, and subtract the Correction Term. NR refers to the number of rows.

The Degrees of Freedom for Rows and Columns are exactly analogous to Formula 13-4 of the last chapter:

$$DFR = NR - 1 = 2 - 1 = 1 \qquad \textbf{14-6}$$

$$DFC = NC - 1 = 2 - 1 = 1 \qquad \textbf{14-7}$$

The number of formulas is rapidly increasing! But so far there's nothing new in the computations.

THE INTERACTION VALUES

The formula for Interaction is new, and requires a new value:

$$SSI = \frac{1}{NN} \Sigma(CELLSUM)^2 - SSR - SSC - C \qquad \textbf{14-8}$$

This says, simply, "Take the sum in each cell (CELLSUM), square it, and add them all up. Divide this sum by the number in each cell. Subtract from that the Sum of Squares for Rows, the Sum of Squares for Columns, and the Correction Term." Working it out is a good deal easier than you'd think.

$$SSI = \frac{(230^2 + 270^2 + 270^2 + 230^2)}{2} - SSR - SSC - C$$

$$= 125800 - 0 - 0 - 125000 = 800$$

The DF for Interaction is simply the product of the DF for Rows and the DF for Columns:

$$DFI = DFR * DFC = 1 * 1 = 1 \qquad \textbf{14-9}$$

All of these values have been entered in the summary table.

THE WITHIN VALUES

The Within or Error values, as they're called in these complex problems, are simply what's left over from the Total values.

$$SSW = SST - SSR - SSC - SSI \qquad \textbf{14-10}$$

$$= 1000 - 0 - 0 - 800 = 200$$

$$DFW = DFT - DFR - DFC - DFI \qquad \textbf{14-11}$$

$$= 7 - 1 - 1 - 1 = 4$$

And, thus, we have most of the major numbers needed.

THE MEAN SQUARES AND F's

The Mean Squares are calculated almost exactly as shown in Formulas 13-7 and 13-8 of the last chapter:

$$MSR = SSR/DFR = 0/1 = 0 \qquad \textbf{14-12}$$
$$MSC = SSC/DFC = 0/1 = 0 \qquad \textbf{14-13}$$
$$MSI = SSI/DFI = 800/4 = 200 \qquad \textbf{14-14}$$
$$MSW = SSW/DFW = 200/4 = 50 \qquad \textbf{14-15}$$

The F's are obtained in this simple example by dividing everything else by MSW:

$$FR = MSR/MSW = 0/50 = 0 \qquad \textbf{14-16}$$
$$FC = MSC/MSW = 0/50 = 0 \qquad \textbf{14-17}$$
$$F\,I = MSI/MSW = 800/50 = 16 \qquad \textbf{14-18}$$

And those are all the fomulas! Only eighteen of them. Clearly, no one could keep track of them all.

A TWO-WAY ANOVA COMPUTER PROGRAM

The following computer program will do all the computations for us. It's arranged so that the computations are worked out generally in the same order as the previous formulas.

```
10 REM PROGRAM 14.1. TWO-WAY ANOVA
20 DIM CR(5),CC(5)                  :REM SAVE SPACE
30 READ NR,NC,NN                    :REM GET SIZE
100 FOR JR=1 TO NR                  :REM ROW LOOP
110 FOR JC=1 TO NC                  :REM COLUMN LOOP
120 S=0                             :REM CLEAR CELL SUM
200 FOR JN=1 TO NN                  :REM CELL LOOP
210 READ X                          :REM READ DATA
220 GS=GS+X:GQ=GQ+X*X               :REM CUMULATE SUMS
230 GN=GN+1
240 CR(JR)=CR(JR)+X                 :REM GET ROW TOTALS
250 CC(JC)=CC(JC)+X                 :REM GET COL TOTALS
260 S=S+X                           :REM GET CELL SUM
270 NEXT JN                         :REM END CELL LOOP
280 II=II+S*S                       :REM CUM INT SUMS
290 NEXT JC                         :REM END COL LOOP
300 NEXT JR                         :REM END ROW LOOP
500 C=GS*GS/GN                      :REM CORRECTION TERM
510 ST=GQ-C:DT=GN-1                 :REM TOTAL VALUES
600 FOR JR=1 TO NR                  :REM SSR LOOP
610 TR=TR+CR(JR)*CR(JR)             :REM LH SIDE 14.4
620 NEXT JR                         :REM END SSR LOOP
630 TR=TR/(NN*NC)                   :REM FINISH LH SIDE
640 SR=TR-C:DR=NR-1                 :REM GET ROW VALUES
650 MR=SR/DR                        :REM FORMULA 14.12
700 FOR JC=1 TO NC                  :REM SSC LOOP
710 TC=TC+CC(JC)*CC(JC)             :REM LH SIDE 14.5
720 NEXT JC                         :REM END SSC LOOP
730 TC=TC/(NN*NR)                   :REM FINISH LH SIDE
740 SC=TC-C:DC=NC-1                 :REM GET COL VALUES
750 MC=SC/DC                        :REM FORMULA 14.13
800 SI=II/NN -TR -TC +C             :REM 14.8-- GET INT
810 DI=DR*DC                        :REM GET DF INT
820 MI=SI/DI                        :REM FORMULA 14.4
830 SW=ST -SR -SC -SI               :REM 14.10-- WITHIN
840 DW=DT -DR -DC -DI               :REM DF WITHIN
850 MW=SW/DW                        :REM 14.15-MS WITHIN
900 FR=MR/MW                        :REM GET ROW F
910 PRINT "FR=",FR                  :REM PRINT IT
920 FC=MC/MW                        :REM GET COL F
930 PRINT "FC=",FC                  :REM PRINT IT
940 FI=MI/MW                        :REM GET INT F
950 PRINT "FI=",FI                  :REM PRINT IT
1000 DATA 2,2,2                      :REM DATA TO USE
1010 DATA 110,120,130,140
1020 DATA 130,140,110,120
```

The program is a reasonably long one, although not really any longer than some of the less efficient ones in the last chapter. However, the logic of the program is much more complex, and it will require several flowcharts to explain it. The flowchart in Fig. 14-4 gives an overall view of the program.

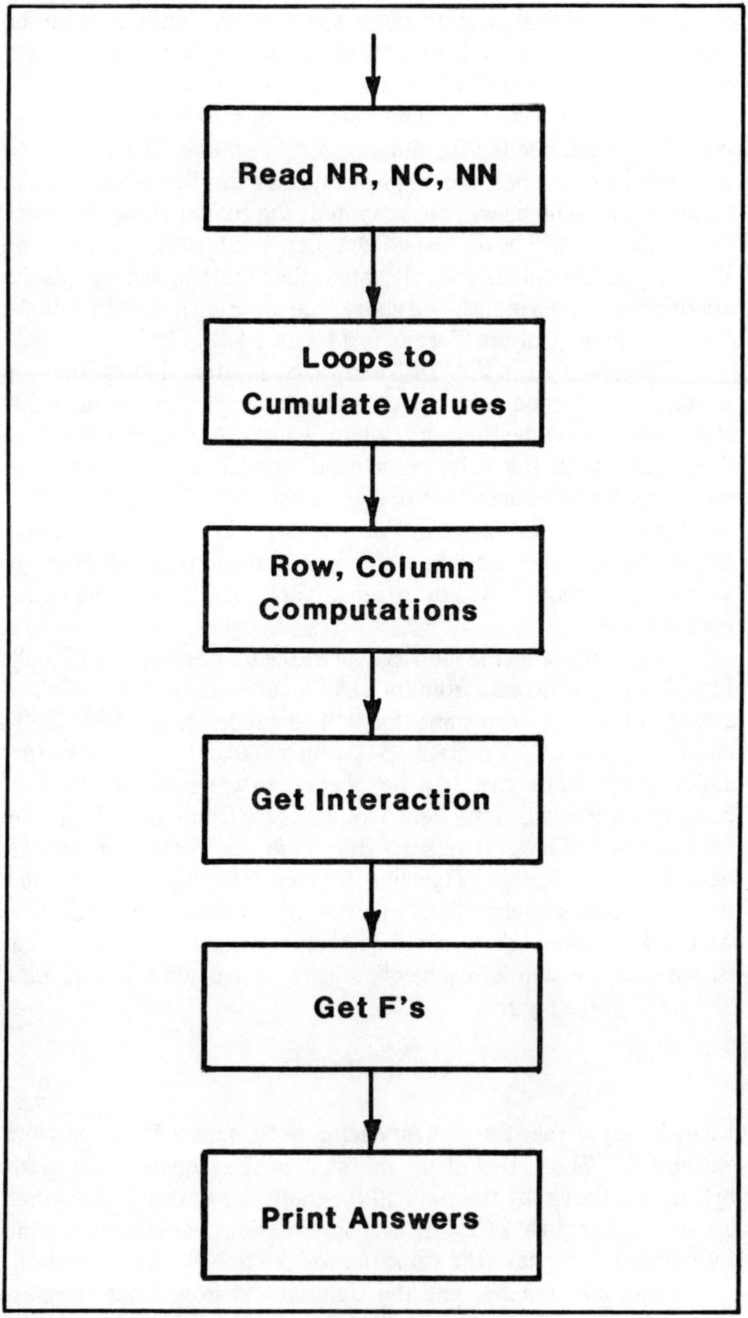

Fig. 14-4. Two-way ANOVA program flowchart.

Line 20 of the program saves space in which to cumulate the Row and Column sums. Line 30 reads in the number of Rows (NR), numbers of Columns (NC), and the number per cell (NN).

Then, from Fig. 14-4, a number of loops are needed to cumulate the values needed for our eighteen formulas. These will be discussed next. Then the Sum of Squares for Rows and Sum of Squares for Columns will be computed; the computations for these are similar and will be discussed together. Next, the Interaction and Within or Error values are calculated, then the F's, and the results are printed out. As mentioned above, the program generally follows the arrangement of the Formulas 14-1 to 14-18.

The flowchart in Fig. 14-5 shows, in shortened form, the way in which the needed sums are cumulated. Loops begin in lines 100 and 110 to read in the Row and Column values. An entire row will be done first, with the columns second. In line 120, we begin to cumulate the values needed to calculate the SS for Interaction. We'll need the sum for each cell in Formula 14-8. Consequently, in line 120 we clear out variable S, which will be used to get this sum for each cell; variable S will be reset to zero each time a new cell is encountered.

Line 200 begins the loop that sums the values within each cell. Line 210 reads the data from the DATA statements. Lines 220 and 230 calculate the grand sums that will be needed to get SST, DFT, and so on, just as in Program 13-3. Lines 240 and 250 get the Row and Column sums that will be needed to calculate the Sum of Squares for Rows and the Sum of Squares for Columns in Formulas 14-4 and 14-5. Line 260 gets the sum, S, for each cell; remember it from the last paragraph? Then line 270 ends the individual cell loop.

Line 280 is tricky. To obtain the Interaction Sum of Squares, we need the sum of the squared cell sums (that's a mouthful!). That is, we take the sum for each cell, square it, and add them up. Line 280 does precisely that:

```
280 II=II+S*S
```

Line 280 says, Take the sum for each cell (S), square it, and add it to variable II. Thus, II will be the sum of the squared cell sums mentioned above, by the time all the loops are finished. And when we get to Formula 14-8, the one for the Sum of Squares for the Interaction, we'll have the value needed for II/NN in that equation.

Lines 290 and 300 end the Column and Row loops, respectively. By line 300, we have all of the sums that we need.

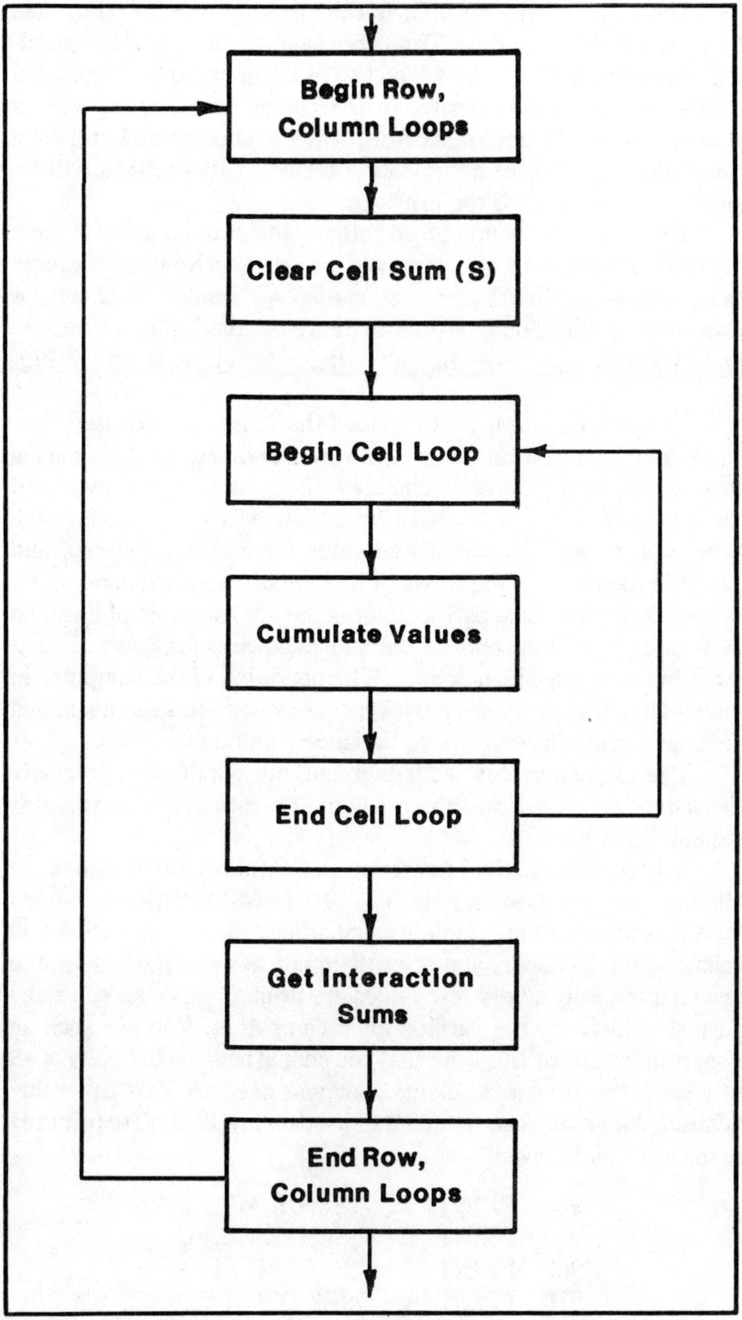

Fig. 14-5. Flowchart for calculation of sums in a two-way ANOVA.

The Correction Term, C, is calculated in line 500 (Formula 14-1), while the SS and DF Total are calculated in line 510, according to Formulae 14-2 and 14-3 just as in Program 13-3. You should be able to follow this easily. In the computer program, SST is abbreviated to ST, since our BASIC will accept only two letters at a time. This practice of abbreviating terms to two letters will be followed in the rest of the program.

The Row and Column SS and other values are obtained in lines 660-750. Program 14-1 is arranged so that the Row and Column calculations are almost identical, and so we'll talk only about the Row values; you should be able to figure out the Column values in lines 700-750 with little difficulty. Refer to the flowchart in Fig. 14-6.

To solve Equation 14-4 we need the sum of the squared Row sums (read that several times until you understand it). That means that we're going to have to cumulate things across the rows, and thus, in line 610 a Row Loop is begun. In line 610, the sum for each Row, CR, (stands for cumulated value for Row) is squared, and added to the total for the Rows, TR. Line 620 ends the loop to get the needed value. Line 630 completes the left hand side of Formula 14-4, and line 640 and 650 get the Sum of Squares for Rows, the DF for Rows, and the Mean Square. The operation of the computer in lines 600-650 is very, very tricky. Such loops are used again and again and again in Analysis of Variance Programs.

The Column values, as mentioned, are obtained in precisely the same way as the Row values in lines 700-750. There's no need to explain them here.

In lines 800-850 the Interaction and Within or Error values are obtained just as shown in Formulas 14-8 to 14-11. There's no need for a flowchart for these simple calculations. Finally, lines 900-950 calculate the F-values, and print them out. As was the case in the last chapter, only a very few values are printed, since screen sizes vary so widely on the various microcomputers. You are such an expert in BASIC by this time that you should have no difficulty at all in adding the print statements that you need to give any value desired. As a hint, the statements I used on my TRS-80 to print the results are as follows:

```
955 PRINT "R  ";SR,DR,MR,FR
960 PRINT "C  ";SC,DC,MC,FC
965 PRINT "I  ";SI,DI,MI,FI
970 PRINT "W  ";SW,DW,MW
975 PRINT "T  ";ST,DT
```

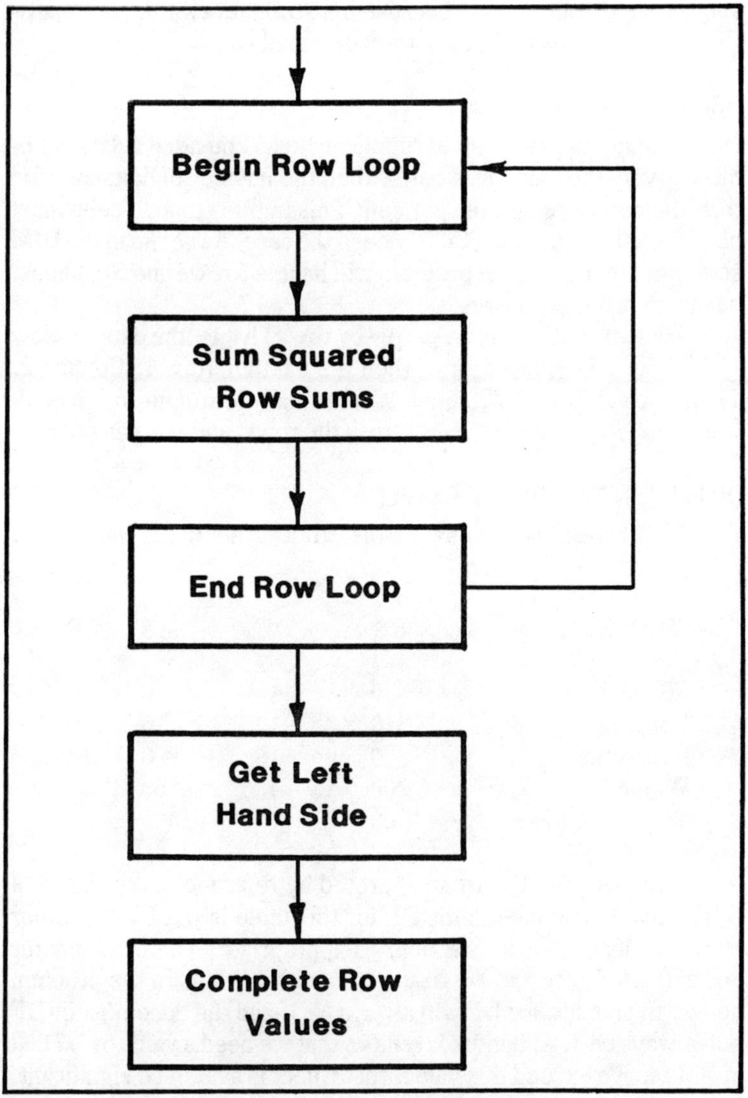

Fig. 14-6. Flowchart for calculation of row sum of squares.

but these won't necessarily work with computers with different size screens.

Program 14-1 is only one approach to programming a Two-Way ANOVA. There are many other ways to do it, some of them using completely different formulas than the ones we have here. I like the approach in Program 14-1 because it's straightforward, and repetitive enough so that it's easy to understand.

USING THE PROGRAM

Running the program is simple enough. You need a data statement giving the number of rows, then the number of columns, and then the number of people per cell. This implies that all cells must be of the same size, and that is indeed the case. As set up in the DIM statement of line 20, the program will handle 5 rows and 5 columns, but that's easily changed.

Then the data is entered, row by row. That is, the data for Row 1, Column 1 is entered first, then the data for Row 1, Column 2. Then comes Row 2, Column 1, and finally the data for Row 2, Column 2. Just enter the data across the rows, and you can't miss!

INTERPRETATION OF THE RESULTS

We'll repeat the summary table, from earlier in the chapter, for reference.

Source	SS	DF	MS	F
Rows	0	1	0	0
Columns	0	1	0	0
Interaction	800	1	800	16
Within	200	4	50	
Total	1000	7		

As mentioned, the F's are interpreted by reference to the table of Appendix E. The numerator DF for this table is the DF for either Rows, Columns, or Interaction, as appropriate. The denominator DF will always be the DF associated with Within. In the problem above, the numerator DF will always be 1, and the denominator DF will always be 4. Appendix E tells us that we need a value of 7.71. If any of the F's exceed this value, that Effect is said to be significant. In the table above, only the Interaction effect exceeds 7.71, and thus we have a significant Interaction.

The significant Interaction could have been inferred from Fig. 14-3, the plot of the means. If the two lines are parallel in this plot, then there is no Interaction. If the lines are not precisely parallel, that is, if they will intersect at any point, including points not on the page, there's an Interaction. Clearly, our lines intersect, and therefore there is an Interaction. Frequently, however, the lines will be almost parallel, but not quite, and then we can't be sure about the Interaction just from the plot of the means. Only the summary table, previously, and the Interaction F can tell the true story.

MORE EXAMPLES

Figure 14-7 presents the same data as that in Fig. 14-1, but the numbers are arranged quite differently. The Means for these data are shown in Fig. 14-8, and a plot of the Means in Fig. 14-9. Put the data into Program 14-1. It should provide you with the summary table that follows:

Source	SS	DF	MS	F
Rows	0	1	0	0
Columns	800	1	800	16
Interaction	0	1	0	0
Within	200	4	50	
Total	1000	7		

From an inspection of Fig. 14-8 and 14-9, we can learn a lot about

	Therapy	Control
Old	110 120	130 140
Young	110 120	130 140

Fig. 14-7. 2 × 2 ANOVA: only a column effect.

MEANS

	Therapy	Control	
Old	115	135	125
Young	115	135	125
	115	135	

Fig. 14-8. 2 × 2 ANOVA: column effect only Means.

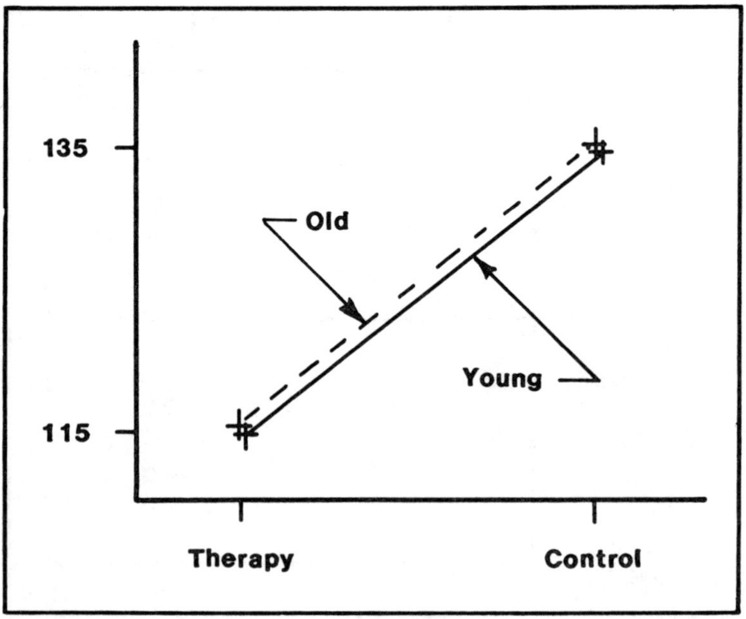

Fig. 14-9. 2 × 2 ANOVA: column effect only plot.

the data. First, notice that the lines for both the Old and Young groups are superimposed on top of each other (they're drawn slightly apart in Fig. 14-6 only so that you can see both). Since they're in the same place, they're parallel, and that means that there's no Interaction. This can clearly be seen from the summary table, where the F for Interaction is Zero. From Fig. 14-8, we notice that the Means for the Rows are identical, and that means that there shouldn't be any difference between the Rows. This situation is reflected in the fact that the F for Rows is again zero. Figure 14-9 does show a difference in Means between the Columns—115 versus 135. That's a substantial difference, and it's reflected in the F for Columns of 16. With 1 and 4 DF, the value needed for significance at the 95 percent level is again 7.71, and the Column F of 16 exceeds this value. Therefore, the Column effect is significant at the 95 percent level. Thus, we may conclude that there is a real difference in IQ between people who have had therapy and people who haven't. People who haven't are smarter, but age doesn't make a difference, and there's no Interaction.

Notice that, in this example, the Sum of Squares has shifted from Interaction, in the first example, to Columns, just as you'd expect. The value of 800 moves from one place to another. In this case, by the way, the result would be precisely the same as that obtained from a simple, one-way ANOVA. That's always the case when the Interaction is zero.

Figure 14-10 shows an example in which there are no significant findings whatever. The Means are shown in Fig. 14-11, and a plot of them in Fig. 14-12. Notice that the Means for all the cells are the same. All of the Averages in the entire table are the same! We'd expect no effects at all, and that is just what is found in the summary table:

Source	**SS**	**DF**	**MS**	**F**
Rows	0	1	0	0
Columns	0	1	0	0
Interaction	0	1	0	0
Within	200	4	50	
Total	200	7		

As expected, none of the F's are significant. Indeed, they're all zero. The Interaction is zero because the lines in Fig. 14-12 are parallel, and the Row and Column F's are zero because the Row and

	Therapy	Control
Old	110 120	110 120
Young	110 120	110 120

Fig. 14-10. 2 × 2 ANOVA: no effect at all.

MEANS

	Therapy	Control	
Old	115	115	115
Young	115	115	115
	115	115	

Fig. 14-11. 2 × 2 ANOVA: no effect Means.

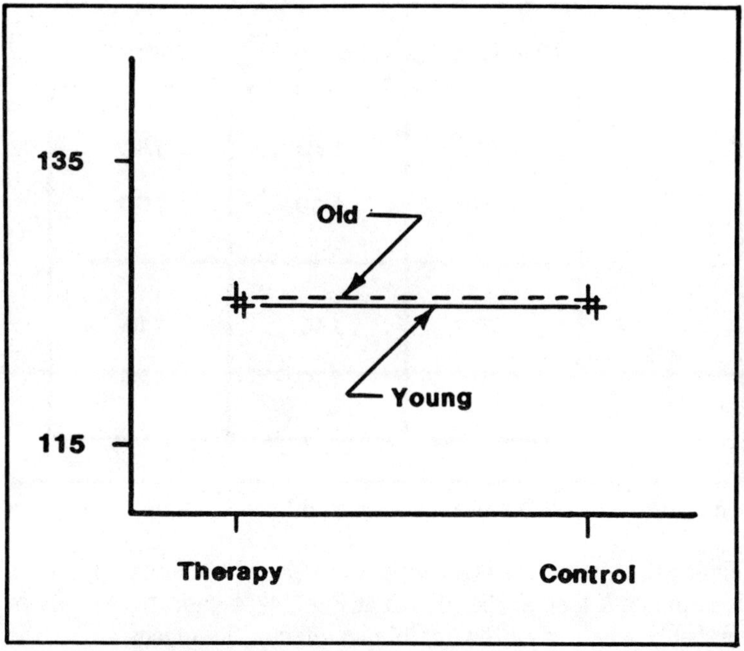

Fig. 14-12. 2 × 2 ANOVA: no effect plot.

Column means are identical. Notice that the Within sum of squares remains precisely the same as it was in the last two examples—200. That's because the variability in the cells (SD) has not changed in any of the examples.

If you study the sequences of figures that I have presented closely and repeatedly and work the problems through on your microcomputer, you'll understand the ideas behind the Two-Way Analysis of Variance far better than most people. Pure examples of this sort help understanding immensely.

A LARGER TABLE

It is impossible to have too many examples of this sort of thing. We'll branch out now to some more complex problems. First of all, by studying the figures, you might have decided that the technique applies only to 2 by 2 tables. That's not the case (the program will do up to a 5 by 5 table), although perhaps 2 by 2 tables make the best examples. Figures 14-13 through 14-15 provide an example of a pure Interaction in a 2 Row by 3 Column table. In this set of fictitious data, we've added a Drug group. Presumably, these people are

	Therapy	Drug	Control
Old	110 120	130 140	150 160
Young	150 160	140 130	110 120

Fig. 14-13. 2 × 3 ANOVA with only interaction.

given drugs, while the other groups are given psychotherapy and no treatment. A look at the Means of Fig. 14-14 indicates that there probably aren't any Row or Column effects (the means are all the same!). Figure 14-15 shows that the lines cross, so that there may

MEANS

	Therapy	Drug	Control	
Old	115	135	155	135
Young	155	135	115	135
	135	135	135	

Fig. 14-14. 2 × 3 ANOVA Means.

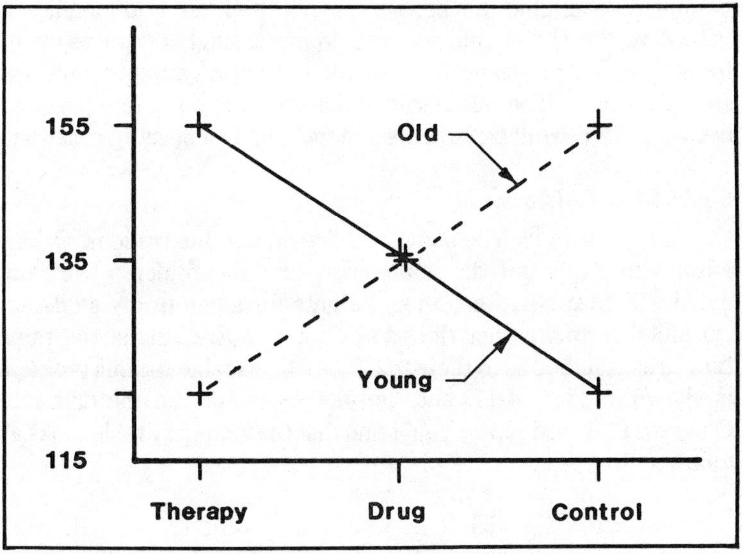

Fig. 14-15. 2 × 3 ANOVA plot.

very well be an Interaction. The summary table bears this out:

Source	SS	DF	MS	F
Rows	0	1	0	0
Columns	0	2	0	0
Interaction	3200	2	1600	32
Within	300	6	50	
Total	3500	11		

Notice that in the summary table, the F for Rows and Columns is indeed zero. Any F of less than one is not significant, and thus nothing is going on there. Do notice, though, that the DF for Columns is now 2, rather than the 1 of the last examples. We now have three Columns, rather than two, and thus the DF should increase.

The Interaction F is, as we suspected, large—32 in this case. With 2 (Interaction) and 6 (Within) DFs, Appendix E says that we need a value of 5.14 for significance at the 95 percent level. Thirty two is clearly higher than 5.14, and thus there really is an Interaction.

Again, we have a pure example. There is an Interaction, and

nothing else at all in this problem. What the Interaction implies is the following: Old people seem to do increasingly better as we go from Therapy to Drugs to Control; however, young people get steadily worse. The net result is that there is no difference at all between the overall treatments and the Old-Young categorization.

SOME REAL DATA

It is wise to include some real data in our illustrations. Otherwise, you might get the impression that all problems work out evenly! Figure 14-16 presents weights for some of my students, classified as to whether they do or don't smoke, and at the same time are classified as to their sex. The Means table for this problem is shown in Fig. 14-17, and the plot in 14-18. Put the data into Program 14-1, and run it. You'll find that the summary table looks as follows:

Source	SS	DF	MS	F
Rows	8789.06	1	8789.06	21.58
Columns	826.56	1	826.56	2.03
Interaction	95.06	1	95.06	.23
Within	4886.25	12	407.19	
Total	14596.90	15		

	Smoke	Don't Smoke
Male	165 195 175 170	180 155 190 142
Female	121 127 132 157	85 115 110 150

Fig. 14-16. Sex-smoking weight ANOVA.

MEANS

	Smoke	Don't Smoke	
Male	176.25	166.75	171.5
Female	134.25	115	124.63
	155.25	140.88	

Fig. 14-17. Sex-smoking weight Means.

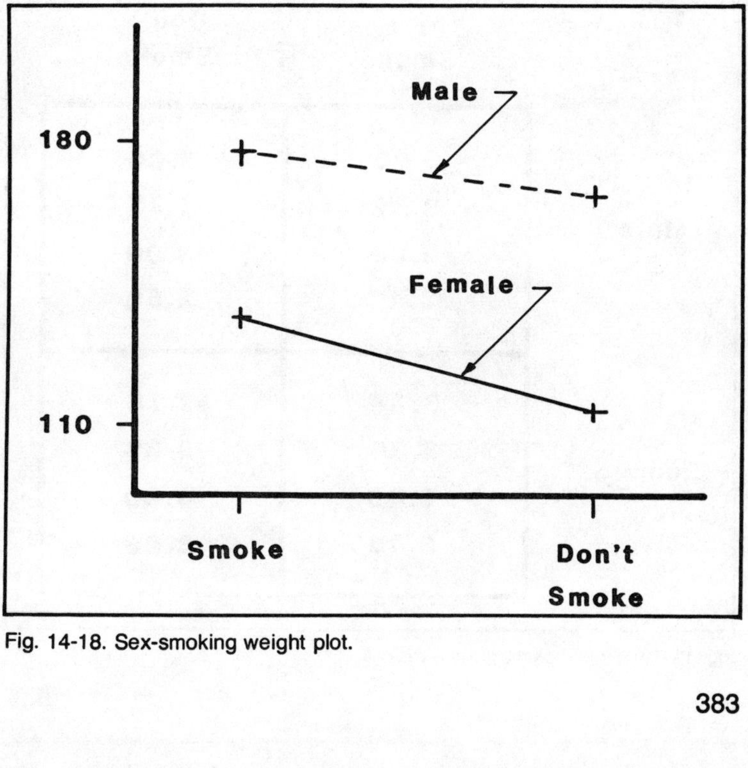

Fig. 14-18. Sex-smoking weight plot.

In this problem we have 16 people, and thus there are 15 total DF. All of the effects will involve 1 and 12 DF, and Appendix E tells us that we'll need a value of about 4.90 for significance.

The summary table tells us, then, that there is a significant Row effect, since the F of 21.58 exceeds 4.90. That means that there is a weight difference between males and females; we would expect this to be the case from casual observation.

However, the F for the Column (Smoke-Don't Smoke) effect is only 2.03, which is not significant. Looking at Fig. 14-17, we note that people who smoke weigh a good deal more than people who don't smoke (155 pounds versus 141 pounds), but with our small sample this is not greater than what could be expected by chance. (Actually, I'd think that people who smoke would weigh less than those who do, but that doesn't seem to be the case.)

Looking at Fig. 14-18, we see that the lines for males and females are very nearly parallel. Thus, we'd expect no Interaction, and the F for Interaction is only .23—clearly not significant. As an aside, I've done many, many Analyses of Variance of this type, and have found very few significant Interactions in real data. They just don't seem to occur very often.

	Smoke	Don't Smoke
Male	3.20 3.52 3.00 3.10	3.00 2.90 3.00 2.50
Female	3.77 3.30 3.45 2.40	2.78 3.39 3.00 3.68

Fig. 14-19. Sex-smoking grade ANOVA.

MEANS

	Smoke	Don't Smoke	
Male	3.20	2.85	3.02
Female	3.23	3.21	3.22
	3.22	3.03	

Fig. 14-20. Sex-smoking grade Means.

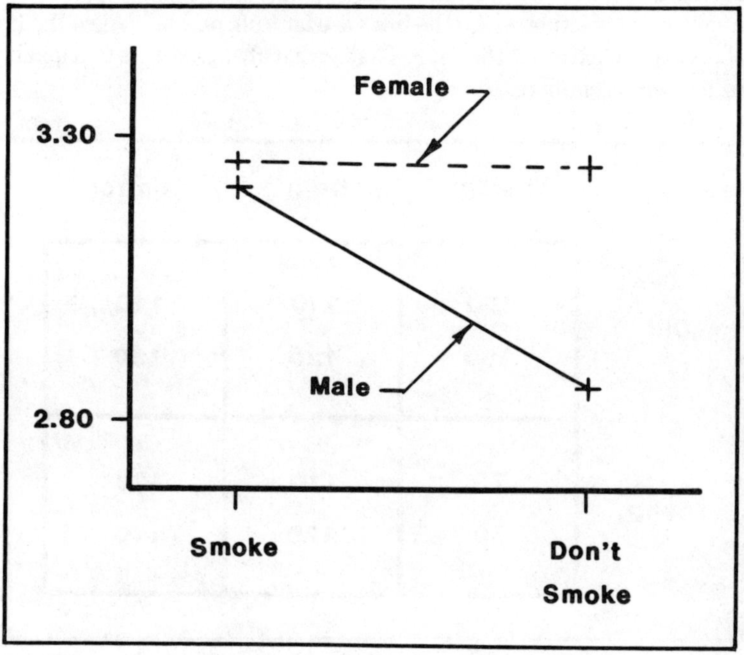

Fig. 14-21. Sex-smoking grade plot.

Another set of real data is shown in Figs. 14-19, 14-20, and 14-21. I had the idea that perhaps intelligence might be related to smoking, and so obtained College Grade Point Average (GPA) on some students, along with information about their smoking habits. At our school, someone who gets all E's has a GPA of zero, and someone who gets all A's has a GPA of four. As you can see from the Summary Table, all of my hypotheses were wrong:

Source	**SS**	**DF**	**MS**	**F**
Rows	.1502	1	.1502	.98
Columns	.1387	1	.1387	.91
Interaction	.1139	1	.1139	.74
Within	1.8384	12	.1532	
Total	2.2412	15		

Whenever an F is less than one, it is clearly not significant, and thus we find that none of the F's in the above table are significant.

Contrary to what could be expected from Fig. 14-20, males don't really have a lower GPA than females, people who don't smoke aren't less bright than those who do. And Fig. 14-21 might imply an Interaction, since the lines are far from parallel. Alas, the F shows that such is not the case. That's what always seems to happen to me with real data.

	Therapy	**Drug**	**Control**
Old	150 160	110 120	110 120
Young	70 80	110 120	120 110

Fig. 14-22. What one hopes for.

MEANS

	Therapy	Drug	Control	
Old	155	115	115	128.33
Young	75	115	115	101.67
	115	115	115	

Fig. 14-23. What one hopes for (Means).

A FINAL FICTITIOUS EXAMPLE

When doing research with Two-Way Analysis of Variance procedures, one secretly hopes that the data will come out as shown in Fig. 14-22. The Means are given as Fig. 14-23, and the plot is shown in Fig. 14-24. Program 14-1 will provide the summary table, as follows:

Source	SS	DF	MS	F
Rows	2133.33	1	2133.33	42.66
Columns	0.00	2	0.00	0.00
Interaction	4266.67	2	2133.34	42.67
Within	300.00	6	50	
Total	6700.00	11		

The F for Rows (Age) is 42.66. With 1 and 6 DF, Appendix E tells us that a value of 5.99 is needed for significance, and 42.66 clearly exceeds that. Thus, we can be 95 percent sure that older people really are smarter than younger people in our very fictitious data. The F for Columns (type of treatment) is zero. Figure 14-23 makes it rather clear that there really is no difference between people undergoing Therapy, Drugs, or no treatment, because the means

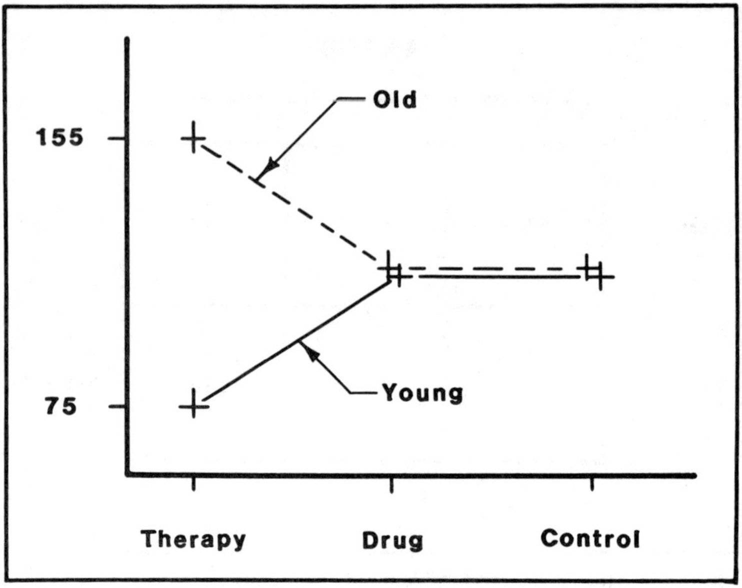

Fig. 14-24. What one hopes for (plot).

are all the same, and the F shows that clearly.

Figure 14-24, however, is very, very interesting. There's no difference between Old and Young people while taking drugs, or in the control group. However, at the left side of the diagram, we note that old people have a much higher IQ with therapy than do young people. This is precisely the sort of effect that one hopes for when calculating an Interaction, and it is very clear in this example. The F of 42.67 shows clearly that the Interaction is significant, for with 2 and 6 DF Appendix E shows that an F of only 5.14 is needed for significance.

Thus, in this example, there is a significant Row or Age effect, and a significant Interaction. The information provided by the Interaction tells us that the Row effect is caused largely by things going on within the therapy group—nothing happens in the drug or control groups. When a researcher calculates an Interaction, he or she hopes that the the Interaction will show something that will not be obvious from the simple Row and Column effects.

COMPARISION WITH ONE-WAY ANOVA

In the last example, the Row effect (Age) was significant. It would be tempting to think that the same conclusion could be reached about the Row effect if a simple One-Way ANOVA were used. It would look as follows:

	Old	**Young**
	150	70
	160	80
	110	110
	120	120
	110	110
	120	120
	------	------
Average	128.33	101.67
SD	19.51	19.51

Here, the data for the Rows in the last example has simply been written as columns (be sure that you understand that!). The summary table for the One-Way Anova looks as follows:

Source	**SS**	**DF**	**MS**	**F**
Between	2133.33	1	2133.33	4.67
Within	4566.67	10	456.67	
Total	6700	11		

Notice that in the case the F, 4.67 is not signficant! Appendix E says that for 1 and 10 DF we need a value of 4.96 for significance at the 95 percent level, and 4.67 doesn't reach that level, but it was significant in the Two-Way ANOVA. How can the two procedures differ so much? You'd expect that they'd yield similar results.

If we compare the summary table for the One-Way Anova with that for the Two-Way, we immediately see the difference. The SS for Interaction in the Two-Way Anova is merged with the SS for Within to make up SS Within in the One-Way Anova. In the simpler ANOVA there's no way to separate out the effect of the Interaction, and thus it winds up in the Within category. If there were a Column effect in the Two-Way Anova, it would also be subtracted from the Within SS.

The result of all this is that Two-Way ANOVAs are almost always more sensitive than One-Way designs. In statistical terminology, we say that they're generally more *powerful*. More complex designs, such as Three- and Four- way designs, have many,

many more terms, and are thus even more sensitive. That's one of the reasons they're so popular in advanced work.

TWO-WAY ANOVA WITH AUTOMATIC SIGNIFICANCE LEVELS

It would be nice if our little computers could calculate the significance levels for us automatically. Nothing could be easier. We simply tack an F-value significance program, such as that in Program 13-4, onto Program 14-1, as shown: (This program just barely fits into my Sharp PC-1500; you may have to eliminate all blanks to make it fit on a small machine.)

```
10 REM PROGRAM 14.2. TWO-WAY ANOVA WITH SIGN
20 DIM CR(5),CC(5)          :REM SAVE SPACE
30 READ NR,NC,NN            :REM GET SIZE
100 FOR JR=1 TO NR          :REM ROW LOOP
110 FOR JC=1 TO NC          :REM COLUMN LOOP
120 S=0                     :REM CLEAR CELL SUM
200 FOR JN=1 TO NN          :REM CELL LOOP
210 READ X                  :REM READ DATA
220 GS=GS+X:GQ=GQ+X*X       :REM CUMULATE SUMS
230 GN=GN+1
240 CR(JR)=CR(JR)+X         :REM GET ROW TOTALS
250 CC(JC)=CC(JC)+X         :REM GET COL TOTALS
260 S=S+X                   :REM GET CELL SUM
270 NEXT JN                 :REM END CELL LOOP
280 II=II+S*S               :REM CUM INT SUMS
290 NEXT JC                 :REM END COL LOOP
300 NEXT JR                 :REM END ROW LOOP
500 C=GS*GS/GN              :REM CORRECTION TERM
510 ST=GQ-C:DT=GN-1         :REM TOTAL VALUES
600 FOR JR=1 TO NR          :REM SSR LOOP
610 TR=TR+CR(JR)*CR(JR)     :REM LH SIDE 14.4
620 NEXT JR                 :REM END SSR LOOP
630 TR=TR/(NN*NC)           :REM FINISH LH SIDE
640 SR=TR-C:DR=NR-1         :REM GET ROW VALUES
650 MR=SR/DR                :REM FORMULA 14.12
700 FOR JC=1 TO NC          :REM SSC LOOP
710 TC=TC+CC(JC)*CC(JC)     :REM LH SIDE 14.5
720 NEXT JC                 :REM END SSC LOOP
730 TC=TC/(NN*NR)           :REM FINISH LH SIDE
740 SC=TC-C:DC=NC-1         :REM GET COL VALUES
750 MC=SC/DC                :REM FORMULA 14.13
800 SI=II/NN -TR -TC +C     :REM 14.8-- GET INT
810 DI=DR*DC                :REM GET DF INT
820 MI=SI/DI                :REM FORMULA 14.4
830 SW=ST -SR -SC -SI       :REM 14.10-- WITHIN
840 DW=DT -DR -DC -DI       :REM DF WITHIN
850 MW=SW/DW                :REM 14.15-MS WITHIN
900 FR=MR/MW                :REM GET ROW F
```

```
905 K1=DR:K2=DW:KØ=FR:GOSUB 3ØØØ :REM ROW SIGN
91Ø PRINT "FR=",FR,PP          :REM PRINT IT
92Ø FC=MC/MW                   :REM GET COL F
925 K1=FC:K2=DW:KØ=FC:GOSUB 3ØØØ :REM COL SIGN
93Ø PRINT "FC=",FC,PP          :REM PRINT IT
94Ø FI=MI/MW                   :REM GET INT F
945 K1=DI:K2=DW:KØ=FI:GOSUB 3ØØØ :REM INT SIGN
95Ø PRINT "FI=",FI,PP          :REM PRINT IT
96Ø END
3ØØØ REM F VALUE SIGNIFICANCE ROUTINE
3Ø4Ø GOSUB 4ØØØ                :REM GOTO DISTRIB
3Ø5Ø PP=1-H                    :REM ADJUST H
3Ø6Ø RETURN                    :REM GO BACK TO PGM
4ØØØ REM APPROXIMATION ROUTINE :REM BEGIN APPROX
4Ø1Ø IF KØ>=1 THEN 4Ø4Ø        :REM BIG ENOUGH?
4Ø2Ø H=1                       :REM CORRECT H
4Ø3Ø RETURN
4Ø4Ø K5=Ø.2222222              :REM SET ONE CONST
4Ø5Ø K6=K5/K1:K7=K5/K2         :REM SET MORE
4Ø6Ø K8=KØ^Ø.3333333           :REM SET MORE
4Ø7Ø K9=(1-K7)*K8-1+K6         :REM TOP
4Ø8Ø Z=K9/SQR(K7*K8*K8+K6)     :REM GET Z
4Ø9Ø IF Z>3 THEN 411Ø          :REM ADJUST IF <3
41ØØ Z=Z*(1.+Ø.Ø8*Z^4/K2^3)    :REM ADJUST
411Ø GOSUB 5ØØØ                :REM NORMAL APPROX
412Ø RETURN
5ØØØ REM NORMAL CURVE APPROXIMATION
5Ø1Ø H=1+Ø.Ø49867347Ø*Z        :REM CALCULATE
5Ø2Ø H=H+.Ø21141ØØ61*Z*Z       :REM  NORMAL CURVE
5Ø3Ø H=H+.ØØ32776263*Z^3       :REM  APPROXIMATION
5Ø4Ø H=H+.ØØØØ38ØØ36*Z^4       :REM  FROM PROGRAM
5Ø5Ø H=H+.ØØØØ4889Ø6*Z^5       :REM  6.2.
5Ø6Ø H=H+.ØØØØØ5383Ø*Z^6
5Ø7Ø H=H^-16/2               :REM FINISH APPROXIMAT.
5Ø8Ø RETURN
8ØØØ DATA 2,2,2
8Ø1Ø DATA 11Ø,12Ø,13Ø,14Ø
8Ø2Ø DATA 13Ø,14Ø,11Ø,12Ø
```

This program is exactly like Program 14-1, with the significance level routine tacked on. If you run it, you'll find that it gives you the significance level for each of the F's.

Its operation is detailed in the flowchart in Fig. 14-25 which shows only the section which calculates and prints the significance levels. Notice that in line 905 the proper DFs are put into Variables K1 and K2, while the F to be evaluated is inserted into Variable K0. The program then branches to subroutine 3000, which calculates the significance level, and is then printed out in line 910. The same thing is done for Columns in line 925, and for the Interaction in line 945. Notice how much effort the subroutine saves! Without the

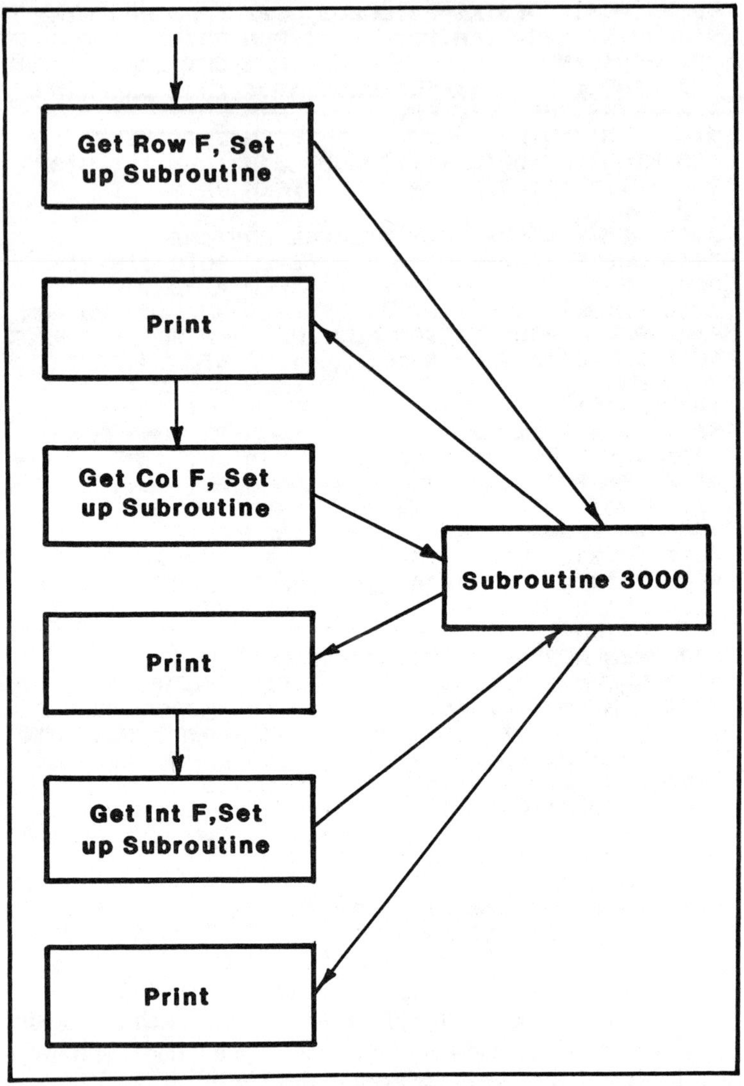

Fig. 14-25. Flowchart for two-way ANOVA with automatic significance levels.

concept of subroutines, all of the programming between lines 3000 and 5080 would have to be repeated over and over again.

UNDERSTANDING MORE COMPLEX DESIGNS

As I've mentioned before, there are hundreds, perhaps thousands, of possible experimental designs. The Two-Way

ANOVA that we've covered in this chapter offers only a glimpse into the world of more complex designs. Thus, one could have a Two-Way design in which measurements were repeated on the same subjects, just like the Paired t-test described in Chapter 11, but of course harder to calculate.

Analysis of these more complex designs almost necessitate the use of a computer, and your microcomputer will do the job very well indeed. There are computer programs for large computers which will do the job for you. The famous Biomedical Computer programs (Dixon, 1983) are an example. However, these programs are frequently very difficult to use, and of course, they require access to a large computer. Use of these number crunchers can be very expensive indeed.

With your knowledge of BASIC, there is no reason why you can't write the programs yourself. You will learn a great deal about the techniques in the process, and save yourself a great deal of money if you don't have access to a very large computer.

To do this, you'll need a book that describes the computational procedures for such complex designs. You won't find a better reference in this regard than Geoffrey Keppel's *Design and Analysis: A Researcher's Handbook*, (1982). It's the only really clear book that I've found on the topic, and, indeed, Keppel all but writes the computer programs for you. There are much more elegant books on the topic, but Keppel can be understood by the average person with little trouble. I cannot recommend it too highly.

We've finished another chapter, and, not too surprisingly, we haven't added any new concepts to our understanding of BASIC. The few concepts that we have used are the essential ones for most statistical programming, and, of course, these are the ones which are common to most microcomputers.

PROBLEMS

1. Figure 14-26 presents some more fictitious data on IQs of Old-Young and Tall-Short people. The summary table looks like this:

Source	SS	DF	MS	F
Age	1200	1	1200	12
Height	300	1	300	3
Interaction	0	1	0	0
Within	800	8	100	
Total	2300	11		

	Tall	Short
Old	110 120 130	120 130 140
Young	90 100 110	100 110 120

Fig. 14-26. Chapter 14, problem 1.

With 1 and 8 DF, we need an F of 5.32 for significance at the 95 percent level, and from Fig. 14-26, we'd conclude that Age makes a real difference in IQ, while Height does not. There is no Interaction at all.

2. Figure 14-27 shows fictitious scores on a learning test for a 2 by 4 table. Programs 14-1 or 14-2 yield the following summary table:

SOURCE	SS	DF	MS	F
Discipline	8	3	2.67	5.33
Height	256	1	256	512.00
Interaction	72	3	24	48.00
Within	4	8	.50	
Total	340	15		

I'll let you figure out the significance levels. Notice that all effects are significant. However, Height and the Interaction are by far the strongest. Plot the means, and you'll see why this should be the case.

3. Perhaps you're getting tired of my simple psychological examples. Let's return to the Ajax Bolt Company. The firm makes some bolts from Iron, and others from Steel. Some of the bolts are nickel plated, and others are chromium plated, yielding results as shown in Fig. 14-28. Using Programs 14-1 or 14-2, we obtain the following results:

Source	**SS**	**DF**	**MS**	**F**
Metal	.003	1	.003	2.26
Plating	.005	1	.005	3.91
Interaction	.002	1	.002	1.81
Within	.011	8	.002	
Total	.021	11		

	Tall	**Short**
Engineering	1 2	15 16
Business	3 4	11 12
Medical	5 6	13 14
Law	7 8	9 10

Fig. 14-27. Chapter 14, problem 2.

	Nickel	Chrome
Iron	.98 .99 1.09	1.05 1.03 1.02
Steel	1.00 .95 .93	1.04 1.02 1.03

Fig. 14-28. Chapter 14, problem 3.

Notice that none of the effects are significant. Thus, Ajax Bolts are fairly uniform. It doesn't make any difference whether they're made of iron or steel, and the plating doesn't seem to affect them either. Of course, in realıty, one would want to take much larger samples in a real study.

I'M AN OPTIMIZED
COMPUTER!

15
Miscellaneous Topics

In this chapter we shall take up two additional matters—the calculation of tables, and ways of converting BASIC programs to FORTRAN. Both subjects are of great potential interest to statistical programmers.

CALCULATING TABLES

Computers are very frequently used to calculate large tables of values. The tables in the Appendices of this book are a good example of such calculations, but in this chapter we'll talk about another sort of table. To a great extent, these calculations are the province of the very big IBM, Control Data, Amdahl, and Cray number crunchers. This is simply because the programs may require awesome amounts of storage, and may require equally awesome amounts of computation. Microcomputers typically have limited storage, and at the same time do computations rather slowly. Still, we can learn the fundamentals of computations of this sort on our small computers, and if the tables are not very large, the tables can be calculated satisfactorily even on our small machines. We shall go through a number of programs, each one more efficient than the last, as a demonstration of the proper techniques in making the programs more efficient, or optimized, as computer specialists say.

As an example, let's talk about calculating tables of Correlations. Take the following set of data:

	Variable		
	1	2	3
Observation	**Age**	**Income**	**Status**
1	21	50	1
2	22	40	4
3	23	30	3
4	24	20	2
5	25	10	5
	------	------	------
ΣX	115	150	15
ΣX^2	2655	5500	55

This data is obviously very fictitious! It will be nice, however, to have such ideal data in the computations that follow. Income refers to income in tens of thousands of dollars per year, and Status refers roughly to your social rank, or socioeconomic status, with 1 as the highest and 5 as the lowest (it's usually scored that way). Plug the data into Program 7-2, and you'll find that variables 1 and 2 have a Correlation of −1, and variables 1 and 3 have a Correlation of +0.6. Variables 2 and 3 have an r of −0.6. Data sets of this sort were used in Chapter 7, the chapter on Correlation. Plots of the data would look a lot like Figs. 7-2, 7-3, and a backwards 7-4.

The problem here is to calculate all of the possible Correlations between variables 1, 2, and 3. This seems like it would be a rather simple task, and yet look at Fig. 15-1. This is a smaller version of Fig. 12-7, from the chapter on Chi-Square. How many possible Correlations are there? Well, there is a Correlation between variables 1 and 2, and one between variables 1 and 3. Then there's one between variables 2 and 1, and another between variables 2 and 2, and so on. Inspection of Fig. 15-1 shows that there are nine possible Correlations between the three variables.

That doesn't seem so bad, but what if there were 100 different variables? This situation is depicted in Fig. 15-2. If you look at this figure, you'll notice that there are 10,000 possible Correlations. That's a lot of r's! Just to store 10,000 Correlations, assuming that each number takes 4 bytes, we'd need 40,000 bytes merely to store the table. Actually, each number may take up to 7 bytes, depending on the sort of BASIC you're using. Thus, storage of the table would require from 40,000 to 70,000 storage locations, more than almost any microcomputer has. If we had 500 variables, we'd have 250,000 Correlations, and perhaps need a million bytes of storage. In such a

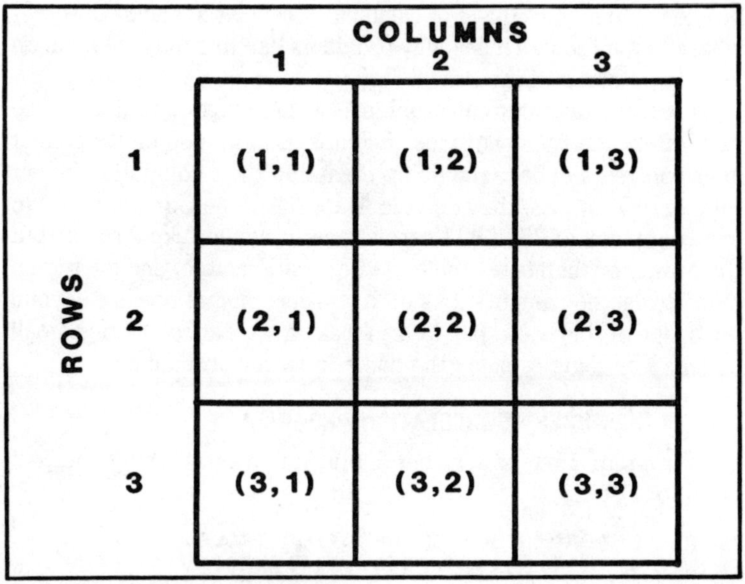

Fig. 15-1. A 3 × 3 Correlation table.

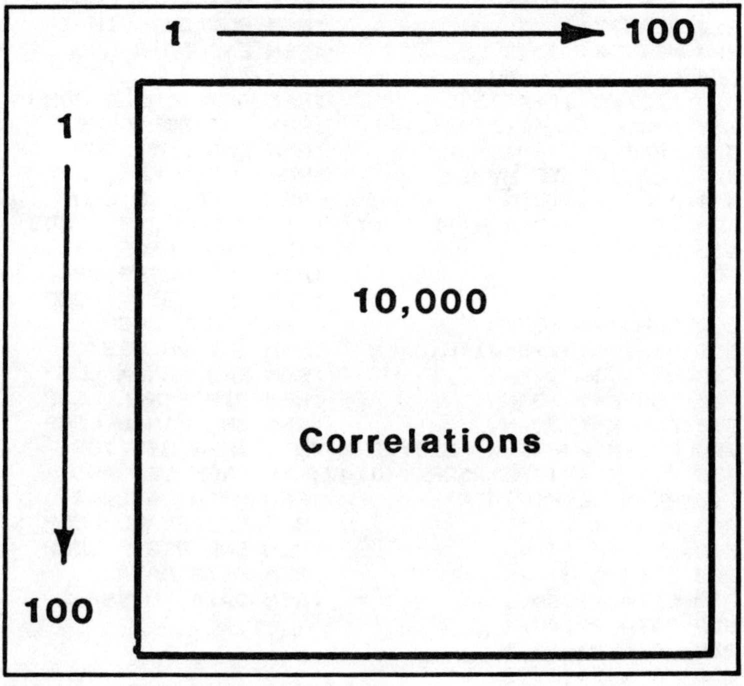

Fig. 15-2. A 100 × 100 Correlation table.

problem, then, you'd have to calculate 250,000 r's, and have somewhere to put them. This is why problems like this must be done on very large computers.

You might think that problems of this magnitude don't arise very often, but that's not true. Scientists, more and more, collect enormous sets of data and dump them into their computers in just this way. Of course, they don't look at each of the Correlations. No one could look at 250,000 Correlations; it would take forever, but the computer then takes the huge Correlation table, and does even more awesome computations on it. Such computations are beyond the scope of this book, but we can make a start at calculating small tables. The logic remains the same in larger problems.

A VERY INEFFICIENT CORRELATION PROGRAM

Program 15-1 is a rather inefficient way of doing the job: Let's look at it.

```
10 REM PROGRAM 15.1 CORRELATION TABLE.
20 DIM XP(4,4),R(4,4),S(4),Q(4),X(4),V(4)
30 READ N,NV                       :REM GET PBM SIZE
100 FOR I=1 TO N                   :REM LOOP FOR OBS
200 FOR J=1 TO NV                  :REM GET DATA LOOP
210 READ X(J)                      :REM PUT DATA IN X
220 NEXT J                         :REM END DATA LOOP
300 FOR J=1 TO NV                  :REM BGN SUM LOOP
310 S(J)=S(J)+X(J)                 :REM GET SIMPLE SUMS
320 Q(J)=Q(J)+X(J)*X(J)            :REM GET SQR SUMS
330 NEXT J                         :REM END SUM LOOP
400 FOR J=1 TO NV                  :REM OUTER CP LOOP
410 FOR K=1 TO NV                  :REM INNER CP LOOP
420 XP(J,K)=XP(J,K)+X(J)*X(K)  :REM CUM CRS PRODS
430 NEXT K                         :REM END INNER CP
440 NEXT J                         :REM END OUTER CP
500 NEXT I                         :REM END OBS   LOOP
600 FOR J=1 TO NV                  :REM DENOM LOOP
610 V(J)=Q(J)-S(J)*S(J)/N          :REM GET VALUES
620 NEXT J                         :REM END DENOM LOOP
700 FOR J=1 TO NV                  :REM BIG FINAL LOOP
710 FOR K=1 TO NV                  :REM SML FINAL LOOP
720 R(J,K)=XP(J,K)-S(J)*S(K)/N     :REM GET TOP
730 R(J,K)=R(J,K)/SQR(V(J)*V(K))   :REM GET ANS
740 PRINT J;K,R(J,K)               :REM PRINT ANSWER
750 NEXT K                         :REM END SMALL LOOP
760 NEXT J                         :REM END BIG   LOOP
800 DATA 5,3                       :REM SIZE DATA
810 DATA 21,50,1                   :REM DATA TO USE
820 DATA 22,40,4
830 DATA 23,30,3
840 DATA 24,20,2
850 DATA 25,10,5
```

First, recall Formula 7-2 for the Correlation Coefficient.

$$r = \frac{\Sigma XY - \dfrac{(\Sigma X)(\Sigma Y)}{N}}{\sqrt{\left(\Sigma X^2 - \dfrac{(\Sigma X)^2}{N}\right)\left(\Sigma Y^2 - \dfrac{(\Sigma Y)^2}{N}\right)}} \qquad \textbf{7-2}$$

To calculate the Correlation, we must know N, the number of observations, the sums of the two variables (ΣX and ΣY), the sums of the variables squared before summing (ΣX^2 and ΣY^2), and the sum of the Cross Products (ΣXY), that is the sum of X and Y multiplied together. Thus, each the calculation of each Correlation requires five quantities.

We want to do the job as efficiently as possible, of course. We could calculate each r separately by reading in the data over and over using the RESTORE statement, and then printing out the value after each calculation. But this would be terribly inefficient. Thus, the Correlation of variable 1 with variable 2 requires ΣX_1, ΣX^2_1, ΣX_2, and $\Sigma X_2^{\,2}$, in addition to ΣXY. The Correlation of Variable 1 with Variable 3 requires ΣX_1, $\Sigma X_1^{\,2}$, ΣX_3, and $\Sigma X_3^{\,2}$, and ΣXY. However, notice that in doing these two separate Correlations by reading in the data over and over, we'd have to calculate ΣX_1 and ΣX^2_1 twice. Indeed, all of these sums would have to be done over and over again with this approach.

Program 15-1 is somewhat more efficient than this crude approach. A very rough overview of the program is provided by the flowchart in Fig. 15-3.

First, we're going to have to use dimensioned variables in this problem; hence line 20. This statement will allow us to calculate a 4 × 4 table. The size of the problem, that is, the number of observations and the number of variables, is read in line 30. The program, must, of course, know these things before it can begin its work. The program is set up to handle a 4 × 4 table. This should fit in most small computers. You can increase the numbers in the DIM statement if your computer has the room.

Line 100 begins the big loop which reads in the observations; the index of this loop, variable I, is not used in the program directly. The loop

```
100   FOR I=1 TO N                :REM  LOOP FOR OBS
```

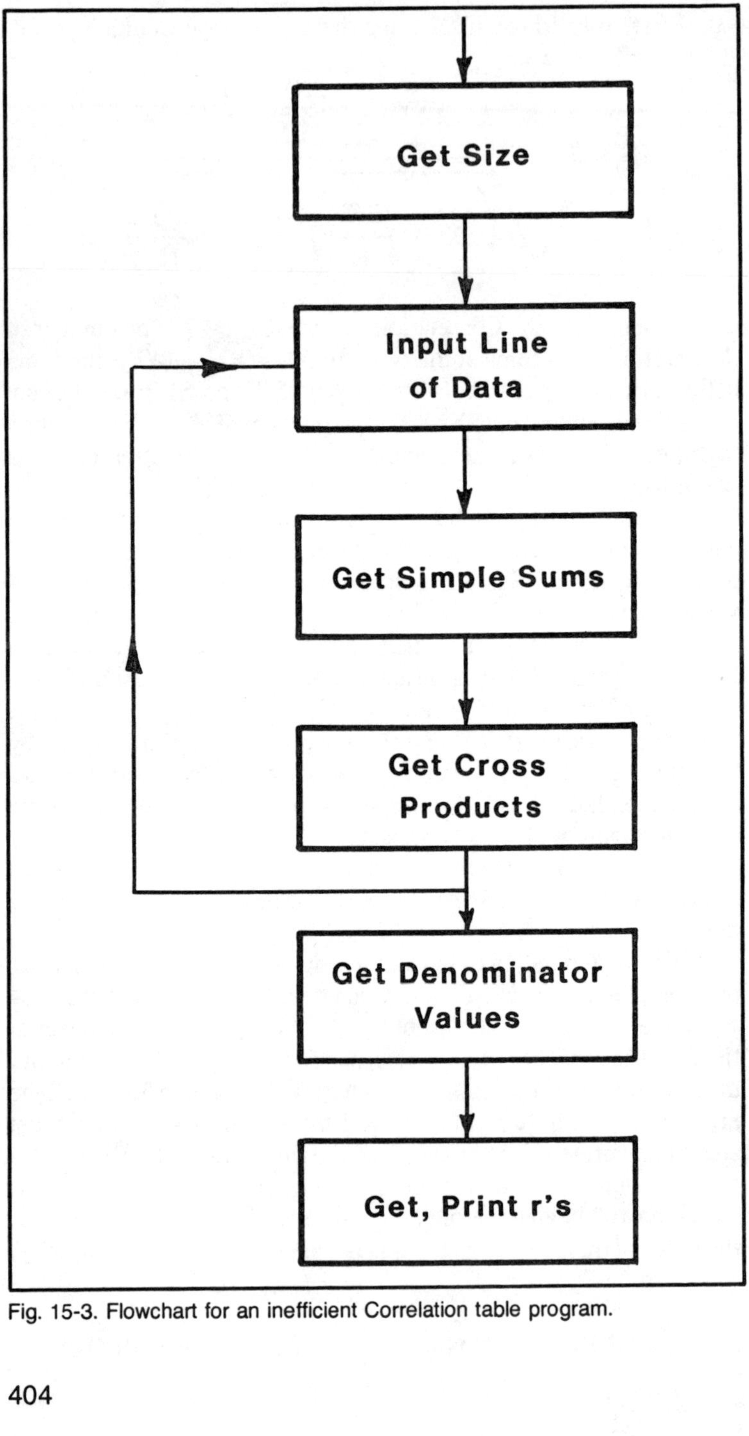

Fig. 15-3. Flowchart for an inefficient Correlation table program.

```
. . .
500  NEXT I                    :REM END OBS  LOOP
```

simply cycles the statements between lines 100 and 500 over and over, going through loop once for each observation of three values.

We must have all the data for each observation all at one time. That's the purpose of the FOR-NEXT loop in lines 210 to 220. Each line of data is read and put into the array, X. Thus, for the first observation, the loop will make X(1) equal to 21, X(2) equal to 50, and X(3) equal to 1. This mirrors the first real DATA statement line:

```
510  DATA 21,50,1
```

You may have to think about this a bit. You see, in the computations that follow, we have to have one whole line of data at once. We can't read each one separately—well, actually, you might be able to do it, but it would take forever.

The loops from lines 300 to 330 get the simple sums needed once the data has been put in X:

```
300  FOR J=1 TO NV           :REM  BGN  SUM LOOP
310  S(J)=S(J)+X(J)          :REM  GET SIMPLE SUMS
320  Q(J)=Q(J)+X(J)*X(J)     :REM  GET SQR SUMS
330  NEXT J                  :REM  END SUM LOOP
```

As you can see from these statements, line 310 will cumulate the sums for each variable in the array called S, and line 320 will cumulate the sums of the same numbers, squared before summing in the array called Q. We've used these designations many times before; they correspond to lines 40 and 50 in Program 7-2. Now we've got many variables, and they have to be an array (that is, dimensioned) in this case. For the Correlation of variable 1 with variable 2, the sum of X in Formula 7-2 corresponds to S(1), the sum of Y corresponds to S(2), and so on. That's easy enough.

Lines 400-440 correspond to line 60 of Program 7-2:

```
400  FOR J=1 TO NV                 :REM  OUTER CP LOOP
410  FOR K=1 TO NV                 :REM  INNER CP LOOP
420  XP(J,K)=XP(J,K)+X(J)*X(K)     :REM  SUM CRS PRODS
430  NEXT K                        :REM  END INNER CP
440  NEXT J                        :REM  END OUTER CP
```

Be very careful about the NEXT statements. If you erroneously typed:

```
430  NEXT J
440  NEXT K
```

The program would get totally confused, and your computer might object, depending upon its sophistication. You see, you must finish up the K loop before the J loop. This is the sort of error that everyone (including the writer) makes all the time.

These two loops operate as follows (see Fig. 15-4). The J loop will start out with the value of one. Then the K loop will become 1, 2, and finally 3. Thus, the sums of XY, or the Cross Products (called XP in the program) as they're called, will be calculated for X(1) and X(1), and summed into XP(1,1). Then the sum of XY—X(1) and X(2)—will be cumulated into XP(1,2). Finally, the sum of XY—X(1) and X(3)—will be cumulated into XP(1,3).

Variable J will then become 2, and XP(2,1), XP(2,2), and XP(2,3) will be cumulated. Finally, J will become 3, and XP(3,1), XP(3,2), and XP(3,3) will be summed.

This is very tricky. The K variable moves the most rapidly in the inner loop, and the J variable moves the least rapidly. The result of the process is shown in Fig. 15-15, omitting the diagonal values for the moment. These are the sums of XY for each possible combination. Thus, for the Correlation of 1 with 2, the sum of XY is 3,350, for the Correlation of 1 with 3, the sum of XY is 351, and for the Correlation of 2 with 3 the sum of XY is 390. Notice that the top half of the table is a mirror image of the bottom half.

Thus, by the end of line 500, we have all of the sums needed. If you run Program 15-1, you can see the various sums by using the Direct mode that most microcomputers have. For example, type:

```
PRINT S(1)
```

and the computer will say 115, the sum for the first variable. Type:

```
PRINT Q(2)
```

and your computer will say 2,655, the sum of the squared numbers for the second variable. Also,

```
PRINT XP(2,3)
```

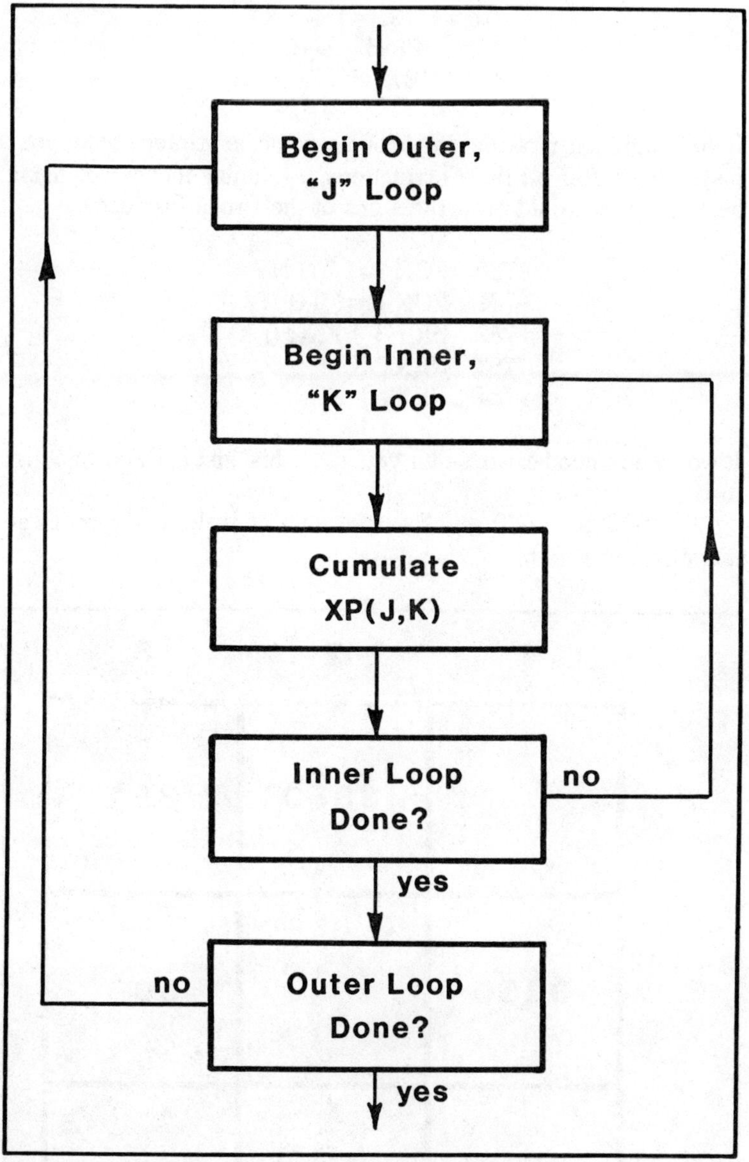

Fig. 15-4. Flowchart for summation of the cross Products.

will tell you the sum of the Cross Products for variables 2 and 3. With this technique, you can look at them all to see whether they're correct. Alternatively, at the end of Program 15-1, you could add the lines:

```
770  FOR J=1 TO NV
771  PRINT J,S(J),Q(J)
772  NEXT J
```

If the program is run, the simple sums would be printed out for your inspection, after all the Correlations. A similar technique, using nested loops, would print the sums of the Cross Products:

```
780  FOR J=1 TO NV
781  FOR K=1 TO NV
782  PRINT J,K,XP(J,K)
783  NEXT K
784  NEXT J
```

It would be a good exercise for you to try this, just to be certain that it works.

Lines 600 to 620 get Σx^2 (the sum of little x squared, remember?) in variable V:

	1	2	3
1		3350	351
2	3350		390
3	351	390	

Fig. 15-5. Sums of the Cross Products.

```
600  FOR J=1 TO NV          :REM DENOM LOOP
610  V(J)=Q(J)-S(J)*S(J)/N  :REM GET VALUES
620  NEXT J                 :REM END DENOM LOOP
```

The V's are the top of the Standard Deviation formula (see Formula 3-5), and are also the two quantities in the bottom of Formula 7-2. They'll be needed for the final r calculation.

FINAL R CALCULATIONS

The loops in lines 700 to 760 finish up the calculations. Let's repeat them here:

```
700  FOR J=1 TO NV                 :REM BIG FINAL LOOP
710  FOR K=1 TO NV                 :REM SML FINAL LOOP
720  R(J,K)=XP(J,K)-S(J)*S(K)/N    :REM GET TOP
730  R(J,K)=R(J,K)/SQR(V(J)*V(K))  :REM GET R
740  PRINT J;K,R(J,K)              :REM PRINT ANSWER
750  NEXT K                        :REM END SMALL LOOP
760  NEXT J                        :REM END BIG   LOOP
```

The structure of the loops is precisely like that of the loops in lines 400-440. We'll get R(1,1), R(1,2), R(1,3), then R(2,1), R(2,2), and so on, just as we did with the Cross Products.

The calculations are made in lines 720 and 730, and are a straightforward application of Formula 7-2, the Correlation formula. In line 720, we get the top of the formula, that is,

$$\Sigma XY - \frac{(\Sigma X)\ (\Sigma Y)}{N}$$

The sum for the first variable is multiplied by the sum for the second variable, and that product is divided by N. That answer is then subtracted from the appropriate Cross Product.

The calculations are then finished up in line 730. The top of the formula (now in R(J,K)) is divided by the square root of the product of the two V values. Each V value is one of the two quantities in parentheses in the term in the bottom of Formula 7-2. The result is then put back into R(J,K). You'll probably have to think about that for a long time before you figure it out! We could calculate the denominator values directly here, but it's faster overall to do them first in lines 600 to 610. That way, they must be calculated only once.

		Age 1	Income 2	Status 3
Age	1	1.00	−1.00	−.60
Income	2	−1.00	1.00	.60
Status	3	.60	−.60	1.00

Fig. 15-6. The final Correlation table.

Run the program. You'll find that the Correlation table calculated by the program is that of Fig. 15-6. Notice that the program seems to run a good while longer than most of our previous programs. It's doing a lot of computation. If we had many variables, say 20, assuming that your little computer had enough memory, it would take very long indeed.

Of course the way in which the computer prints the data leaves a lot to be desired. It's admittedly not very elegant. Commercial programs format the data in a nice, readable way, but such elegant printing programs frequently require more programming than the Correlation program itself, so we'll forego discussing them here. Besides, the precise way in which it's done depends heavily on the screen size of your computer, or the width of your printer.

A SELF-COUNTING CORRELATION PROGRAM

As always, I'm lazy, and I hate to count the number of observations. Let's take Program 15-1 and make it do the job by adding a couple of lines. The result in Program 15-2:

```
10 REM PROGRAM 15.2 SELF-COUNTING R TABLE
20 DIM XP(4,4),R(4,4),S(4),Q(4),X(4),V(4)
30 READ NV                        :REM N VARIABLES
100 FOR I=1 TO 10000              :REM LOOP FOR OBS
200 FOR J=1 TO NV                 :REM GET DATA LOOP
210 READ X(J)                     :REM PUT DATA IN X
215 IF X(1)=-1 THEN 600           :REM SKIP IF DONE
220 NEXT J                        :REM END DATA LOOP
225 N=N+1                         :REM GET # OBS
300 FOR J=1 TO NV                 :REM BGN SUM LOOP
310 S(J)=S(J)+X(J)                :REM GET SIMPLE SUMS
320 Q(J)=Q(J)+X(J)*X(J)           :REM GET SQR SUMS
330 NEXT J                        :REM END SUM LOOP
400 FOR J=1 TO NV                 :REM OUTER CP LOOP
410 FOR K=1 TO NV                 :REM INNER CP LOOP
420 XP(J,K)=XP(J,K)+X(J)*X(K)  :REM CUM CRS PRODS
430 NEXT K                        :REM END INNER CP
440 NEXT J                        :REM END OUTER CP
500 NEXT I                        :REM END OBS   LOOP
600 FOR J=1 TO NV                 :REM DENOM LOOP
610 V(J)=Q(J)-S(J)*S(J)/N         :REM GET VALUES
620 NEXT J                        :REM END DENOM LOOP
700 FOR J=1 TO NV                 :REM BIG FINAL LOOP
710 FOR K=1 TO NV                 :REM SML FINAL LOOP
720 R(J,K)=XP(J,K)-S(J)*S(K)/N      :REM GET TOP
730 R(J,K)=R(J,K)/SQR(V(J)*V(K))  :REM BET BOT
740 PRINT J;K,R(J,K)              :REM PRINT ANSWER
750 NEXT K                        :REM END SMALL LOOP
760 NEXT J                        :REM END BIG   LOOP
800 DATA 3                        :REM GET N VARIABLES
810 DATA 21,50,1                  :REM DATA TO USE
820 DATA 22,40,4
830 DATA 23,30,3
840 DATA 24,20,2
850 DATA 25,10,5
860 DATA -1,0,0
```

The approach in this program is exactly the same one that we've used again and again in this book—so much so that we won't discuss it at any length. Line 100 is changed so that a large number of observations is to be read in. However, line 215 checks to see whether the first observation in the line is a −1, just as before. If so, the program jumps to line 600. Line 225, then, counts the observations, since the program won't know how many there are beforehand. Your programming skills should be sufficiently developed by this time so that you can understand this program without difficulty.

A BETTER WAY

Program 15-1 works well, but as we have mentioned, it's very inefficient. The DIM statement

20 DIM XP(4,4),R(4,4),S(4),Q(4),X(4),V(4)

demands space for 48 numbers. However, the dimension statement only allows sixteen Correlations. Surely we can cut things down a bit.

Program 15-1 was written mainly as a way of demonstrating the underlying principles of the process. Program 15-3 is a considerable improvement from the standpoint of memory usage:

```
10 REM PROGRAM 15.3. BETTER CORRELATION TABLE.
20 DIM R(4,4),S(4),X(4)          :REM SAVE SPACE
30 READ N,NV                     :REM GET PBM SIZE
100 FOR I=1 TO N                 :REM LOOP FOR OBS
200 FOR J=1 TO NV                :REM GET DATA LOOP
210 READ X(J)                    :REM PUT DATA IN X
220 NEXT J                       :REM END DATA LOOP
400 FOR J=1 TO NV                :REM OUTER CP LOOP
405 S(J)=S(J)+X(J)               :REM GET SIMPLE SUMS
410 FOR K=1 TO NV                :REM INNER CP LOOP
420 R(J,K)=R(J,K)+X(J)*X(K)      :REM CUM CRS PRODS
430 NEXT K                       :REM END INNER CP
440 NEXT J                       :REM END OUTER CP
500 NEXT I                       :REM END OBS LOOP
600 FOR J=1 TO NV                :REM DENOM LOOP
610 X(J)=R(J,J)-S(J)*S(J)/N      :REM GET VALUES
620 NEXT J                       :REM END DENOM LOOP
700 FOR J=1 TO NV                :REM BIG FINAL LOOP
710 FOR K=1 TO NV                :REM SML FINAL LOOP
720 R(J,K)=R(J,K)-S(J)*S(K)/N        :REM GET TOP
730 R(J,K)=R(J,K)/SQR(X(J)*X(K))     :REM GET ANS
740 PRINT J;K,R(J,K)             :REM PRINT ANSWER
750 NEXT K                       :REM END SML LOOP
760 NEXT J                       :REM END BIG LOOP
800 DATA 5,3                     :REM SIZE DATA
810 DATA 21,50,1                 :REM DATA TO USE
820 DATA 22,40,4
830 DATA 23,30,3
840 DATA 24,20,2
850 DATA 25,10,5
```

This program is very similar to Program 15-1, but notice that the DIM statement:

20 DIM R(4,4),S(4),X(4)

requires only 24 numbers. We've saved half of the storage in this program. Furthermore, the program is somewhat shorter than

Program 15-1. The price for this, however, is exacted in the fact that the program becomes somewhat more difficult to understand.

Let's look through the program. Lines 10-220 are basically the same as those in the former program. Nothing has changed here.

Aha, but notice that there are no lines 300-330. Does that mean that we're not calculating the simple sums? Not at all! We're just doing them more efficiently. The cumulation of the sums of the variables is done by tucking the line 405 into the first part of the Cross Product loops.

```
405   S(J)=S(J)+X(J)
```

You'll have to think about it for a bit, but this will achieve the same result as lines 300-330 in Program 15-1.

"Wait!!!!!" you say. "You're not cumulating the Q's, the sums of the squared numbers. Something's wrong!"

In a way, you're right—but the sums of the squared numbers are being computed. Figure 15-7 is another view of the Cross

	1	2	3
1	2655	3350	351
2	3350	5500	390
3	351	390	55

Fig. 15-7. The hidden numbers!

Products table. In Fig. 15-5, earlier, we didn't show you the whole table, for it would have ruined our surprise. Notice in Fig. 15-7 that the diagonal of the Cross Products table contains the sums of the squared numbers! How can that be?

Very simple—the Cross Products are the sum of the X's multiplied by the Y's. In the diagonal of the table, X and Y are the same variable. Thus the diagonal contains the sum of X squared. You'll have to think about *that* a bit. In short, in Program 15-1 we were calculating the sum of the squared numbers twice! Since XP(1,1) was precisely the same thing as Q(1), we don't need the Q values at all. That's a nice savings. Noticė, also, that in Program 15-3 the Cross Products are placed immediately in variable R, rather than a separate variable, XP. Generally, we don't need to print the Cross Products, so we'll just replace them later in line 720, where they become the r.

Lines 600 through 620 are virtually the same as the ones in the previous program. However, instead of Q(J), we insert R(J,J), now that we've figured out that they're the same quantity. To save even more space, instead of putting the denominator values into a separate variable, V, we'll put them into the array X. This may confuse you. Notice, however, that by line 600 variable X is not needed anymore. It was used to store the data temporarily while the various sums were being calculated. Consequently, we can use it to hold what we used to call V.

The remainder of the program is practically the same as Program 15-1. Line 730 has the variable X replacing the variable V, but that's not a major change.

The program is how much shorter, and the storage is greatly reduced. This sort of optimization is precisely what programming is all about. The problem is, however, that of knowing when to optimize! For example, you might think that you could save lines 600-620 by tucking a line into the final loops.

```
700  FOR J=1 TO NV                :REM  BIG FINAL LOOP
705  X(J)=R(J,J)-S(J)*S(J)/N
710  FOR K=1 TO NV                :REM  SML FINAL LOOP
```

Try it, but **don't forget and leave it in the program.** It doesn't work, for the program needs all the denominator values before it can enter the final calculations—but it looks as though it should work! This one baffled me for several days once, when I was a student.

MORE OPTIMIZATION

The alert reader may have noticed that one half of the Correlation table is a mirror image of the other half. Look at Fig. 15-4 again. The Correlations above the diagonal are just the same as Correlations in corresponding positions below the diagonal. That seems wasteful. Further, notice that all of the Correlations in the diagonal are +1.00. That should be obvious, for the Correlation of anything with itself is, of course, one. Thus, we could eliminate the computations for the top half of the table (or the bottom half, if you like). Program 15-4 will do the job for us, at least in part:

```
10 REM PROGRAM 15.4. STILL BETTER R TABLE.
20 DIM R(4,4),S(4),X(4)        :REM SAVE SPACE
30 READ N,NV                   :REM GET PBM SIZE
100 FOR I=1 TO N               :REM LOOP FOR OBS
200 FOR J=1 TO NV              :REM GET DATA LOOP
210 READ X(J)                  :REM PUT DATA IN X
220 NEXT J                     :REM END DATA LOOP
400 FOR J=1 TO NV              :REM OUTER CP LOOP
405 S(J)=S(J)+X(J)             :REM GET SIMPLE SUMS
410 FOR K=1 TO J               :REM INNER CP LOOP
420 R(J,K)=R(J,K)+X(J)*X(K)    :REM CUM CRS PRODS
430 NEXT K                     :REM END INNER CP
440 NEXT J                     :REM END OUTER CP
500 NEXT I                     :REM END OBS LOOP
600 FOR J=1 TO NV              :REM DENOM LOOP
610 X(J)=R(J,J)-S(J)*S(J)/N    :REM GET VALUES
620 NEXT J                     :REM END DENOM LOOP
700 FOR J=2 TO NV              :REM BIG FINAL LOOP
710 FOR K=1 TO (J-1)           :REM SML FINAL LOOP
720 R(J,K)=R(J,K)-S(J)*S(K)/N     :REM GET TOP
730 R(J,K)=R(J,K)/SQR(X(J)*X(K))  :REM GET ANS
740 PRINT J;K,R(J,K)           :REM PRINT ANSWER
750 NEXT K                     :REM END SML LOOP
760 NEXT J                     :REM END BIG LOOP
800 DATA 5,3                   :REM SIZE DATA
810 DATA 21,50,1               :REM DATA TO USE
820 DATA 22,40,4
830 DATA 23,30,3
840 DATA 24,20,2
850 DATA 25,10,5
```

The differences are minor. In line 410 we have:

410 FOR K=1 TO J

instead of FOR K=1 to NV. The computations will be done only as far as the diagonal of the Cross Products table. We must include the diagonal, since that's where the sums of the squared numbers are being stored. Line 700 says:

```
700   FOR J=2 TO NV
```

This statement will start the Correlations with variable 2, thus eliminating the Correlation of variable 1 with variable 1. Line 710,

```
710   FOR K=1 TO (J-1)
```

limits the calculation of the Correlations to the lower half of the table.

The upshot of the changes in Program 15-4 is simply that not as much computation will have to be done, and no redundant answers will be printed out.

AN EFFICIENT CORRELATION TABLE PROGRAM

The programs discussed in this chapter up to this point are all right as far as they go, but they have one great disadvantage: they all use memory very inefficiently. To demonstrate this, let's examine Fig. 15-4 again. Notice that roughly half the table simply duplicates what is found in the other half. Each Correlation appears twice. Thus, the Correlation between variable 1 and variable 2 is −1.00, and the Correlation between variable 2 and variable 1 is also −1.00. They both have the same value, but each has its own memory location.

To see the problem even more acutely, look at Fig. 15-8. This demonstrates a 1,000 × 1,000 r table. Notice that 1,000,000 storage locations are needed. These are broken down as shown in Fig. 15-7. Notice that 499,500 locations are needed for storage of the Correlations. Another 1,000 locations are needed for the sums of the squared numbers. They can be either in the diagonal of the table or in another array, but we have to have them. Thus we simply must have 500,500 storage locations. However, 499,500 storage locations are redundant, simply replications of the other half of the table.

This isn't much of a problem with a 3 × 3 table. In a tiny table of this sort, six storage locations are absolutely necessary, and only three are wasted. However, as the tables get larger and larger, unacceptable amounts of storage are wasted. Storage on large computers costs something like $1 per word. Consequently, in a

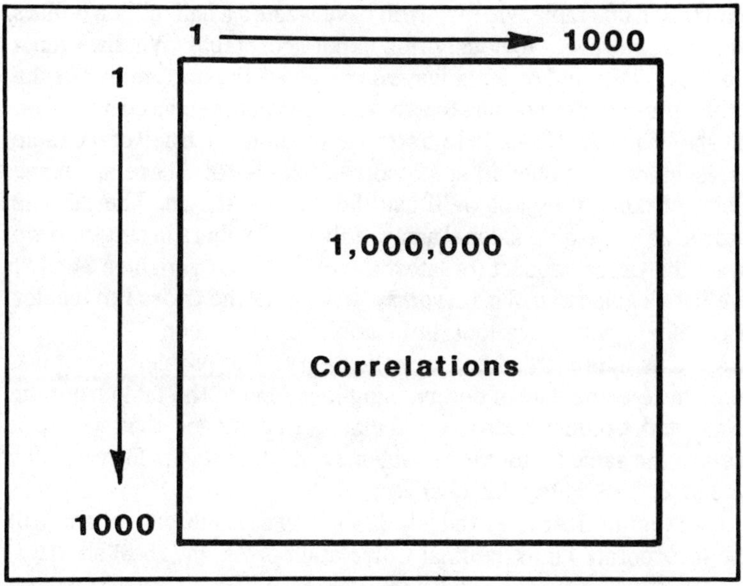

Fig. 15-8. A 1,000 × 1,000 Correlation table.

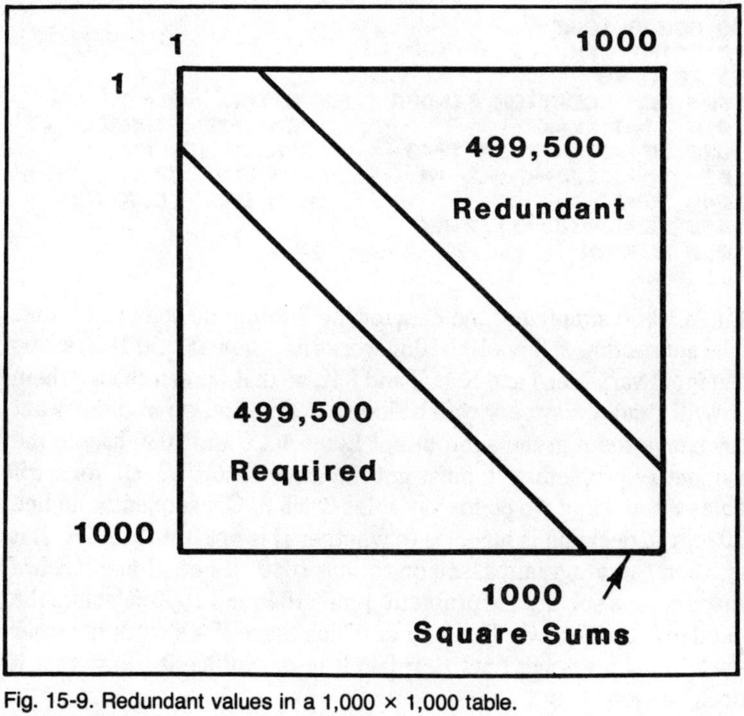

Fig. 15-9. Redundant values in a 1,000 × 1,000 table.

1,000 × 1,000 table, we may well have wasted a half million dollars worth of storage. Obviously this is not acceptable. What we must do, then, is to figure out a way of using just the bottom half of the table, plus enough storage to save the sums of the squared numbers, as shown in Fig. 15-10. In this storage scheme, each cell of the table is assigned a number in some sort of sequence. There are many ways of doing this, but we'll use the method shown. The table is stored as a one-dimensional array, only ten numbers in length. If we want the Cross Product (or later, Correlation) for variables 2 and 3, we'll look at location 5 in this array. If we want the Cross Product for variables 4 and 4, we look in location 10, and so on.

This approach adds a complication to the problem: we must now have some way of determining locations in the table from the Row and Column addresses. Further, the means that we must return the same result for variables 2 and 3 that it gets for variables 3 and 2. This gets a bit tricky.

Program 15-5 does the job. It's set up as a subroutine, and will be incorporated into the final Correlation program (15-6) shortly.

```
10 REM PROGRAM 15.5. TEST OF LOCATION SUBROUTINE
20 INPUT J,K
30 GOSUB 1000
40 PRINT J;K, LC
50 GOTO 20
1000 REM LOCATION FINDER SUBROUTINE
1010 JJ=J:KK=K                :REM TEMP STORAGE
1020 IF JJ>KK THEN 1040       :REM IS JJ>KK?
1030 TM=JJ:JJ=KK:KK=TM        :REM FLIP JJ AND KK
1040 JJ=JJ-1                  :REM CALC LOCATION
1050 LC=JJ*(JJ+1)/2+KK
1060 RETURN
```

Lines 10-50 simply get the data for the Subroutine and print it out. The subroutine in lines 100-1060 works as follows. Line 1010 saves the input variables J and K as JJ and KK, so that we can change them at will. J and K may very well be loop indexes, and we wouldn't want to change them in the subroutine! Lines 1020 and 1030 handle the symmetry problem. We must get the same result, recall, for variables 2 and 3 that we do for variables 3 and 2. Consequently, in line 1020, the decision is made as to whether JJ is greater than KK. If it is, then the program passes on to line 1040. If not, JJ and KK are flipped, thus solving the problem. Lines 1040 and 1050 calculate the needed location, LC. The way in which these lines work is rather tricky, and we won't try to explain it here. Suffice it to say that it does the job!

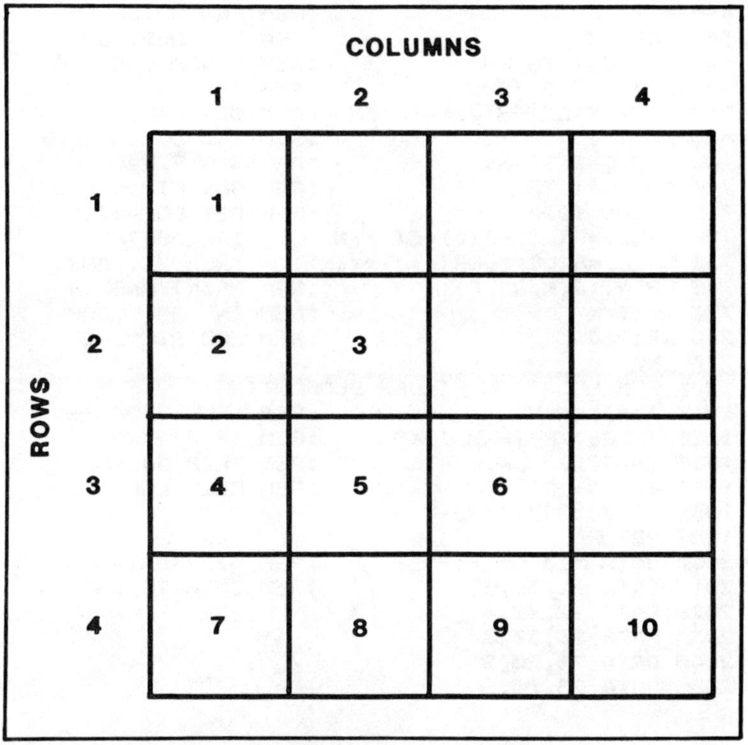

Fig. 15-10. How the r's will be stored.

Run Program 15-5, and refer to Fig. 15-10. If you give it 2 and 3 for the Row and Column addresses, it will say 5, as it should. If given 3 and 2, it also says 5. Try all of the possible locations in the table. The Subroutine works! We now have the means to write an efficient Correlation table program.

The final r table program is shown as Program 15-6.

```
10 REM PROGRAM 15.6. SUPER CORRELATION PROGRAM.
20 DIM R(28),S(7),X(7)          :REM SAVE SPACE
30 READ N,NV                    :REM GET PRM SIZE
100 FOR I=1 TO N                :REM LOOP FOR OBS
200 FOR J=1 TO NV               :REM GET DATA LOOP
210 READ X(J)                   :REM PUT DATA IN X
220 NEXT J                      :REM END DATA LOOP
400 FOR J=1 TO NV               :REM OUTER CP LOOP
405 S(J)=S(J)+X(J)              :REM GET SIMPLE SUMS
410 FOR K=1 TO J                :REM INNER CP LOOP
415 GOSUB 1000                  :REM GET LOCATION
420 R(LC)=R(LC)+X(J)*X(K)       :REM CUM CRS PRODS
430 NEXT K                      :REM END INNER CP
```

```
440 NEXT J                         :REM END OUTER CP
500 NEXT I                         :REM END OBS LOOP
600 FOR J=1 TO NV                  :REM DENOM LOOP
605 K=J:GOSUB 1000                 :REM GET LOCATION
610 X(J)=R(LC)-S(J)*S(J)/N         :REM GET VALUES
620 NEXT J                         :REM END DENOM LOOP
700 FOR J=2 TO NV                  :REM BIG FINAL LOOP
710 FOR K=1 TO (J-1)               :REM SML FINAL LOOP
715 GOSUB 1000                     :REM GET LOCATION
720 R(LC)=R(LC)-S(J)*S(K)/N           :REM GET TOP
730 R(LC)=R(LC)/SQR(X(J)*X(K))        :REM GET ANS
740 PRINT J;K,R(LC)                :REM PRINT ANSWER
750 NEXT K                         :REM END SML LOOP
760 NEXT J                         :REM END BIG LOOP
770 END
1000 REM LOCATION FINDER SUBROUTINE
1010 JJ=J:KK=K                     :REM TEMP STORAGE
1020 IF JJ>KK THEN 1040            :REM IS JJ>KK?
1030 TM=JJ:JJ=KK:KK=TM             :REM FLIP JJ AND KK
1040 JJ=JJ-1                       :REM CALC LOCATION
1050 LC=JJ*(JJ+1)/2+K
1060 RETURN
2000 DATA 5,3                      :REM SIZE DATA
2010 DATA 21,50,1                  :REM DATA TO USE
2020 DATA 22,40,4
2030 DATA 23,30,3
2040 DATA 24,20,2
2050 DATA 25,10,5
```

The operation of the program is almost exactly like that of Program 15-4, and the line numbers have been made almost identical. However, whenever the big table must be used, a call is made to subroutine 1000, which calculates the location needed, and puts it in LC. Then the table is addressed with a statement like R(LC).

The result of this system is that roughly half the storage space is saved. Notice that Program 15-6 is dimensioned to handle a 7 × 7 table, and we need to store only 42 numbers for that purpose. With much larger tables, you'll find that an enormous savings has been effected.

There are certain disadvantages to the procedure. Some time is wasted in calculating LC again and again, and some computer memory is needed for storing subroutine 1000 and the statements that are needed to call it. But the saving in memory is enormous.

CALCULATING MANY t-TESTS

A similar approach may be used if we have to calculate a number of t-tests on several groups. Take the following data:

	1 Dogs	2 Cats	3 Gnus
	110	130	150
	120	140	160
	130	150	170
	140	160	180
	150	170	190
	------	------	------
Average	130	150	170

Using the t-test procedures of Chapter 10, we'd like to calculate t-tests between all possible combinations of the three groups. Recall from Chapter 10 that Formulas 10-2 and 10-3 require, however, only the sums for each group, and the sums of the squared numbers for each group. The Averages, sums of little x squared, and so on, can be derived from these values. Consequently, we don't have to store a big table of values such as that needed for the Cross Products in the earlier part of the chapter. Program 15-7 will do the job for us:

```
10 REM PROGRAM 15.7. SIMPLE t-TEST TABLE.
20 DIM X(7),S(7),Q(7),A(7),V(7)  :REM GET SPACE
30 READ N,NV                     :REM GET PBM SIZE
100 FOR I=1 TO N                 :REM LOOP FOR OBS
200 FOR J=1 TO NV                :REM SUM LOOP
210 READ X(J)                    :REM GET DATA
220 S(J)=S(J)+X(J)               :REM GET SUMS
230 Q(J)=Q(J)+X(J)*X(J)          :REM GET SQRD SUMS
240 NEXT J                       :REM END SUM LOOP
500 NEXT I                       :REM END OBS LOOP
600 FOR J=1 TO NV                :REM DENOM LOOP
610 A(J)=S(J)/N                  :REM GET AVERAGE
620 V(J)=Q(J)-S(J)*S(J)/N        :REM GET LTL X SQRD
630 NEXT J                       :REM END DENOM LOOP
700 FOR J=2 TO NV                :REM BIG FINAL LOOP
710 FOR K=1 TO (J-1)             :REM SML FINAL LOOP
720 H=A(J)-A(K)                  :REM TOP OF FRM 10.3
730 SE=SQR(((V(J)+V(K))/(N+N-2))*(1/N+1/N))
740 T=H/SE                       :REM GET T
750 PRINT J;K,T                  :REM PRINT IT
760 NEXT K                       :REM END SML LOOP
770 NEXT J                       :REM END BIG LOOP
800 DATA 5,3                     :REM SIZE DATA
810 DATA 110,130,150                :REM DATA TO USE
820 DATA 120,140,160
830 DATA 130,150,170
840 DATA 140,160,180
850 DATA 150,170,190
```

This program has been numbered to correspond, as much as possible to the previous programs in the chapter. Line 20 saves the needed space. Enough is saved for a 7 × 7 table. Since we don't have to save the entire 7 × 7 table, however, we can be a bit more generous in putting things in different variables. Thus, we have more things in the DIM statement. This makes debugging a good deal easier.

Line 100 begins the loop to cycle through the observations. Lines 200 to 240 cumulate the sums and the sums of the squared numbers, but since we aren't collecting a huge table, we can read the data in at the same time that we obtain the sums, thus saving some loops.

Lines 600-630 calculate the Averages (A's) and the sums of little x squared (V's). They could be done later, but it's more time-effective to do them all at once.

Finally, lines 700-770 calculate the t-tests, using Formulas 10-2 and 10-3. The loops that cycle through the various t's are precisely the same as those in Program 15-6; again there's no point in calculating redundant t's. The difference between the Averages is calculated in line 720, and the Standard Error of the Difference Between Means in line 730. Things are simplified a good deal by the fact that the group sizes are the same. Therefore N1 = N2 = N, and the formula is somewhat simpler than that used in Program 10-1. The t is calculated in line 740, and printed in line 750. Lines 760 and 770 finish up the loops.

Run the program, and you'll find that you get all the t's at once, as shown in Fig. 15-11 (only the nonredundant t's are shown). With 8 Degrees of Freedom, we need values of 2.31 and 3.36 for significance, so we find that only the t between Dogs and Gnus (4.00) is significant. Notice that the program runs a good deal faster than the Correlation table program. There's not as much computation involved. There's really no need to save the t's at the end, but if you wanted to you could save them in an array T(J, K), just as you did with the Correlations. Of course, you'd have to add T(7,7) to the DIM statement.

Statistical workers frequently calculate large numbers of t's in precisely this manner, and the program will do a 20 × 20 table of t's just as easily as the three in the example (of course, it will take longer, and you'll have to change the DIM statement).

CALCULATION OF PAIRED-t TABLES

It is also possible to calculate tables of matched or paired t's,

		Dogs	Cats	Gnus
		1	2	3
Dogs	1			
Cats	2	2.00		
Gnus	3	4.00	2.00	

Fig. 15-11. A t-test table.

just as we did in Chapter 11. Assume that you have the following fictitious IQ scores in a fictitious experiment. Notice that each person is tested under three separate conditions. This is precisely the situation needed for the matched or paired t-test.

		Condition	
Person	**Control**	**Tea**	**Coffee**
1	120	160	220
2	100	120	140
3	80	125	180
	-----	-----	-----
Average	100	135	180

Now, let's determine how to calculate the paired t's between the various Means.

This can get more complicated than you might think. Formulas 11-1, 11-2, and 11-3, which are so simple and handy for the paired t-test, become rather awkward when calculating tables. This is simply because the difference approach used in Chapter 11 would require the storage of say, a 4 × 4 table of the differences, and another 4 × 4 table of the squared differences. That's very inefficient. For this purpose, an alternate, but mathematically equivalent formula is better:

$$t = \frac{\overline{X}_1 - \overline{X}_2}{\sqrt{s^2_{x_1} + s^2_{x_2} - 2 r s_{x_1} s_{x_2}}} \qquad \textbf{15-1}$$

Notice that this formula involves the Standard Error of the Mean (Formula 9-1) for each variable, the Average for each variable, and the Correlation between them. This formula is less popular in elementary statistics courses because so much computation is involved, but of course, your microcomputer doesn't care about the amount of computation that it has to do.

Let's implement Formula 15-1 in a program. We'll add it to Program 15-1, since that's the one that's the easiest to understand.

```
10 REM PROGRAM 15.8. PAIRED T TABLE.
20 DIM XP(4,4),R(4,4),S(4),Q(4),X(4),V(4)
25 DIM A(4),SE(4)
30 READ N,NV                    :REM GET PBM SIZE
100 FOR I=1 TO N                :REM LOOP FOR OBS
200 FOR J=1 TO NV               :REM GET DATA LOOP
210 READ X(J)                   :REM PUT DATA IN X
220 NEXT J                      :REM END DATA LOOP
300 FOR J=1 TO NV               :REM BGN SUM LOOP
310 S(J)=S(J)+X(J)              :REM GET SIMPLE SUMS
320 Q(J)=Q(J)+X(J)*X(J)         :REM GET SQR SUMS
330 NEXT J                      :REM END SUM LOOP
400 FOR J=1 TO NV               :REM OUTER CP LOOP
410 FOR K=1 TO NV               :REM INNER CP LOOP
420 XP(J,K)=XP(J,K)+X(J)*X(K)  :REM CUM CRS PRODS
430 NEXT K                      :REM END INNER CP
440 NEXT J                      :REM END OUTER CP
500 NEXT I                      :REM END OBS   LOOP
600 FOR J=1 TO NV               :REM DENOM LOOP
605 A(J)=S(J)/N                 :REM GET AVERAGE
610 V(J)=Q(J)-S(J)*S(J)/N       :REM GET VALUES
615 SE(J)=SQR(V(J)/(N*(N-1)))  :REM GET STD ERROR
620 NEXT J                      :REM END DENOM LOOP
700 FOR J=2 TO NV               :REM BIG FINAL LOOP
710 FOR K=1 TO (J-1)            :REM SML FINAL LOOP
720 R(J,K)=XP(J,K)-S(J)*S(K)/N    :REM GET TOP
730 R(J,K)=R(J,K)/SQR(V(J)*V(K))  :REM GET R
735 BT=SE(J)*SE(J)+SE(K)*SE(K):REM 1ST PT OF BTM
736 BT=BT-2*R(J,K)*SE(J)*SE(K)   :REM CONT BOTTOM
```

```
737 BT=SQR(BT)                 :REM FINISH BOTTOM
738 T=(A(J)-A(K))/BT           :REM GET T
740 PRINT J;K,R(J,K),T         :REM PRINT ANSWERS
750 NEXT K                     :REM END SMALL LOOP
760 NEXT J                     :REM END BIG   LOOP
800 DATA 3,3                   :REM SIZE DATA
810 DATA 120,160,220           :REM DATA TO USE
820 DATA 100,120,140
830 DATA 80,125,180
```

This program is exactly like Program 15-1, with only the necessary additions. In fact, I simply took Program 15-1 and tacked the needed statements onto it. You can see the additions easily if you look for the lines that aren't multiples of 10! Line 25 adds another DIM statement, so that we'll have a place to put the Averages and Standard Errors. Line 605 calculates the Averages we'll need, and line 615 the needed Standard Errors of the Mean. Lines 735 through 738 calculate Formula 15-1; the calculations are straightforward.

There is one change from Program 15-1. The limits of the loops in lines 700 and 710 have been changed so that we don't get t's between, say, variable 1 and variable 1. The program will blow up in such circumstances, for the denominator of Formula 15-1 will become zero. Consequently, an approach like that of Program 15-4 is used in setting up the loops.

If you run the program, you'll find that the paired t's are as shown in Fig. 15-12. With two Degrees of Freedom, we need values of 4.30 and 9.92, so that only the difference between the Control condition and the Tea condition is significant, and only at the 95 percent level. Notice that the computer does this problem in no time at all! Take it from an old timer (me) that the calculation of Formula 15-1 is not easy. There's something about the hand calculation of that formula that seems to lead to errors. But, of course, computers don't make mistakes!

Since it is based on Program 15-1, Program 15-8 is very inefficient in terms of storage. However, with your newly acquired programming skills, you can easily change it so that it will use the strategy of Program 15-6, and thus be just as efficient, but you'll find that it will be much more difficult to understand.

USE OF THIS BOOK WITH FORTRAN

It is conceivable, but not very likely, that you might have a computer available to you that doesn't offer BASIC. Many large computers focus on FORTRAN, the scientific programming language that was developed by IBM in the later 1950s.

		Control 1	Tea 2	Coffee 3
Control	1			
Tea	2	4.58		
Coffee	3	4.00	3.58	

Fig. 15-12. A paired t-test table.

It might surprise you to learn that conversion of the simple programs in this book to FORTRAN is very simple. It's likely that you'd need the help of someone who knows FORTRAN well, but the two languages are so similar that you might be able to make the conversion yourself.

Let's take, as an example, Program 3-6, the efficient Standard Deviation program. As you recall, it looks like this in BASIC:

```
10 REM PROGRAM 3.6. BETTER STANDARD DEVIATION
20 READ X                        :REM GET DATA
30 IF X=-1 THEN 80               :REM TEST FOR END
40 S=S+X                         :REM SUM NUMBERS
50 Q=Q+X*X                       :REM SUM SQUARED NOS
60 N=N+1                         :REM GET COUNT
70 GOTO 20                       :REM CONTINUE
80 V = Q - S*S/N                 :REM TOP OF FORMULA
90 D=SQR(V/N)                    :REM GET STD DEV
100 PRINT "STANDARD DEVIATION IS",D:REM PRINT IT
120 DATA 80,90,100,110,120,-1  :REM DATA TO USE
```

A FORTRAN equivalent of this program is as follows:

```
C PROGRAM 3.6F. FORTRAN STANDARD DEVIATION
C
   20 READ(5,1) X
    1 FORMAT(F10.4)
      IF(X.EQ.-1) GO TO 80
      S=S+X
      Q=Q+X*X
      N=N+1
      GO TO 20
   80 V= Q - S*S/N
      D=SQRT(V/N)
      WRITE (6,2) D
    2 FORMAT(6F10.4)
      END
```

There are obvious differences. For example, the main part of FORTRAN statements must begin in column 7, and the line numbers (if any) end by column 5. FORTRAN programming harks back to the days when people punched all of their programs on IBM cards, and it was relatively easy to make sure that things were in the proper columns on these cards simply by thumbing through them. Statement numbers are not required in FORTRAN unless needed to keep track of a GO TO or a similar statement. Only one FORTRAN statement may be put on each line. Any comments that occur must be on a separate line, and start with the letter C in the first column. Finally, FORTRAN reads only columns 1-72. The remaining columns are ignored. Any mark in column 6 means that that line is a continuation of the previous line.

Notice that many of the statements are identical, except for the presence or absence of Line Numbers. Thus, in BASIC, we have

```
40  S=S+X
50  Q=Q+X*X
60  N=N+1
70  GOTO 20
```

while in FORTRAN, we have

```
S=S+X
Q=Q+X*X
N=N+1
GOTO 20
```

The statements are identical. The real differences in FORTRAN programs lie in the READ and WRITE statements. Input/output statements in FORTRAN are much more flexible than those in BASIC, in many ways. You have rather complete control over the way things are printed out. FORMAT statements are much like the PRINT USING statements in many BASICs, but I like FORMAT statements better.

However, FORMAT statements are very difficult and tricky things to understand. They're not consistent in their syntax, and there are seemingly thousands of rules to recall while using them. Students seem to have more difficulty with FORMAT statements than with any other feature of FORTRAN.

For example, the statement,

```
20   READ (5,1) X
```

in FORTRAN means: "Read the value for variable X from device 5, according to the pattern set up in FORMAT statement 1." Device 5 is usually the input device on IBM equipment. When this statement is encountered, the monster computer will then look on the disks for the file that contains the data (it can have any name you like, usually), and get the value 80.

The FORMAT statement,

```
1 FORMAT (F10.4)
```

means, "expect a decimal point number 10 columns wide, with the decimal point 4 columns to the left of the right margin." FORTRAN will put in the decimal points for you; that's a nice feature.

The WRITE statement,

```
WRITE (6,2) D
```

means "write D on device 6, the way you're told in FORMAT statement 2."' The FORMAT statement,

```
2   FORMAT(6F10.4)
```

simply says to print the Standard Deviation in six fields 10 wide, using 4 decimal places for each. We have only one thing to print, not six, but FORTRAN will figure out the situation and print only one. Alphabetic information can be included in FORMAT statements also, but you have enough difficulties without worrying about that!

Finally, the symbol for square root is SQR in BASIC, and SQRT in FORTRAN. I like SQRT better. I still occasionally think that SQR means square the number

A table of BASIC-FORTRAN equivalencies for the simple BASIC statements used in Chapter 3, and, frankly, for most of the rest of the book, is as follows:

BASIC		FORTRAN
ABS(−1)		ABS(−1)
DATA		(no exact equivalent)
FOR J=1 TO 3		DO 1000 J=1,3
NEXT J	1000	CONTINUE
GOTO 100		GOTO 100
INPUT		READ (more complex)
IF J=1 THEN 100		IF (J.EQ.1) GOTO 100
PRINT		WRITE (more complex)
READ		READ (more complex)
REM		C (in Column 1)
RESTORE		REWIND
SQR		SQRT

FORTRAN is a very complex language, and has many, many statements. Thus, there is a DATA statement in FORTRAN, but it works in a different fashion than the one in BASIC. Further, there is a PRINT statement in most FORTRAN installations, but it's a bit confusing to use.

The mathematical and logical operations are either identical, or obvious.

BASIC	FORTRAN
=	=
+	+
−	−
*	*
/	/
>	.GT.
<	.LT.
<>	.NE.
<=	.LE.
=>	.GE.
↑	**

These should give you no difficulty.

Let's convert Program 3-6 so that it uses a FOR-NEXT loop.

```
10 REM PROGRAM 3.6A. FOR-NEXT STD DEVIATION
20 FOR J=1 TO 5                :REM START LOOP
30 READ X                      :REM GET DATA
40 S=S+X                       :REM SUM NUMBERS
50 Q=Q+X*X                     :REM SUM SQUARED NOS
60 N=N+1                       :REM GET COUNT
70 NEXT J                      :REM END LOOP
80 V = Q- S*S/N                :REM TOP OF FORMULA
90 D=SQR(V/N)                  :REM GET STD DEV
100 PRINT "STANDARD DEVIATION IS",D:REM PRINT IT
110 DATA 80,90,100,110,120,-1  :REM DATA TO USE
```

A FORTRAN equivalent of this program is as follows:

```
C PROGRAM 3.6AF. DO LOOP STANDARD DEVIATION
      DO 1000 J=1,5
      READ (2,1) X
    1 FORMAT(F10.4)
      S=S+X
      Q=Q+X*X
      N=N+1
 1000 CONTINUE
      V = Q - S*S/N
      D=SQRT(V/N)
      WRITE (6,2) D
    2 FORMAT(6F10.4)
      END
```

The DO LOOP in FORTRAN does precisely the same job as the FOR-NEXT loop in BASIC. Everything between the DO statement and the line marked 1000 will be repeated 5 times. The CONTINUE statement is a dummy statement. It doesn't do anything. It's just there as a place to hang the 1000. When I was a student, I first thought that it meant go on to the next statement, but that's not true. FORTRAN has many inconsistencies of this sort. The data, by the way, will be in an external file for this program

We won't belabor the point, but you should realize, now, that it is perfectly possible for someone with a little knowledge of FORTRAN to make the conversion from BASIC to FORTRAN programs.

PROBLEMS

1. The following data has been used in previous chapters, and is data collected many years ago in one of my classes in statistics.

The scores are scores on the Scholastic Aptitude Test (SAT), the total number of points in the class, the score on another aptitude test, and finally data on the student's tendency toward extraversion (this latter bit of data was constructed from the SAT scores to yield very nearly a zero Correlation).

SAT	Grade	Aptitude	Extraversion
460	44	39	62
350	34	18	88
530	24	16	55
520	42	37	55
700	68	42	72
670	74	42	66

Using the Correlation programs in the previous sections of this chapter, calculate the Correlations between the various variables. (Answers are in Fig. 15-13.)

		SAT 1	Grade 2	Aptitude 3	Extrav 4
SAT	1				
Grade	2	.78			
Aptitude	3	.64	.84		
Extrav	4	−.31	.13	−.21	

Fig. 15-13. Problem number 1, Chapter 15.

2. The following data is thoroughly fictious data on schizophrenia scores. In this problem groups of schizophrenics are given four separate treatments: a control treatment, in which nothing is done, one involving staying in a warm room, one involving staying in a cool room, and one in which the treatment is done through hydrotherapy, or the use of water:

Control	Heat	Cold	Water
50	70	20	51
53	68	25	63
58	72	33	69
48	69	30	52
58	77	21	52
53	72	28	53

		Control 1	Heat 2	Cold 3	Water 4
Control	1				
Heat	2	8.50			
Cold	3	10.17	18.33		
Water	4	.96	4.40	8.23	

Fig. 15-14. Problem number 2, Chapter 15.

		Control 1	Heat 2	Cold 3	Water 4
Control	1				
Heat	2	15.59			
Cold	3	10.55	16.57		
Water	4	1.27	4.12	12.14	

Fig. 15-15. Problem number 3, Chapter 15.

Your job in this problem is that of calculating simple t-tests between all possible combinations of the groups, using one of the programs from earlier in the chapter. The results that you should obtain are shown in Fig. 15-14. With 10 D.F., values of 2.23 and 3.17 are needed for significance.

3. This problem is the same as Problem 2. However, in this case assume that the same person appears in all of the groups. Thus, a matched or paired t-test is possible between each group. The results that should be obtained are shown in Fig. 15-15. Here we have 5 D.F., and values of 2.57 and 4.03 will be needed for significance.

Note on Appendices

When I began this book, I thought that it would be fun to use the microcomputer to calculate the tables that follow. People generally don't think tiny computers are capable of such tasks.

It turned out to be more of a job than I had anticipated, and my poor little TRS-80 spent many, many hours grinding out these tables; one-half of the F table took about four hours.

The formulas used are found in Abramowitz and Stegun (1964). You could do the programming yourself, although the operations are a bit tricky.

If you compare these tables with the tables found in most statistics books, you'll occasionally find a difference of one one-hundredth of a point in the values, especially in the F table. The tables found in most statistics texts were calculated in the 1920s and 1930s, or earlier, and it amuses me to think that perhaps my little computer may have produced more accurate results than the workers who originally produced the tables. Calculation of such tables was an enormous job in the old days, using mechanical calculators, tables of sines, cosines, and logarithms, etc. And perhaps, just perhaps, my results are more accurate! Hurray for Minnie Computer!

Appendix A
Normal Curve Table

Z	P	Z	P	Z	P
0.01	0.0040	0.51	0.1950	1.01	0.3438
0.02	0.0080	0.52	0.1985	1.02	0.3461
0.03	0.0120	0.53	0.2019	1.03	0.3485
0.04	0.0160	0.54	0.2054	1.04	0.3508
0.05	0.0199	0.55	0.2088	1.05	0.3531
0.06	0.0239	0.56	0.2123	1.06	0.3554
0.07	0.0279	0.57	0.2157	1.07	0.3577
0.08	0.0319	0.58	0.2190	1.08	0.3599
0.09	0.0359	0.59	0.2224	1.09	0.3621
0.10	0.0398	0.60	0.2257	1.10	0.3643
0.11	0.0438	0.61	0.2291	1.11	0.3665
0.12	0.0478	0.62	0.2324	1.12	0.3686
0.13	0.0517	0.63	0.2357	1.13	0.3708
0.14	0.0557	0.64	0.2389	1.14	0.3729
0.15	0.0596	0.65	0.2422	1.15	0.3749
0.16	0.0636	0.66	0.2454	1.16	0.3770
0.17	0.0675	0.67	0.2486	1.17	0.3790
0.18	0.0714	0.68	0.2517	1.18	0.3810
0.19	0.0753	0.69	0.2549	1.19	0.3830
0.20	0.0793	0.70	0.2580	1.20	0.3849
0.21	0.0832	0.71	0.2611	1.21	0.3869
0.22	0.0871	0.72	0.2642	1.22	0.3888
0.23	0.0910	0.73	0.2673	1.23	0.3907
0.24	0.0948	0.74	0.2703	1.24	0.3925
0.25	0.0987	0.75	0.2734	1.25	0.3944
0.26	0.1026	0.76	0.2764	1.26	0.3962
0.27	0.1064	0.77	0.2793	1.27	0.3980
0.28	0.1103	0.78	0.2823	1.28	0.3997
0.29	0.1141	0.79	0.2852	1.29	0.4015
0.30	0.1179	0.80	0.2881	1.30	0.4032
0.31	0.1217	0.81	0.2910	1.31	0.4049
0.32	0.1255	0.82	0.2939	1.32	0.4066
0.33	0.1293	0.83	0.2967	1.33	0.4082
0.34	0.1331	0.84	0.2995	1.34	0.4099
0.35	0.1368	0.85	0.3023	1.35	0.4115
0.36	0.1406	0.86	0.3051	1.36	0.4131
0.37	0.1443	0.87	0.3078	1.37	0.4147
0.38	0.1480	0.88	0.3106	1.38	0.4162
0.39	0.1517	0.89	0.3133	1.39	0.4177
0.40	0.1554	0.90	0.3159	1.40	0.4192
0.41	0.1591	0.91	0.3186	1.41	0.4207
0.42	0.1628	0.92	0.3212	1.42	0.4222
0.43	0.1664	0.93	0.3238	1.43	0.4236
0.44	0.1700	0.94	0.3264	1.44	0.4251
0.45	0.1736	0.95	0.3289	1.45	0.4265
0.46	0.1772	0.96	0.3315	1.46	0.4279
0.47	0.1808	0.97	0.3340	1.47	0.4292
0.48	0.1844	0.98	0.3365	1.48	0.4306
0.49	0.1879	0.99	0.3389	1.49	0.4319
0.50	0.1915	1.00	0.3413	1.50	0.4332

Z	P	Z	P	Z	P
1.51	0.4345	2.01	0.4778	2.51	0.4940
1.52	0.4357	2.02	0.4783	2.52	0.4941
1.53	0.4370	2.03	0.4788	2.53	0.4943
1.54	0.4382	2.04	0.4793	2.54	0.4945
1.55	0.4394	2.05	0.4798	2.55	0.4946
1.56	0.4406	2.06	0.4803	2.56	0.4948
1.57	0.4418	2.07	0.4808	2.57	0.4949
1.58	0.4429	2.08	0.4812	2.58	0.4951
1.59	0.4441	2.09	0.4817	2.59	0.4952
1.60	0.4452	2.10	0.4821	2.60	0.4953
1.61	0.4463	2.11	0.4826	2.61	0.4955
1.62	0.4474	2.12	0.4830	2.62	0.4956
1.63	0.4484	2.13	0.4834	2.63	0.4957
1.64	0.4495	2.14	0.4838	2.64	0.4959
1.65	0.4505	2.15	0.4842	2.65	0.4960
1.66	0.4515	2.16	0.4846	2.66	0.4961
1.67	0.4525	2.17	0.4850	2.67	0.4962
1.68	0.4535	2.18	0.4854	2.68	0.4963
1.69	0.4545	2.19	0.4857	2.69	0.4964
1.70	0.4554	2.20	0.4861	2.70	0.4965
1.71	0.4564	2.21	0.4864	2.71	0.4966
1.72	0.4573	2.22	0.4868	2.72	0.4967
1.73	0.4582	2.23	0.4871	2.73	0.4968
1.74	0.4591	2.24	0.4875	2.74	0.4969
1.75	0.4599	2.25	0.4878	2.75	0.4970
1.76	0.4608	2.26	0.4881	2.76	0.4971
1.77	0.4616	2.27	0.4884	2.77	0.4972
1.78	0.4625	2.28	0.4887	2.78	0.4973
1.79	0.4633	2.29	0.4890	2.79	0.4974
1.80	0.4641	2.30	0.4893	2.80	0.4974
1.81	0.4649	2.31	0.4896	2.81	0.4975
1.82	0.4656	2.32	0.4898	2.82	0.4976
1.83	0.4664	2.33	0.4901	2.83	0.4977
1.84	0.4671	2.34	0.4904	2.84	0.4977
1.85	0.4678	2.35	0.4906	2.85	0.4978
1.86	0.4686	2.36	0.4909	2.86	0.4979
1.87	0.4693	2.37	0.4911	2.87	0.4979
1.88	0.4699	2.38	0.4913	2.88	0.4980
1.89	0.4706	2.39	0.4916	2.89	0.4981
1.90	0.4713	2.40	0.4918	2.90	0.4981
1.91	0.4719	2.41	0.4920	2.91	0.4982
1.92	0.4726	2.42	0.4922	2.92	0.4982
1.93	0.4732	2.43	0.4925	2.93	0.4983
1.94	0.4738	2.44	0.4927	2.94	0.4984
1.95	0.4744	2.45	0.4929	2.95	0.4984
1.96	0.4750	2.46	0.4931	2.96	0.4985
1.97	0.4756	2.47	0.4932	2.97	0.4985
1.98	0.4761	2.48	0.4934	2.98	0.4986
1.99	0.4767	2.49	0.4936	2.99	0.4986
2.00	0.4772	2.50	0.4938	3.00	0.4987

Appendix B
t-Table

DF	95%	99%
1	12.71	63.66
2	4.30	9.92
3	3.18	5.84
4	2.78	4.60
5	2.57	4.03
6	2.45	3.71
7	2.36	3.50
8	2.31	3.36
9	2.26	3.25
10	2.23	3.17
11	2.20	3.11
12	2.18	3.05
13	2.16	3.01
14	2.14	2.98
15	2.13	2.95
16	2.12	2.92
17	2.11	2.90
18	2.10	2.88
19	2.09	2.86
20	2.09	2.85
21	2.08	2.83
22	2.07	2.82
23	2.07	2.81
24	2.06	2.80
25	2.06	2.79
26	2.06	2.78
27	2.05	2.77
28	2.05	2.76
29	2.05	2.76
30	2.04	2.75
31	2.04	2.74
32	2.04	2.74
33	2.03	2.73
34	2.03	2.73
35	2.03	2.72
36	2.03	2.72
37	2.03	2.72

DF	95%	99%
38	2.02	2.71
39	2.02	2.71
40	2.02	2.70
41	2.02	2.70
42	2.02	2.70
43	2.02	2.70
44	2.02	2.69
45	2.01	2.69
46	2.01	2.69
47	2.01	2.68
48	2.01	2.68
49	2.01	2.68
50	2.01	2.68

Appendix C
Correlation Significance Table

Number Of Pairs	95%	99%
3	0.997	1.000
4	0.950	0.990
5	0.878	0.959
6	0.811	0.917
7	0.754	0.875
8	0.707	0.834
9	0.666	0.798
10	0.632	0.765
11	0.602	0.735
12	0.576	0.708
13	0.553	0.684
14	0.532	0.661
15	0.514	0.641
16	0.497	0.623
17	0.482	0.606
18	0.468	0.590
19	0.456	0.575
20	0.444	0.561
21	0.433	0.549
22	0.423	0.537
23	0.413	0.526
24	0.404	0.515
25	0.396	0.505
26	0.388	0.496
27	0.381	0.487
28	0.374	0.479
29	0.367	0.471
30	0.361	0.463
31	0.355	0.456
32	0.349	0.449
33	0.344	0.442
34	0.339	0.436
35	0.334	0.430
36	0.329	0.424
37	0.325	0.418

Number Of Pairs	95%	99%
38	Ø.32Ø	Ø.413
39	Ø.316	Ø.4Ø8
4Ø	Ø.312	Ø.4Ø3
41	Ø.3Ø8	Ø.398
42	Ø.3Ø4	Ø.393
43	Ø.3Ø1	Ø.389
44	Ø.297	Ø.384
45	Ø.294	Ø.38Ø
46	Ø.291	Ø.376
47	Ø.288	Ø.372
48	Ø.285	Ø.368
49	Ø.282	Ø.365
5Ø	Ø.279	Ø.361
51	Ø.276	Ø.358
52	Ø.273	Ø.354

Appendix D
Chi-Square Table

DF	95%	99%
1	3.84	6.64
2	5.99	9.21
3	7.81	11.34
4	9.49	13.28
5	11.07	15.09
6	12.59	16.81
7	14.07	18.47
8	15.51	20.09
9	16.92	21.67
10	18.31	23.21
11	19.68	24.73
12	21.03	26.22
13	22.36	27.69
14	23.68	29.14
15	25.00	30.58
16	26.30	32.00
17	27.59	33.41
18	28.87	34.81
19	30.14	36.19
20	31.41	37.57
21	32.67	38.93
22	33.92	40.29
23	35.17	41.64
24	36.42	42.98
25	37.65	44.31
26	38.88	45.64
27	40.11	46.96
28	41.34	48.28
29	42.56	49.59
30	43.77	50.89

Appendix E
F/Table

DF Between (Numerator)

DF Within (Denominator)	1	2	3	4	5	6
1	161.45	199.50	215.71	224.58	230.16	233.99
2	18.51	19.00	19.16	19.25	19.30	19.33
3	10.13	9.55	9.28	9.12	9.01	8.94
4	7.71	6.94	6.59	6.39	6.26	6.16
5	6.61	5.79	5.41	5.19	5.05	4.95
6	5.99	5.14	4.76	4.53	4.39	4.28
7	5.59	4.74	4.35	4.12	3.97	3.87
8	5.32	4.46	4.07	3.84	3.69	3.58
9	5.12	4.26	3.86	3.63	3.48	3.37
10	4.96	4.10	3.71	3.48	3.33	3.22
15	4.54	3.68	3.29	3.06	2.90	2.79
20	4.35	3.49	3.10	2.87	2.71	2.60
30	4.17	3.32	2.92	2.69	2.53	2.42
40	4.08	3.23	2.84	2.61	2.45	2.34
50	4.03	3.18	2.79	2.56	2.40	2.29
60	4.00	3.15	2.76	2.53	2.37	2.25
70	3.98	3.13	2.74	2.50	2.35	2.23
80	3.96	3.11	2.72	2.49	2.33	2.21
90	3.95	3.10	2.71	2.47	2.32	2.20
100	3.94	3.09	2.70	2.46	2.31	2.19

DF Within (Denominator)	DF Between (Numerator) 7	8	9	10	11	12
1	236.77	238.88	240.54	241.88	242.98	243.91
2	19.35	19.37	19.38	19.40	19.41	19.41
3	8.89	8.85	8.81	8.79	8.76	8.75
4	6.09	6.04	6.00	5.96	5.94	5,91
5	4.88	4.82	4.77	4.73	4.70	4.68
6	4.21	4.15	4.10	4.06	4.03	4.00
7	3.79	3.73	3.68	3.64	3.60	3.57
8	3.50	3.44	3.39	3.35	3.31	3.28
9	3.29	3.23	3.18	3.14	3.10	3.07
10	3.14	3.07	3.02	2.98	2.94	2.91
15	2.71	2.64	2.59	2.54	2.51	2.47
20	2.51	2.45	2.39	2.35	2.31	2.28
30	2.33	2.27	2.21	2.16	2.13	2.09
40	2.25	2.18	2.12	2.08	2.04	2.00
50	2.20	2.13	2.07	2.03	1.99	1.95
60	2.17	2.10	2.04	1.99	1.95	1.92
70	2.14	2.07	2.02	1.97	1.93	1.89
80	2.13	2.06	2.00	1.95	1.91	1.88
90	2.11	2.04	1.99	1.94	1.90	1.86
100	2.10	2.03	1.97	1.93	1.89	1.85

Bibliography

Abramowitz, M., and Stegun, I. *Handbook of Mathematical Functions*. Washington, D.C.: National Bureau of Standards, 1964.

Dixon, W. *BMDP Statistical Software*. Berkeley, California: University of California Press, 1983.

Federer, W. *Experimental Design: Theory and Applications*. New York: MacMillan, 1955.

Fisher, R.A. and Cornish, E. A. The percentile points of distributions having known cumulants. *Technometrics*, 1960, *2*, 209-225.

Hastings, C. *Approximations for Digital Computers*. Princeton, New Jersey: Princeton University Press, 1955.

Jaspen, N. The calculation of probabilities corresponding to values of *z*, t, F, and chi-square. *Education and Psychological Measurement,* 1965, *25*, 877-880.

Kelly, F., Beggs, D., and McNeil, K. *Research Design in the Behavioral Sciences*. Carbondale, Illinois: Southern Illinois University Press, 1969.

Keppel, G. *Design and Analysis*. Englewood Cliffs, New Jersey: Prentice-Hall, 1973.

Veldman, D. *Fortran Programming for the Behavioral Sciences*. New York: Holt, Rinehart, Winston, 1967.

Index

Edited by Brint Rutherford